International Conference on Futuristic Trends in Mechanical Engineering (ICOFTIME-2020)

IOP Conference Series: Materials Science and Engineering
Volume 1013

Bengaluru, India
24 - 25 April 2020

ISBN: 978-1-7138-3320-8
ISSN: 1757-8981

Printed from e-media with permission by:

Curran Associates, Inc.
57 Morehouse Lane
Red Hook, NY 12571

proceedings.com

Some format issues inherent in the e-media version may also appear in this print version.

Institute of Physics
Dirac House, Temple Back
Bristol BS1 6BE UK

Phone: 44 1 17 929 7481
Fax: 44 1 17 920 0979

techtracking@iop.org

Additional copies of this publication are available from:

Curran Associates, Inc.
57 Morehouse Lane
Red Hook, NY 12571 USA
Phone: 845-758-0400
Fax: 845-758-2633
Email: curran@proceedings.com
Web: www.proceedings.com

TABLE OF CONTENTS

MANUFACTURING

MATERIALS

COMPOSITE

DESIGN

HEAT TRANSFER

ENERGY

BIO FUEL

IOP Conf. Series: Materials Science and Engineering 1013 (2021) 011001 doi:10.1088/1757-899X/1013/1/011001

International Conference on Futuristic Trends in Mechanical Engineering (ICOFTIME2020)

PRESIDENCY UNIVERSITY, Bengaluru, India.

11th - 12th September, 2020

in association with

UCSI UNIVERSITY, Kuala Lumpur, Malaysia

The Conference Report

Day 1 (Friday, 11-09-20):

The conference is inaugurated by beloved Honorable Chancellor, Sri. Nissar Ahmed, Dr. Radha Padmanabhan, Vice Chancellor; Dr. A M Surendra Kumar, Pro Vice Chancellor & Dean In charge SOE; Mr. A Abdul Bari, Registrar; Dr. Rajesh, Director, Chancellor's Office; The Chief Guest Dr. Elango Natarajan, UCSI University, Kuala Lumpur, Malaysia & Dr. Udaya Ravi M, Convener ICOFTIME2020 and HOD Mechanical Engg. Book of Abstracts released during the inauguration function by the dignitaries.

Fig 1: Lighting the lamp by H'ble Chancellor and other delegates

After the inaugural function keynote speech delivered by 3 speakers Dr. Elango Natarajan, Professor, UCSI University, Malaysia on **'soft robotics'**, Dr. S Rajkumar, Sr. Vice President, Rane Engine Valve Ltd, Chennai on **'the shift from IC engine to EV'** Dr. Jyotishkumar Parameswara Pillai, Research Professor, King Mongkut's University, Thailand on **'Thermosetting blends and composites for advanced applications'.**

Fig 2: Book of Abstracts released during the inauguration function by the dignitaries.

We have received a total of 119 papers for our conference, after a thorough review process only 72 papers were selected for presentation.

Afternoon 2 PM to 5 PM, 7 parallel sessions were conducted in different streams such as Manufacturing, Composites, Design, Materials, Thermal, Energy. Each session is chaired by the experts from various reputed institution across the country. Dr. Manjaiah M, NIT Warangal, Dr. Mohan Kumar G C NITK, Surathkal, Dr. Wei Hong Lim, UCSI University, Malaysia. Dr. Rakesh S G, Amrita Vishwa Vidya Peetam, Bengaluru Campus. Dr. Amarnath M IIIT Design & Manufacturing, Jabalpur. Dr. V Krishna, PES University, Bengaluru. And Dr. Govinde Gowda A.T.M.E College of Engineering Mysuru. At the end the of the day 1 total of 44 papers were presented by the registered authors from various institution across the globe.

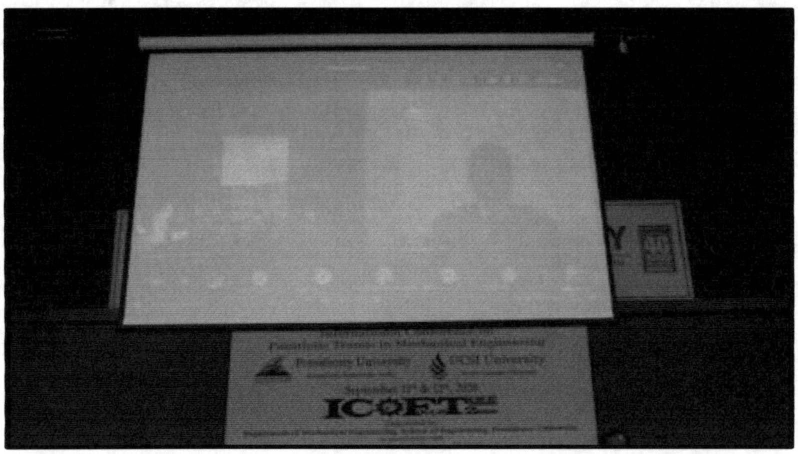

Fig. 3: UCSI Prof. Elango Natarajan attending conference Online

IOP Conf. Series: Materials Science and Engineering 1013 (2021) 011001 doi:10.1088/1757-899X/1013/1/011001

Day-2 (Saturday, 12-09-2020):

Today morning Key note session started at 10 am. Keynote speech delivered by Dr. S Ganesh Sundara Raman, Professor, Metallurgical Engineering, IITM, Chennai on "**Importance of fatigue life in the machine component**"

Fig. No. 4: HoD, ME, Dr. Udaya Ravi M speaking in the valedictory function.

From 11 AM to 1 :15 PM 3 parallel sessions were conducted in different streams such as Manufacturing, Composites, Design, Materials, Thermal, Energy. Each session is chaired by the experts from various reputed institution across the globe. Dr Vasu M, NIT, Tiruchirappalli. Dr Pradyumna Ghosh, Professor, Mech, IIT BHU and Dr B M Girish, Alliance University. Total of 22 papers were presented by the registered authors from various institution across the globe.

Fig. 5: Session Chair, Dr. G C Mohan Kumar, NITK Surathkal, conducting the session No. 2

After the lunch break Key note session started at 2:30 PM Keynote speech delivered by Dr. Zulfiqar A Khan, Professor of Design, Engineering & Computing, Bournemouth University, UK on **"Importance of coating and its application".**

Fig. 6: Keynote speech delivered by Dr. Zulfiqar A Khan, Professor of Design, Engineering & Computing, Bournemouth University, UK.

In the valedictory session Dr. Wei Hong Lim, UCSI University, Malaysia, has expressed the need of research in current scenario with the futuristic area of Mechanical Engineering. He has communicated his satisfaction about quality of paper presented in his session. He has mentioned in his speech about possibilities of collaboration of research, faculty exchange and student exchange program with UCSI University.

Fig. 7: valedictory speech by Dr. Wei Hong Lim, HoD, ME & Mechatronics Dept. UCSI University, Malaysia

Out of total 72 registered papers, 66 papers were presented. 46 papers will be published in the IOP Conference Series and 20 papers will be published in The Springer Lecture Note Series.

The program was come to end with concluding remark by Dr Udaya Ravi M, Convener and HoD of Mechanical Engineering and followed by vote of thanks.

IOP Conf. Series: Materials Science and Engineering 1013 (2021) 011001 doi:10.1088/1757-899X/1013/1/011001

We would like to extend thanks to Mr Govindaraju N, Magnum Engineers , C-272, 7th Cross, 3rd Main, 1st Stage, Peenya Industrial Estate, Bangalore –560 058 for sponsoring the event.

Schedule of Event

Day 1: Friday, 11th September 2020

Event	Time	Speakers
Inauguration	9.30 am to 10.00 am	**Mr. Nissar Ahmed,** Chancellor **Dr. Radha Padmanabhan**, Vice Chancellor **Dr. A M Surendra Kumar**, Pro-Vice Chancellor **Dr. Udaya Ravi M,** Convener and HOD Mechanical **Dr. Elango Natarajan** Professor, UCSI University, Malaysia.
Keynote Speech	10 am to 10.45 am	**Dr. Elango Natarajan**, Professor, UCSI University, Malaysia. Topic of the speech: **Soft Robotics**
Keynote Speech	10.45 am to 11.30 am	**Dr. S Rajkumar**, Sr. Vice President, Rane Engine Valve Ltd, Chennai. Topic of the speech: **The shift from IC Engines to EV**
Keynote Speech	11.30 am to 12.15 pm	**Dr. Jyotishkumar Parameswara Pillai**, Research Professor, King Mongkut's University, Thailand. Topic of the speech: **Thermosetting blends and composites for advanced applications**
Sessions (Parallel Session)	**Paper ID**	**Session Chair** **2 pm to 5.00 pm**
Session 1	**011, 012, 016, 017, 018, 039, 044**	**Manufacturing** **Dr. Manjaiah M, Professor,** Department of Mechanical Engineering NIT Warangal Telangana.
Session 2	**047, 099, 002, 023, 027, 032, 094**	**Manufacturing/Composite** **Dr. Mohan Kumar G C Professor** - HAG, Department of Mechanical Engineering NITK, Surathkal
Session 3	**083, 097, 101, 102, 104, 105, 110**	**Design** **Dr. Wei Hong Lim, Professor and Head,** Mechanical and Mechatronics Department, UCSI University, Malaysia.
Session 4	**057, 062, 063, 074, 081, 092, 098, 109**	**Composite** **Dr. Rakesh S G, Professor,** Department of Mechanical Engineering, Amrita Vishwa Vidya Peetam, Kasavanahalli Bengaluru Campus.
Session 5 M: 9425805414	**006, 008, 031, 035, 059, 077, 080**	**Materials** **Dr. Amarnath M, Professor,** Dept. of Mechanical Engg IIIT Design & Manufacturing, Jabalpur Madhya Pradesh, India
Session 6	**005, 007, 024, 025, 033, 053, 060, 064,**	**Thermal/Energy** **Dr. V Krishna, Professor & Dean Student Affairs,** Department of Mechanical Engineering, PES University, Bengaluru
Session 7	**078, 084, 086, 087, 088, 089, 096, 113**	**Thermal/Energy** **Dr. Govinde Gowda Professor & Dean – Academics,** Department of Mechanical Engineering A.T.M.E College of Engineering Mysuru

Day 2: Saturday, 12th September 2020

Event	Time	Speakers

IOP Conf. Series: Materials Science and Engineering 1013 (2021) 011001 doi:10.1088/1757-899X/1013/1/011001

Keynote Speech	10 am to 10.45 am	**Dr. S Ganesh Sundara Raman**, Professor, Metallurgical Engineering, IITM, Chennai
Sessions (Parallel Session)	**Paper ID**	**Session Chair** **11.15 am to 1.15 pm**
Session 8	**037, 038, 045, 051, 052, 055, 056,**	**Composite** **Dr Vasu M, Professor,** NIT Campus, Tiruchirappalli
Session 9	**010, 020, 022, 043, 085, 100,**	**Session Chair (General)** **Dr Pradyumna Ghosh, Professor**, Mech, IIT BHU.
Session 10	**095, 093, 115, 111, 112, 114, 116**	**Materials/Design** **Dr B M Girish, Professor,** HOD Mech, Alliance Univ.
Lunch	1.30 pm to 2.15 pm	
Keynote Speech	2.30 pm to 3.15 pm	**Dr. Zulfiqar A Khan,** Professor of Design, Engineering & Computing, Bournemouth University, UK.
Valedictory Session	3.30 pm -3.45 pm	**Dr. Wei Hong Lim**, Professor and Head, Mechanical and Mechatronics Department, UCSI University, Malaysia.
	4.00 pm 4.15 pm	**Best Paper award:** • Manufacturing (Supply chain, Industrial Engineering), • Design (Robotics, Automation, Additive Manufacturing) • Materials • Analysis • Thermal
	4.15 pm to 4.30 pm	**Vote of thanks**

IOP Conf. Series: Materials Science and Engineering 1013 (2021) 011001 doi:10.1088/1757-899X/1013/1/011001

Chief Patrons

Mr. Nissar Ahmed, Chancellor

Mr. Salman Ahmed, Vice President

Patrons

Dr. Radha Padmanabhan, Vice Chancellor

Dr A M Surendra Kumar, Pro-Vice Chancellor

Organizing Committee:

Convener

Dr. Udaya Ravi M. Professor & Head, Dept. of Mechanical Engineering, Presidency University.

Organizing Secretary

Dr. Bhaskar Pal, Professor Dept. of Mechanical Engineering, Presidency University.

Joint Secretary

Dr. Ramachandra C. G, Associate Professor Dept. of Mechanical Engineering, Presidency University

TREASURER

Dr. Mahesha K. Professor Dept. of Mechanical Engineering. Presidency University.

Technical Committee

Dr. C S Ramesh, Professor & Dean R&IC

Dr. B V Prabhu, Professor and Asso. Dean

Dr. Ramesh S, Professor

Dr. Akshay Nanjangud, Associate Professor

THEMATIC AREAS

Automation and Robotics and allied areas

AUTOMATION, MECHATRONICS AND ROBOTICS

CAD/CAM/CAE/CIM/FMS IN MANUFACTURING

IOT IN MANUFACTURING

ARTIFICIAL INTELLIGENCE AND MACHINE LEARNING

FEA / FEM TECHNIQUE

MEMS AND NANOTECHNOLOGY

IOP Conf. Series: Materials Science and Engineering 1013 (2021) 011001 doi:10.1088/1757-899X/1013/1/011001

Additive Manufacturing and allied areas

MATERIALS FOR 3D PRINTING

RAPID PROTOTYPING

SUSTAINABLE MANUFACTURING

Nano and Smart Materials and allied areas

COMPOSITE AND SMART MATERIALS

ADVANCED MATERIAL PROCESSING AND TECHNIQUES

MODELING AND OPTIMISATION TECHNIQUES

NANO AND MICRO-MACHINING

Energy, Environment and Sustainability

BIOFUELS / LUBRICANTS

COMPUTATIONAL FLUID DYNAMICS

REFRIGERATION AND AIR CONDITIONING

GREEN ENERGY

ELECTRIC VEHICLES

COMBUSTION AND IC ENGINES

Other Areas

LOGISTICS AND SUPPLY CHAIN MANAGEMENT

PRODUCT DESIGN AND DEVELOPMENT

BIO MECHANICS

AVIATION AND UMAVs

VIBRATION AND ACCOUSTICS

SHOCKWAVES & HYENTHALPY AERODYNAMICS

IOP Conf. Series: Materials Science and Engineering 1013 (2021) 011001 doi:10.1088/1757-899X/1013/1/011001

Advisory Board

IOP Conf. Series: Materials Science and Engineering 1013 (2021) 011002 doi:10.1088/1757-899X/1013/1/011002

Peer review declaration

All the papers published in this *IOP Conference Series: Materials Science and Engineering* (MSE) have been peer-reviewed through processes administered by the Editors. All Reviews conducted by expert referees to the professional and scientific standards expected of a proceedings journal published by IOP Publishing.

Type of peer review: All the conference papers used a double-blind review process.

Conference submission management system: All the submitted materials, timely received in easy chair portal (https://easychair.org) and a copy to the conference e-mail icoftime2020@presidencyuniversity.in, were considered by the program organizing committee.

Each paper has undergone plagiarism detection. If the content similarity is more than acceptable condition, the author has asked to rewrite the paper in such a way that the submitted article should be in the limit of plagiarism.

The program committee identified the reviewers who received the authors' articles through the conference email along with a review format. The article has submitted to two independent peer reviewers with a scientific degree. After receiving the reviewer comments, the corresponding author has informed to incorporate the changes as per the reviewer suggestion and submit the article with publishing format as prescribed by the IOP.

All articles that have passed the peer review procedure are original works that have not been previously published in other publications in their current or similar form; they have not been reviewed in other editions. Materials of low scientific level are not accepted for publication.

Number of submissions received: The number of applications received is 117, of which 43 were accepted for work.

Number of submissions sent for review: The number of applications sent for review process is 98

Number of submissions accepted: 43 were accepted for the publication in IOP Conference series.

Acceptance Rate (Number of Submissions Accepted / Number of Submissions Received X 100): 36.75

Average number of reviews per paper: The article has submitted to two independent peer reviewers with a scientific degree.

Total number of reviewers involved: The total number of reviewers involved is 58 experts in this particular subject area.

Any additional info on review process: The authors of which had corrected the substantiated comments of the reviewers, were accepted; the articles that received a negative review from the reviewer were excluded from the conference materials.

Contact person for queries: Dr Bhaskar Pal, Professor, Department of Mechanical Engineering, Presidency University, Itgalpur, Rajanakunte, Yelahanka, Bengaluru 560 064 Karnataka, India.

Optimization of surface roughness of Titanium Gr-9Alloy Turning using Taguchi method

Raghavendra M J [1], Dr. Ramachandra C G[2], Dr. T R Srinivas[3], and M Prashanth Pai[4]
1Department of Mechanical Engineering, Srinivas Institute of Technology, Mangaluru, India.
2Department of Mechanical Engineering, Presidency University, Bengaluru, India
3Department of Industrial & Production Engineering, SJCE, Mysuru, India.
4Department of Mechanical Engineering, P.A College of Engineering, Mangaluru, India.

E-mail: raghavendra.ram@sitmng.ac.in

Abstract: - In this work, the performance of chemical vapors deposition coated carbide insert has been described by Taguchi's manner, a whirling of Titanium(gr-9) alloy, for investigation consider a machine hurtful parameters like machining Feed-rate, Depth of cut and Cutting speed by the help of input evaluation of investigation of surface irregularity and cutting tangential force. The experimental test was carried. Based on the optimization techniques like Taguchi's of an orthogonal array system. The optimization of the output response by the analysis of variants (ANOVA) to estimate the nominal surface roughness and maximum tool life. The test result shows that increases in feed-rate and cutting speed correspondingly, the surface irregularity is higher, varies in cutting tangential forces increases cutting speed force also increased, decrease with feed-rate and speed correspondingly and significantly decrease cutting tangential forces.

Keywords: Titanium, Taguchi optimization technique; Turning; Cutting force; Surface roughness;

1. Introduction

Titanium is one of the best super alloys. Which is highly applicable in aerospace and medical field? Due to its chemical position and mechanical behavior. titanium super alloy is a hard material. Which rust-free alloys because it is not chemical reaction with air and water. the iron percentage in the chemical composition is very less. When aluminum and vanadium is added in the alloys some of the mechanical properties may varies. Titanium alloys is having different types of grades system due to its chemical composition and different application. the titanium is very hard materials but the machining will be done with help of coated carbide inserts like PVD A& CVD. While making the components with the help of of titanium alloys for machining the cutting speed should be high from 50m/min to 300m/min and able to achieve an actual good surface quality when the feed is less than the 0.1mm/rev. the cutting force which acts on the tools and also tool wear is very less when matching action is taken place in above condition. exactly concerning multiple efforts and productivities with non-linear relative's midst themselves, are found extremely effective cheers to the progressive procedures and multiplication science. Hence, optimization of high-speed machining replies of Ti–3Al–2.5V has been scrutinized within this training employing a Taguchi's Method is used.

2.Experimental procedure

The below figure illustrates that experimental setup of research work. Titanium grade 9 alloy as a wok specimen of diameter 50 mm and 160 mm length specimen are used in research work for machine-ability testing. The cutting inserts are CVD coated carbide inserts about 80-degree diamond shape and 0.8 mm nose radius having the geometry minus chip breaker. The pullouts are fasenated into the tool holder of right hand of ISO designation PCLNR 2525-M12. the chemical vapour deposition coated carbide inserts having multi-layer of coating materials Al2o3 and TiN having thickness about 8-16 micro meter thickness layer on inserts. The CNMG 120408 CVD inserts which purely made for machining like hard metals, super alloys like Ti and Ni based alloys The titanium grade 9 alloy is a wide variety of application in sports equipment, dental applications, surgical applications, aerospace and aircraft solicitations due to its high forte and high current resistance power the titanium alloys find application in this area.

Work piece material consist of 3% Al, 0.2% Fe, 0.15% O, 0.035% N, 3% V and 92.75% Ti, 0.015% H and 0.86% others. The input parameters are cutting speed, feed rate and depth of cut or selected every level of machining design pf experimentations are implemented on full factorial method of orthogonal array L9 of taguchi's technique. The experimental trial runs or carried out on machining of Ti grade 9 with CVD inserts. With the help of taguchi's L9 the output response of results is optimized by taguchi's technique. the interference of results is drawn from the ANOVA table with the help of Minitab software, the graphical representation of each trial run are validate by analysis of SN ratio. The effect of each machining process parameters at different levels are analysed by SN ratio. To validate the output response was carried out by SN ratio considering smaller the better criteria each factor of response as shown in delta statistics and the rank table. The SN ratio indicates that when increasing cutting speed and feed rate the surface roughness on material also increases when decreases feed rate and cutting speed the surface roughness also reduces. The trial runs are passed on the conventional engine lathe under dry environment machining condition. The lathe runs between 30 rpm to 1250 rpm, the specimen of trial runs between two centres, the centre which consist of three jaw chuck and dead centre. The main key factor for these trial runs was to evaluate the machining factors which special effects on tool wear, surface roughness and tip temperature on machining of Ti-Gr-9 compound.

.

Figure 1. Experimental setup

3. RESULT AND DISCUSSION

The justification of trial outcomes was accompanied on using the inspiration of taguchi's L9 technique with support of Minitab. The illustration of the table explains the restrictions on Titanium grade 9 alloys through coated CVD inserts. The formulated S/N ratio ideals are assimilated machining process of titanium grade-9 with productivity comeback.

3.1 Analysis of Signal to noise ratio

Signal to noise relation is to quantity castoff in a production structure. That associates flat of the favorite signal to the noise ratio. The Taguchi's system of orthogonal array practices is used to quantity the input ideals for the peak explanations and signal to noise ratio is to be grander values. Their determination be three appearances like i) lesser the better, ii) superior the better and iii) minimal are recovering.in this experimentations, force of cutting, finish of surface is recorded by means of quality appearances like smaller the healthier. Where y is the restrained value and n is the amount investigational trial manner in which contribution constraints consuming tall gesture to noise ratio which gives the smallest adjustment and superiority of organization.

Table 1. CVD coated Cutting insert variables in machining

	Input			Output Response		S/N Ratios	
Sl No	ap	Vc	f	Fz	Ra	Fz	Ra
1	60	0.25	0.015	167.45	6.03	-44.48	-44.48
2	60	0.3	0.03	186.18	6.78	-45.40	-45.40
3	60	0.35	0.045	235.45	6.98	-47.44	-47.44
4	70	0.25	0.03	163.28	5.38	-44.26	-44.26
5	70	0.3	0.045	198.77	5.96	-45.97	-45.97
6	70	0.35	0.015	276.77	6.66	-48.84	-48.84
7	80	0.25	0.045	175.45	7.76	-44.88	-44.88
8	80	0.3	0.015	288.98	7.81	-49.22	-49.22
9	80	0.35	0.03	376.65	8.66	-51.52	-51.52

3.2 Analysis of variance (ANOVA)

The discrete process limitations will be strong-minded by using ANOVA Minitab soft wear. ANOVA is practices which inspect the consequence of machining progression constraints and taguchi's system cannot belief for define the consequence of specific method restrictions will be in calculation involvement and roughly of the features impact the proposal classical.

CVD Coated Insert ANOVA Table

Table 2 Represents investigation of Surface Roughness (R_a)

Source	DoF	Sum S	Mean S	Value-F	Value-P	%
ap	2	9.568	4.568	8.36	0.012	17.51
Vc	2	15.487	7.412	11.36	0.019	**28.35**
f	2	28.369	15.697	14.59	0.043	51.93
Error	2	1.2068	563			2.21
Total	8	54.6308				100

Table 3 Represents investigation of cutting force (Fz)

Source	DoF	Sum S	Mean S	Value-F	Value-P	%
ap	2	18.23	9.630	6.652	0.118	**23.59**
Vc	2	21.637	11.576	7.632	0.133	28.00
f	2	33.56	16.489	16.78	0.078	43.42
Error	2	3.856	1.987			4.99
Total	8	77.283				100

From the analysis of the ANOVA table no3 which gives the overall information reading the output response of the machining parameters on the titanium grade-9 alloys. Considering the input variable in the dry machining all the input variables have the 2 degree of freedom each, ANOVA table represents the overall values of the output response. the sum of square values is 9.568 by the depth of cut, 15.487 are the cutting speed and 28.369 from the feed rate, by conserving significantly the feed rate is more concentrated on the surface roughness on the titanium grade-9 alloys, correspond lying the cutting speed and depth of cut are 15.48 and 9.568 respectively. As per the standard if the probability values which are laws less than the 1, when the p-values is less than or equal to one, whatever the model we

selected for machining is good for machining parameters. In the similar way the percentage contribution feed rate is directly influences on the surface roughness on the titanium garde-9 alloys, the cutting speed is indirectly contributed the surface roughness have the % about 28.35.
From the table no 4 give the brief representation on the cutting force on the alloys titanium grade-9, where the percentage contribution on the cutting force 43.42% which is directly influence by the fee rate on the cutting force on the titanium alloys. Whereas other two parameters like depth of cut and cutting speed are sucesely contributed 23.59 and 28.00 respectively, which means the other two parameters are indirectly contributed on the cutting force.in this model the p-values are less the 1

3.3 Mean effect plots
Aimed at the investigation the statistics to review number of critical limitations and thus the most outcome plots on the surface coarseness stayed evaluated. Through an assistance of soft wear like Minitab expressions, the distinction of disconnected rejoinder per three machining bounds individually. indoors the designs Axis of X signposts the ethics of cutting restrictions and Y-axis represents comeback charge. The intrigues are rummage-sale to treasure the plan complaint for surface roughness. Figure 4 shows that the MEP for surface roughness, the diagram displays that intensification the depth of cut pointedly coarseness also upsurges in smaller rate, but once the speed growths their motivation be unremitting growths in surface roughness. Some feed rate intensifications there's decline surface roughness.

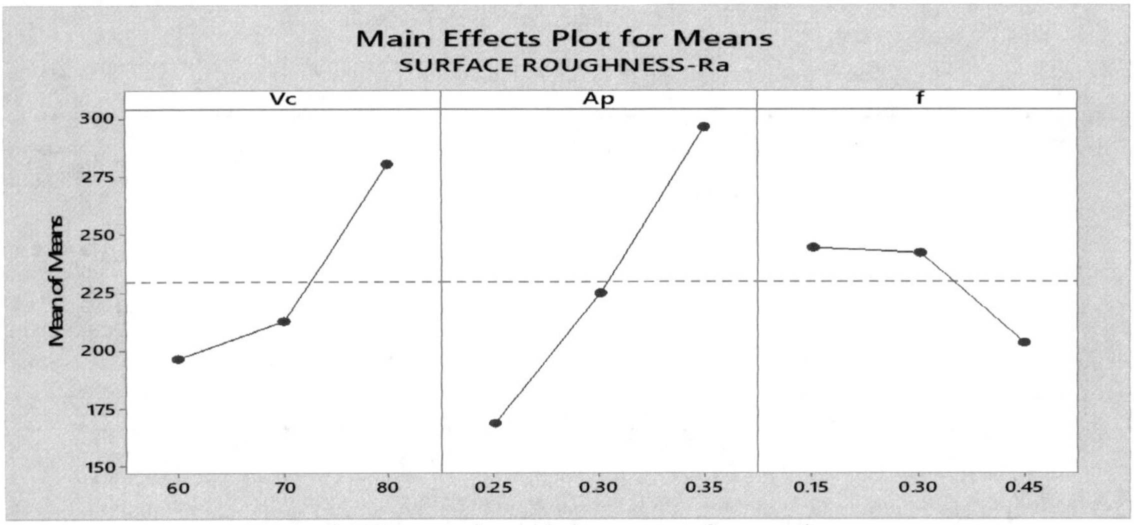

Figure 2 CVD coated carbide insert on surface roughness

Figure 3 expressions that signal to noise Ration on surface roughness diagram confirmations an upsurge the depth of cut suggestively surface irregularity similarly declines, then after nearby willpower be higher speed roughly unceasing dewdrop downhearted in surface roughness. A feed rate rises around is growths a surface roughness. From Figs. 3 and 4 it's pure that diverse depths of cut require unalike speed, feed and spindle speed in command to get a noble exterior roughness.

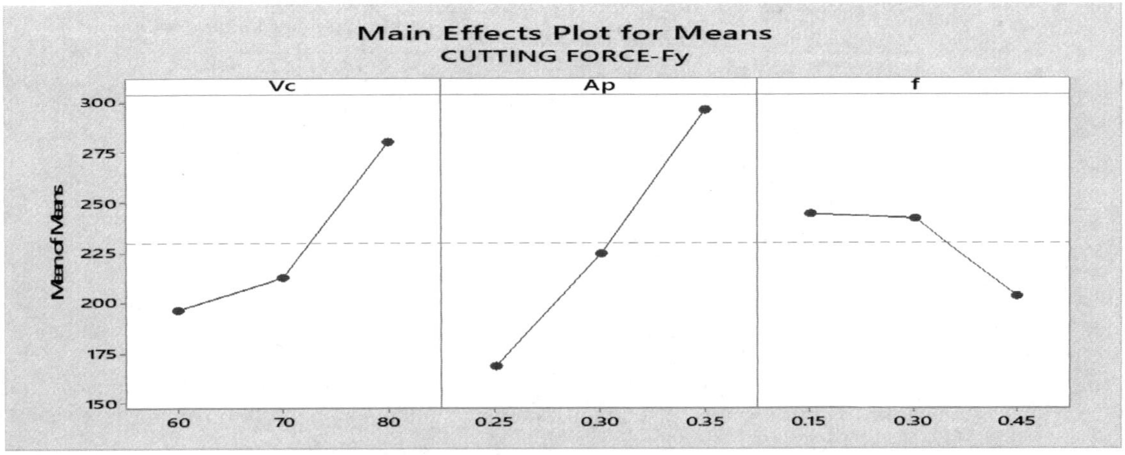

Figure 3 MEP of cutting force on CVD inserts

The optimal principles of cutting speed (60 m/min) besides shaft rapidity (1200 rpm) are repeatedly endangered to high hurry machining thru moral accuracy, lessening in burr formation, stoppage of manufactured up edges and immaterial surface roughness devoid of hurt to insert or specimen.
The table no 4 expressions the answer for signal to noise ratio to dissimilar cutting situation for slighter is better. The reaction table no 4 indications a speed rate is suggestively dictatenmachining concert while associate to agreed dissimilar difficult standards.

Table 4 optimal conditions

Optimal cutting parameters for uncoated inserts			
	a_p	V_c	F
R_a	0.25	60	0.15
F_z	0.30	80	0.15

Table 5 Surface roughness response table of Smaller the best

Level	a_p(depth of cut	Vc(speed)	f(feed rate)
1	-08.326	-7.205	-11.561
2	-11.662	-11.858	-09.65
3	-13.455	-12.562	-12.356
Delta	2.786	1.45	1.42
Rank	3	1	2

The table top no 5 S/N ratio shows that feed rate is expressively underwriting supplementary of turning presentation in the cutting fore disorder and monitored by the Depth of cut and Cutting speed.

Table 6 Response table for S/N ratio for cutting force Smaller the best

Level's	a_p (Depth of cut)	Vc (speed)	F (feed rate)
1	-34.86	-36.25	-52.66
2	-38.16	-35.99	-37.65
3	-36.49	-35.68	-3756
Delta	3.68	3.01	4.54
Rank	2	3	1

The table no 6 demonstrations the reaction of the S/N ratio of various machining complaint for lesser are best. retort table no 7 displays that cutting speed is expressively govern the machining presentation however associate to given unlike complex ethics.
The counter no 6 S/N fraction signposts that feed rate is expressively subsidizing supplementary of the machining presentation within the fore of cutting complaint tailed by the Depth of cut and Cutting speed.

3.4 Approval Test
It is initiate that the optimime illness from the taguchis array L9 gained the principles, constructed the optimal values of conforimation trials and it is accompanied and find that optimum parameters .

Table 7 Trial of Conformation Test

	Input-response			Output-response		Signal to noise Ratios	
Sl No	ap	Vc	f	Ra	Fz	Ra	Fz
1	0.25	60	0.15	1.45	166.45	-07.56	-56.17

4. CONCLUSIONS
In the experinmentataal work, the trail runs are carreied out with FFD of Level-9 orthogonl taguchi's skills was engaged to finding out optimal solution for creadte problem under the machining run at various speed , feed and depth of cut. By using the ANOVA table, feed rade are more infleuence on the cutting forece on the alloys, where as the cutting speed and depth of cut are infleuence comperatelively lesss on cuttin ng force.
Observes that when the cutting speed increases cutting force alos ncreases, while surface roughness, increassing the feed rate and cutting speed the surface roughness also increases correspondingly. While dcreses the feed rate and cutting force surface rouhghness also decresese.
While in machining increases the depth of cut , the cutting tool tip temperature also increases which results in consumption of energy is more on the system, while the tool tip temperature increases the plastic deformation of the alloys which growths the superiority of the surface .
Observed the when the cutting force intensifications with feed rate the deline in surfcae roughness and cutting force , which results need to br remove the materials consuming more energy on the system.

References
[1] H. Chelladurai & V. K. Jain & N. S. Vyas, "Development of a cutting tool condition monitoring system for high speed turning operation by vibration and strain analysis", Int J Adv Manuf Technol (2008) 37:471–485 DOI 10.1007/s00170-007-0986-z
[2] Adam G. Rehorn · Jin Jiang · Peter E. Orban, "State-of-the-art methods and results in tool condition monitoring", Int J Adv Manuf Technol (2005) 26: 693–710 DOI 10.1007/s00170-004-2038-2
[3] N. H. Abu-Zahra and J. H. Lang, "Tool Chatter Monitoring in Turning Operations Using Wavelet Analysis of Ultrasound Waves", Int J Adv Manuf Technol (2002) 20:248–254
[4] Mei Wang & Jie Wang, "CHMM for tool condition monitoring and remaining useful life prediction", Int J Adv Manuf Technol (2012) 59:463–471 DOI 10.1007/s00170-011-3536-7
[5] Alan Hase & Masaki Wada & Toshihiko Koga & Hiroshi Mishina, "The relationship between acoustic emission

signals and cutting phenomena in turning process", Int J Adv Manuf Technol (2014) 70:947–955 DOI 10.1007/s00170-013-5335-9

[6] M. Vasu, and Nayaka H. Shivananda, "Comparative study of coated and uncoated tool inserts with dry machining of EN47 steel using Taguchi L9 optimization techniqueAIP Conference Proceedings 1943, 020063 (2018); doi: 10.1063/1.5029639.

[7] Vasu M, Shivananda Nayaka H, "Investigation of Cutting Force Tool Tip Temperature and Surface Roughness during Dry Machining of Spring Steel, Materials Today: Proceedings 5 (2018) 7141–7149 .

[8] Vasu M and H Shivananda Nayaka, "Investigation of machinability characteristics on EN47 steel for cutting force and tool wear using optimization technique", Mater. Res. Express 5 (2018) 066501 https://doi.org/10.1088/2053-1591/aac67f.

[9] Pradeep V. Badiger, Vijay Desai "Cutting Forces, Surface Roughness and Tool Wear Quality Assessment Using ANN and PSO Approach During Machining of MDN431 with TiN/AlN-Coated Cutting Tool", Arabian Journal for Science and Engineering https://doi.org/10.1007/s13369-019-03783-0

[10] Pradeep V Badiger, Sachin Kumar "Experimental study of machinability factors in machiningof al6061-b4c composites: doe approach", International Journal of Modern Manufacturing TechnologiesISSN 2067–3604, Vol. X, No. 2 / 2018.

[11] Raghavendra M J, Ramachandra C Optimization of surface roughness in turning operation in machining of ti-6al-4v", IOP Conf. Series: Materials Science and Engineering 376 (2018) 012118 doi:10.1088/1757-899X/376/1/012118.

[12] M J Raghavendra, C G Ramachandra, "Influence of machining cutting speed on cutting force, tool flank wear on pvd inserts by turning of ti-6al-4v alloy", AIP Conference Proceedings 2204, 040024 (2020); https://doi.org/10.1063/1.5141597

[13] A. Siddhpura & R. Paurobally, "A review of flank wear prediction methods for tool condition monitoring in a turning process", Int J Adv Manuf Technol (2013) 65:371–393 DOI 10.1007/s00170-012-4177-1

[14] Mozammel Mia1 & Md Awal Khan, "High-pressure coolant on flank and rake surfaces of tool in turning of Ti-6Al-4V: investigations on surface roughness and tool wear", Int J Adv Manuf Technol (2017) 90:1825–1834 DOI 10.1007/s00170-016-9512-5

Optimization of EDM Electrode by Direct Metal Laser Sintering (DMLS) method for SS316L Material

Shashank S [1], Dr. Thomas Pinto[2], Dr. Ramachandra C G[3], Raghavendra M J [4]
1,4 Department of Mechanical Engineering, Srinivas Institute of Technology, Mangaluru, India.
2 Dean, SUCET, Mangaluru, India.
3 Department of Mechanical Engineering, Presidency University, Bengaluru, India.

E-mail: shashank.s844@gmail.com

Abstract: EDM is one of the methods used in processing. It is a process in which the eroded material starts from the piece with the aid of a series of sparks to obtain the desired shape. It is useful for processing hard metals. The work done by several researchers helps in producing EDM electrodes through SLS, as a different method to conventional processing techniques, and to assess the action of electrodes produced with the technique. Due to this reason, experimental research is carried by SLS out on electrodes made from brass, copper alloy and nickel alloy of bronze and nickel, and steel alloy disintegration of a solid substance. EDM recital is evaluated in terms of material deposition (Vw), which represents the volume of extract from the piece per unit of time (cubic meters per second) and correlative magnitude wear (θ), which shows the relationship between the rate of electrode wear and the rate of material withdrawn. A comparative study between the electrodes produced by SLS and the electrodes made in a conventional manner with respect to SS316L material is made.
.**Keywords**: Selective laser sintering, Direct metal laser sintering, EDM Electrode, SS316L;

1. Introduction

Direct laser sintering of metal is a technique just like selective laser sintering, but is only used in the production of a 3D metal prototype. RP and RM technology is a new technology in which 3D pieces are completed directly from CAD data without usage of any customary tools. Complex parts that cannot be built with the customary process can be created with much shorter delivery time. The different RP techniques, like SLS (Selective Laser Sintering), SLA (stereolithography), 3D preprint, FDM and DMLS capable of producing prototypes of different materials. DMLS implies laser sintering using a metal fine particle, and that blende parts are put together by a direct route during the construction proceeding [5]. Since high thermal energy is required to stimulate sanitation and concretion of material during sintering with an abundantly short period of time, complication such as deformation shall be generated [1]. To refrain such issue affecting the precision of the piece create, some specification must be considered, such as laser power, scanning speed, beam displacement, raster model and work piece inclination. It is noted that Direct Metal Laser Sintering technology not only helps to achieve high precise built parts, but also helps to easily upgrade the blooming and study of micro-reactors [3]. In addition, the machine permits parts of the building manipulate incompatible materials like chrome-cobalt nickel alloy, stainless steel and titanium. It requires management of process specifications to create a better mechanism by sintering the powder to achieve an agile structure in the parts. In Fig 1. Below is a simplified diagram of the Direct Metal Laser Sintering. For establishment of any part, the machine enacts the ensuing steps. The building and distributor platform drops one layer thick so that recovery blade can move without collisions. When the corrector is positioned in the correct position, the distributor scaffold rises to provide some amount of powder for the next layer. Thus, the recoil shifts from sideling position, through this process the metallic dust extends from the dispenser to the construction field and the remaining metal dust falls into a manifold [2]. Thereafter the scanning tip moves the laser beam through a 2D section and turns it on and off precisely throughout the exposure of the assigned areas. The blotting up of energy by the metallic dust will initiate the polymerization and the sintering of the already ossify sector. This process continues layer by layer up to the time of completion of all parts of a craft. In this way, within a few hours, the machine can manufacture 3D segments with great entanglement and precision. Moreover, throughout the construction process, the mould parts can achieve roughly final patrimony, but depends on the requisition of the metal being machined, can be completed with a post-treatment treatment, such as hardening or surface treatment.

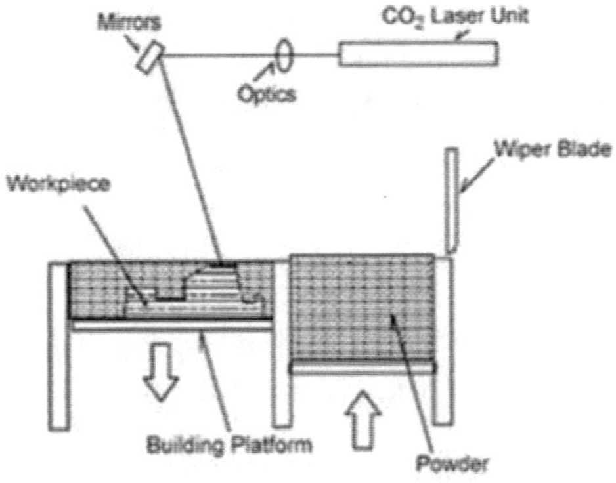

Fig 1: Direct Metal Laser Sintering process

2. Literature review

To understand previous works reported in the literatures, so as to adopt established methods to set a foundation for the present work. To have an insight into the production of EDM electrodes via SLS as an alternative technique to conventional machining processes[1]. To develop intricate shaped EDM tool by SLS technique which is not possible by conventional machining process.

3. Experimental details

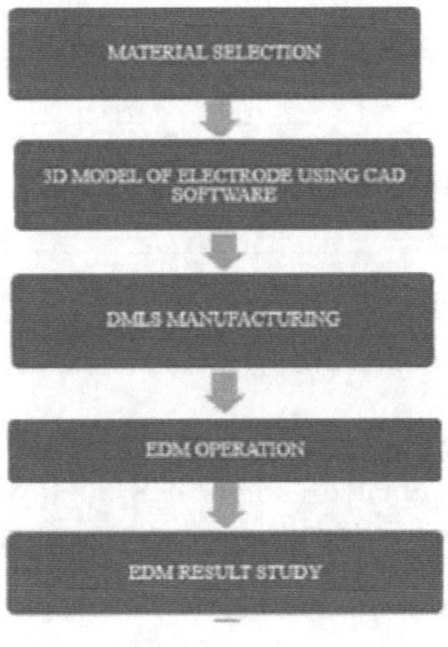

3.1 MATERIAL SELECTION

The material selected for manufacturing of EDM electrode is Stainless steel 316L.Copper is the most widely used EDM electrode material due to the attractive properties like high patrimony and high mechanical strength. The major problem is that copper is difficult to laser sinter due to its high reflectivity. Copper is having high reflectivity of 0.99 to laser light of wavelength 10.6 μm. Because of which copper alloys or materials that give similar results copper is used as non-convectional electrode

3.2 Direct Metal Laser Sintering Machine (DMLS) Work

Laser sintering was done in Intech DMLS, Bangalore, with the EOSINT M 280 Xtended DMLS machine. The machine includes an ytterbium fiber laser power (Yb) of 400 W and a radius diameter of 0.4 mm. In the construction chamber, there is a construction platform through which a dust delivery platform and a recovery blade are used to spray dust in to the construction platform. The 3D CAD model of a 10 mm diameter and 25 mm length sample for the DMLS process is modeled using the "Magics RP" software. The CAD model in STL format will be divided by the "EOS RP Tools" software. The thickness of the layer must be kept at a constant of 0.06 mm. The segmented data is uprooting to the proceeding computer of the DMLS machine where the laser path is initiated with the PSW software. A base plate of the required size of 250 mm X 250 mm X 22 mm in steel is mounted on the building platform with four screws and the base plate is leveled with a 10 μm resolution comparator with respect to the coated sheet

3.3 *Electrical Discharge Machining (EDM) Work*
The machine used for the work is V3525 a 500 X 300 Series Spark Erosion type die sinking EDM machine.

3.4 *Experimental Details*
The hypothesis test was conducted on the V3525 precision electric discharge machine, which consists of a working plane, a servo control system and a dielectric power supply system. For the experimentation the current position was varied from 5 A to 20 A, 9 timed power areas, 9 pulse duration position and a 50-75-micron spark. Grade 30 EDM oil is used as a dielectric fluid and tests have been accomplished for a peculiar set of input variables. The number of hypothesis tests and input levels are decided based on the experiment design and input variables and their stages presented in Table 1. MRR and TWR are calculated using precision digital balancing of 1 mg and the processing time with the digital clock with precision 1 thousandth of a second and the roughness of the surface is measured using a machine that measures the roughness of the surface for a sampling length of 5 mm.

Table 1: Input parameter levels

Input Parameters	Current (amp)	Pulse on Time (μs)	Pulse off Time (μs)
Symbol	A	B	C
Level 1	10	5	4
Level 2	15	6	5
Level 3	20	7	6

4. Results and Discussions
OPTIMIZATION USING TAGUCHI METHOD

The S / N ratios for MRR are premeditated as indicated in equation.1. The Taguchi method is pre-owned for the investigation of the result for the response of the processing parameter for wider and better criteria.

The bigger, the better: S / N ratio = -10log (1 / n.Σn i = 1 1 / yi). . . . . (1)

When the S / N ratios can be purposive from the discern values, yi, which constitute the empirically discern value of dissection i and n = 1 is the frequent number of each appraisal in L-9 OA that is performed.

The scrutiny of the variance for the consideration is determined by the Minitab software and the results are shown below and the table of answers is shown below

Table 2: Signal to noise ratio table

SI NO.	I (Amps)	Pulse ON	Pulse OFF	MRR(g/min)	EWR(g/min) × 〖10〗 ^(−3)	Ra(μm)	S/N Ratio	
1	10	5	4	0.0268	0.514	10.47	-15.6382	3.6702
2	10	6	5	0.0285	0.596	11.09	-16.14	3.9048
3	10	7	6	0.0281	0.519	12.3	-17.0346	4.2823
4	15	5	4	0.0424	0.785	13.4	-17.7858	4.7424
5	15	6	5	0.0455	0.839	14.49	-18.4647	5.1248
6	15	7	6	0.0397	1.071	15.04	-18.7957	5.3835
7	20	5	4	0.0825	1.449	15.7	-19.1837	5.7438
8	20	6	5	0.0811	1.698	16.12	-19.4241	5.9663
9	20	7	6	0.0803	2.565	15.82	-19.3257	6.1551

. *Table 3: Analysis of Variance of MRR*

Source	DF	Seq SS	Contribution	Adj SS	Adj MS	F-value	P-value
Regression	3	15.0605	93.85%	15.0605	5.0202	25.45	0.002
Amps	1	13.8647	86.40%	13.8647	13.8647	70.28	0
Pulse on	1	1.0824	6.74%	1.0824	1.0824	5.49	0.066
Pulse off	1	0.1134	0.71%	0.1134	0.1134	0.58	0.482
Error	5	0.9864	6.15%	0.9864	0.1973		

Table 4: Response for Mean

Level	Amps	Pulse on	Pulse off
1	31.12	26.85	27.09
2	27.45	26.52	26.75
3	21.8	26.99	26.51
Delta	9.32	0.46	0.58
Rank	1	3	2

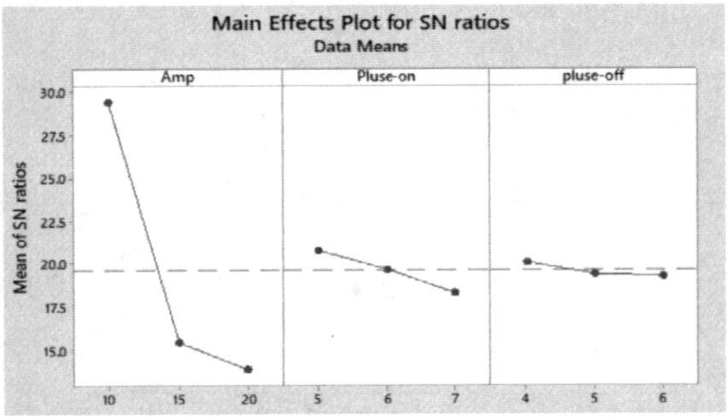

Fig 2: Main effect plot for signal to noise ratio for MRR

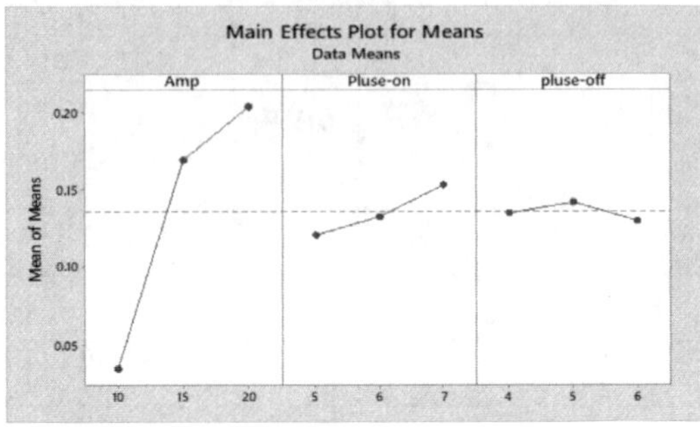

Fig 3: Main effect plot for means for MRR

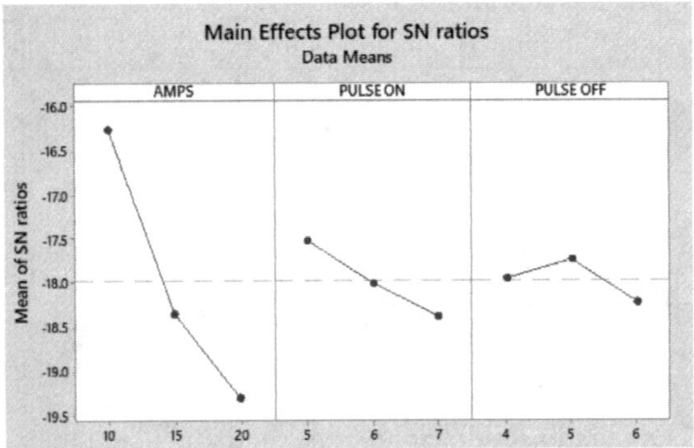

Fig 4: Main effect plot for signal to noise ratio for EWR

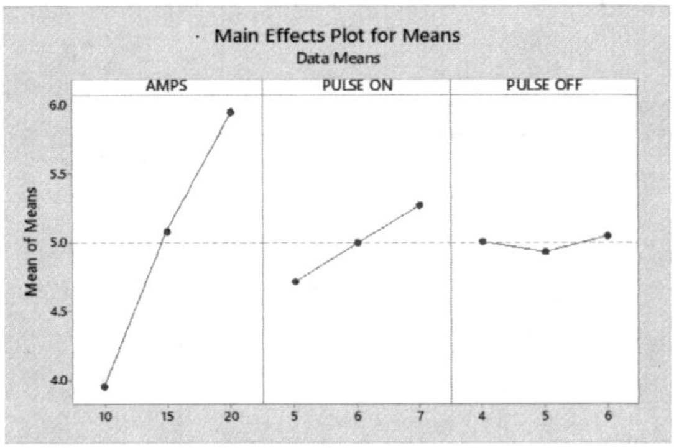

Fig 5: Main effect plot for means for EWR

5. Conclusions

It is determined that current is the vital factor between the different process parameters involved in the EDM process and the contribution is high in both cases. Based on hypothesis testing, the following conjecture are furnished: With the increase in current, the value of MRR and EWR is cautiously expanding. To have a favourable condition, i.e. an increase in MRR and a decrease in EWR, it is necessary to select a nominal current and also the roughness of the surface increases.With magnifying current and to maintain a moderate current that will provide a good finish superficial together with a moderate TWR and MRR

References

[1] Fred L. Amorim , Armin Lohrengel , Norbert Müller , Guenter Schäfer ,Tiago Czelusniak (2013) , "Performance of sinking EDM electrodes made by selective laser sintering technique", International Journal of Advanced Manufacturing Technology , 65:1423–1428 00170-01242670.

[2] Fred L. Amoriml, Walter L. Weingaertner, (2007) "The behavior of graphite and copper electrode on die sinking EDM of AISI P20 tool steel", journal f the Brazilian society of Mechanical science and engineering, volume 29, issue 4, ISSN 1806-3691

[3] AzharEqubal , Nitesh Kumar Dixit, Anoop kumarsood,(2013) "Rapid prototyping application in manufacturing of EDM electrode", International Journal of Scientific & Engineering Research, volume 4, issue8, ISSN 2229-5518.

[4] NorlianaMohd Abbas , Darius G. Solomon, Md. FuadBahari (2006) "A review on current research trends in electrical discharge machining (EDM)" ,Volume 47, Issues 7–8, Pages 1214–1228.

[5] P. Raja, G. Ramamoorthi, Dr. G. Nallakumarasamy (2017) "Experimental Analysis and Optimization of Electrical Discharge Machining on Titanium GR-2 with FEA", International Journal of Emerging Technologies in Engineering Research (IJETER), Volume 5, Issue 7.

[6] [6]Cristian Pisarciuc (2014), "a review of unconventional procedures for Manufacturing tool electrodes used in edm", Volume 15, issue 3(43), pp- 205-208

[7] Brent E. Stucker, Walter L. Bradley, Somchin (Jiab) Norasetthekul , Philip T. Eubank (1995) "The Production of Electrical Discharge Machining Electrodes Using SLS: Preliminary Results".
[8] Chandramouli S, Shrinivas Balraj U and Eswaraiah K (2014) "Optimization of Electrical Discharge Machining Process Parameters Using Taguchi Method", International Journal of Advanced Mechanical Engineering, Volume 4, issue 4, pp. 425-434.
[9] Dr. M.Indira Rani , Ketan (2014) "Optimization of Various Machining Parameters of Electrical Discharge Machining (EDM) Process on AISI D2 Tool Steel Using Hybrid Optimization Method", International Journal of Application or Innovation in Engineering & Management, Volume 3, Issue 9, ISSN 2319 – 4847.
[10] L.A. Dobrzański, M. Musztyfaga, A. Drygała (2010)"Selective laser sintering method of manufacturing front electrode of silicon solar cell", journal of achievement in material and manufacturing engineering ,volume 42 ,issue 1-2,pp 1-9.
[11] Eubank, P.T., and B. Bozkrut,(1992) "Recent Developments in Understanding the Fundamentals of Spark Erosion for Composite Materials, in Machining of Composite Materials , pp 159-166.
[12] Tobin, J.R., B. Badrinarayan, lW. Barlow, J.J. Beaman, and D.L. Bourell, "Indirect Metal Composite Part Manufacture Using the SLS Process", Solid freeform Fabrication Symposium, pp 303-307.

IOP Conf. Series: Materials Science and Engineering 1013 (2021) 012003 doi:10.1088/1757-899X/1013/1/012003

PRODUCTION OF CUTTING TOOL VIBRATION IN TURNING USING ARTIFICIAL NEURAL NETWORK

Murugappan Elango [1], Adithyan Annamalai[2], Praveen Raju A[3],
[1,2,3] Department of Mechanical Engineering, Thiagarajar College of Engineering, Madurai, Tamil Nadu, India 625015.

E-mail: adithyanannamalai@gmail.com

Abstract: - The cutting parameters like cutting speed, feed, and depth of cut can affect the tool life. Also, the vibration of the tool can affect the tool life. The aim of this study is to predict the cutting tool vibration to enhance the tool life. In this paper, vibration of tool is studied with HSS tool (high speed steel) while turning mild steel. The tool vibration is measured using X Viber vibration meter. The experimental data are tabulated and imported to artificial neural network (ANN). A multilayer perceptron model is trained with back-propagation algorithm using the experimental data. The rake angle, cutting speed, and feed are considered as input parameters for training the ANN. The trained ANN is used to predict tool vibration for different conditions. The predicted values are compared for the specific range to justify the tool change.

Keywords: Machine learning, vibration, neural network

1. Introduction

Tool life is the volume of material removed by a machining tool during its total usage span. The volume may be in cubic mm or cubic cm measured at a standard cutting speed. The most dominant factors affecting tool life are work material, tool material, cutting speed and tool shape chosen for a particular machining operation. But the other machining parameters have also been found to have a secondary effect on the tool life. During the process of shearing the metal, high temperatures are generated at the tool cutting edge region, and these temperatures are found to have a controlling influence on the rate of wear of the cutting tool and on the friction between the chip and tool. Tooling cost is an important factor which should be reduced in order to keep the manufacturing cost to bare minimum. Tool failure can be identified by observing the following factors namely, higher power consumption, poor surface finish, dimensional inaccuracy of the machined material, appearance of a burnishing band on machined surface, tool or work piece vibration, among others. The vibration of machine tools during cutting, affects certain parameters such as surface finish of the workpiece, life of the cutting tool, positioning accuracy of slides, life of machine tool parts particularly transmission elements, removal rate of the material, appearance of chatter marks and errors in dimension and form. The vibration occurring in a machine tool may be classified as follows

I. Free vibration induced by a sudden variation in the cutting force due to change in feed or depth of cut (resulting from variation in machining allowance, blow holes, hard spot or entry of tool into the work piece.

II. Self-excited vibration or chatter due to the interaction of cutting process and machine tool dynamics.

III.Stick-slip motion of the sideways at low sliding speeds.

IV. Forced vibrations due to the periodic forces caused by cutting process, unbalance or defects in transmission system elements like, motor rotors, gears, pulleys, bearings, work pieces and rotating fixtures, etc.

2. Literature review

Anurag Naidu and Sanjay Asati [1], reported a remedial idea to increase production rate by improving the tool life. They developed a system to measure the accuracy and surface finish of the work part. The system could also measure the vibrations generated due to variations in cutting parameters using

the intelligent systems. Balla Srinivasa Prasad, et.al [2], reported that the tool vibration and wear condition had a direct effect on the tool failure and decreased the life of tool. So, they found a relationship between the tool vibrations and wear properties of tool by using laser Doppler vibrometer. The tool was provided with carbide coatings so that it could improve its wear resistant properties. Chang Tseng [3], explained that during milling process, the work piece and surface roughness was affected by the tool vibration and its wearing properties. The vibration was measured by four accelerometers joined to the tool and the parameter used to derive a relationship with surface roughness was the MSE of the signal. The tool was provided with carbide coatings to improve its wear resistant properties. Future research would be directed towards active control for decreasing vibration in the process. Filippov, et.al [4], had stated that the tool life was affected by the tool vibration. So, they measured the tool vibration using acoustic emission monitoring system to choose the perfect tool material according to the machining materials. Gouarir, et.al [5], had investigated that the tool wear could account directly for tool life failure. So, they devised a tool wear prediction system, which used a force sensor and machine learning (ML)algorithm to monitor the progression of the tool flank wear. More specifically, a convolutional Neural Network (CNN) was used as a method to predict tool wear. By controlling the tool wear, tool life was improved and this could further be enhanced by selecting proper cutting tool parameters. Krishna kumara, et.al [6], reported that the tool wear affects the tool life. During the machining process, vibration signals were monitored continuously using an accelerometer. The features from the signal were extracted and a set of prominent features were selected using Dimensionality Reduction Technique. Experimental observations indicated that the tool life would decrease with cutting speed, feed rate, depth of cut and material hardness. Using the right tool in the right cutting condition in order to reduce tool wear and fracture, increase machining accuracy and to increase tool life and productivity was very significantly evident
Marcelo Mendes de Aguiar, et.al [7], delineated a problem that the surface finish of the workpiece was affected by the tool vibration and tool wear properties. So, they intended to enhance the surface finish of the work piece by reducing the tool vibration. Milling operations were performed on AISI materials with different hardnesses and the cutting parameters were measured using FRF setup (Frequency Response Function). The workpiece surface finish was determined by cutting parameters namely tool vibration, depth of cut, feed, etc. Nexhat Qehaja, et.al [8], have promulgated that the tool life was dependent on the cutting parameters. So, they constructed a prediction model for a first order tool during turning of hardened 42CrMo4 steel at different levels of hardnesses. Experimental observations indicated that tool life will decrease with feed rate, depth of cut cutting speed, and material hardness. Outeiro, et.al [9], outlined that the tool vibration enhances with increase in temperature. So, they measured the residual stress produced by the tool vibration using neutron diffraction. Controlling the tool vibration disguisedly meant controlling the temperature of the machining process. Sam Paul, et.al [10], had asserted that the presence tool vibrations lead to a poor surface finish, which ultimately resulted in a damaged cutting tool and the production of an irritable and unacceptable noise. Experimental observations indicated that the tool life would decrease with cutting speed, feed rate, depth of cut and material hardness. Siamak Ghorbani, et.al [11], had deduced a relationship between tool vibration and tool life. Static and dynamic characteristics of cutting tools at different machining conditions were analyzed for different cutting tools. Changes in cutting parameters would drastically improve the cutting tool life. Venkata Rao, et.al [12], proved that tool wear affects the tool life. The tool wear and work piece vibrations were recorded by means of a laser vibrometer and surface roughness of work piece was also monitored. The tool will be changed when the particular wear condition occurs, which was found by the neural network software.

Xu Chuangwen, et.al [13], had identified that tool wear and milling depth significantly affected the cutting force and vibration. So, they recorded the cutting force and vibration acceleration signals with which the wear depth on the flank face of the tool was measured. This study observed a tool wear morphology using a scanning electron microscope and investigated the distributions of surface elements on the damaged tools by energy spectrum analysis. The tool and workpiece surfaces were affected by the cutting parameters and this could be considerably avoided by providing proper coating on the tool surface.

The above studies help to conclude that artificial neural network had been used to predict the tool life or tool wear only with respect to machining parameters. The perdition of tool vibration with respect to tool geometry and machining parameter is not addressed. Thus, this paper considers the different

IOP Conf. Series: Materials Science and Engineering 1013 (2021) 012003 doi:10.1088/1757-899X/1013/1/012003

combination of cutting speed, feed, and depth of cut, rake angle and its tool vibration. The aim of this study is to predict the optimal cutting tool vibration at different cutting parameter combinations to enhance the tool life.

The rest of the paper is structured as follows: Section 2 describes the experimental setup, experimental procedure and our observation. Section 3 gives a description of the ANN's structure, results obtained from ANN and the actual vibration values. Section 4 discusses about the validation of the ANN. Finally, the conclusion is discussed in the last section.

3. Experimental details

The single point ¼ inch cutting tool material is grinded to various rake angles such as 7°, 8°, 9°, 10° and 11°s. The other tool nomenclatures are side rake angle 8, end clearance angle 6°, side clearance angle 6°, end cutting edge angle 6°, side cutting edge angle 8° and nose radius 1 mm. The machining parameter cutting speed 30 m/min and feed 0.275 mm is selected. The mild steel round rod specimen having 25 mm diameter workpiece is loaded on to the lathe machine. The transducer of X-viber is placed above the tool post and tool vibrations were analyzed for all the above rake angles. Then the experimental data is collected and imported to the neural network to predict the tool vibration error for turning operation with mild steel as the work piece and high-speed steel as the tool material. Then, the results are analyzed and concluded. Figure 1 and 2 shows the tool before machining and after machining and Figure 3 depicts the tools with various rake angles.

4. Results

The tool vibrations are analysed for tools with different rake angles and at constant a depth of 0.25mm. Tables 1 to 5 gives us the vibration values for tools with rake angles 7, 8, 9, 10 and 11 respectively.

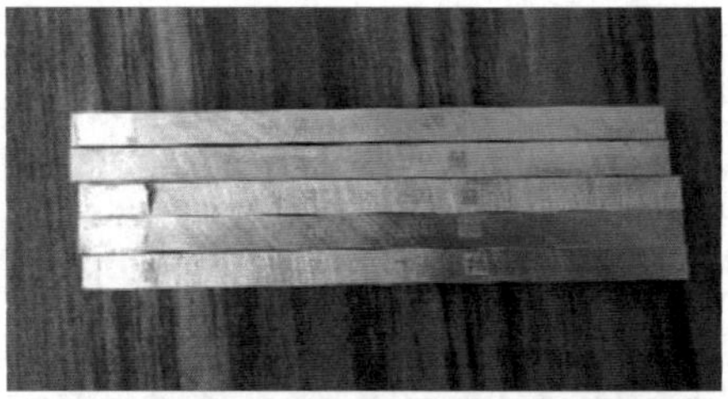

Figure.1. Before machining

Figure.2. After machining

Figure. 3. Various rake angles after machining

4.1 Artificial neural network

Artificial neural networks are also known as neural nets, artificial neural system, parallel distributed processing system and connectionist system. The network is commonly referred to as a directed graph having a set of nodes (vertices) and a set of connections (edges/links/arcs) between nodes.

Each node contributes functions like simple computation and each connection transfers information or signal between nodes. The connections between two nodes are labeled with a number called as connection strength or weight. The weight represents, to what extent the signal is to be amplified or diminished by the connection. A network with single or fewer numbers of nodes cannot solve all the problems. Henceforth, networks which are constructed with large numbers of nodes are used to solve complex problems. Some types of the neural networks are fully connected networks, layered networks, acyclic networks, feed forward networks and modular networks. Figure 4 and 5 shows the architecture of an artificial neural network.

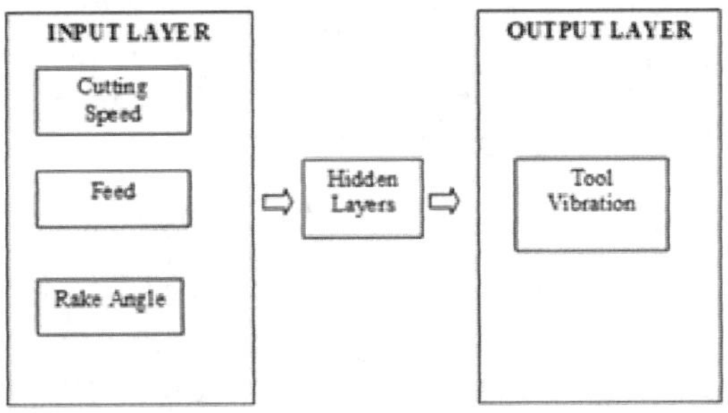

Figure.4. Construction of artificial neural network

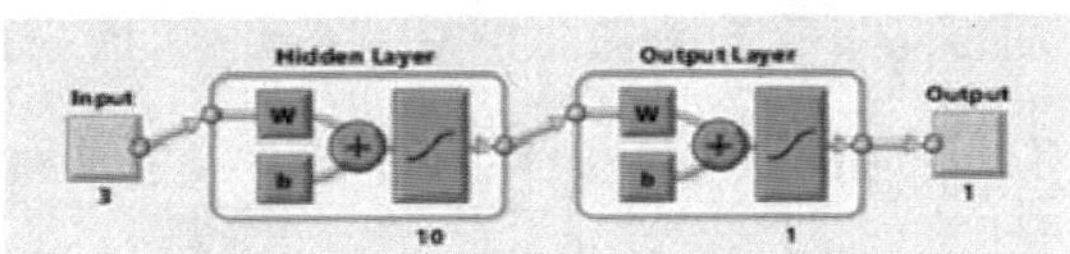

Figure.5. Artificial neural network

4.2 Actual tool vibration

The X-viber apparatus was used to measure the exact tool vibration of the cutting tools with different rake angles, cutting speeds and feeds. Table 1 gives the values of actual tool vibration values at rake angle 7° at different cutting speeds and feeds

Table 1. Tool vibration values at 7° rake angle

Rake Angle (°)	Cutting Speed (m/min)	Feed (mm/rev)	Tool Vibration (mm/sec)
7	15.7	0.336	1.701
	23.5	0.319	1.233
	37.6	0.290	1.345
	56.5	0.227	1.684

Table 2 gives the values of actual tool vibration values at rake angle 8° at different cutting speeds and feeds.

Table 2. Tool vibration values at 8° rake angle

Rake Angle (°)	Cutting Speed (m/min)	Feed (mm/rev)	Tool Vibration (mm/sec)
8	15.7	0.336	1.047
	23.5	0.319	1.585
	37.6	0.290	1.706
	56.5	0.227	1.606

Table 3 gives the values of actual tool vibration values at rake angle 9° at different cutting speeds and feeds.

Table 3. Tool vibration values at 9° rake angle

Rake Angle (°)	Cutting Speed (m/min)	Feed (mm/rev)	Tool Vibration (mm/sec)
9	15.7	0.336	1.114
	23.5	0.319	1.631
	37.6	0.290	1.788
	56.5	0.227	1.746

Table 4 gives the values of actual tool vibration values at rake angle 10° at different cutting speeds and feeds.

Table 4. Tool vibration values at 10° rake angle

Rake Angle (°)	Cutting Speed (m/min)	Feed (mm/rev)	Tool Vibration (mm/sec)
10	15.7	0.336	1.023
	23.5	0.319	1.232
	37.6	0.290	1.562
	56.5	0.227	1.458

Table 5 gives the values of actual tool vibration values at rake angle 11° at different cutting speeds and feeds.

Table 5. Tool vibration values at 11° rake angle

Rake Angle (°)	Cutting Speed (m/min)	Feed (mm/rev)	Tool Vibration (mm/sec)
11	15.7	0.336	1.154
	23.5	0.319	1.212
	37.6	0.290	1.380
	56.5	0.227	1.554

4.3 Predicted tool vibration

The neural network model was used to predict the tool vibration values. Table 6 delineates the values of tool vibration at different rake angles, speeds and feeds. It can be observed that the predicted values closely match with the actual vibration values in most cases

Table 6. Predicted Values For 7° 8° 9° 10° and 11° rake angles

Rake Angle (°)		7	8	9	10	11
Cutting Speed (m/min)	Feed (mm/rev)	Predicted Tool Vibration (mm/sec)				
15.7	0.336	1.5063	1.0481	1.3038	1.0653	0.9308
23.5	0.319	1.0721	1.5655	1.6304	1.2276	1.2021
37.6	0.290	1.3477	1.7064	1.7863	1.5619	1.3803
56.5	0.227	1.6842	1.6073	1.7458	1.4558	1.5523

5. Discussion

A feed-forward four layered neural networks is constructed. The network is constructed with four layers namely input layer, output layer and two hidden layers. Each hidden layer consists of 5 nodes. The ANN with one hidden layer gives errors significantly high. Hence two-layer network is considered. The input neurons are cutting speed, feed and rake angle and output neuron is tool vibration. Neurons in the hidden layers are determined by examining different neural networks. MATLAB software is used for training on this network and the ANN is trained using front propagation algorithm.
The neural network is trained with 14 samples, validated for 3 samples and tested for 3 samples. Percentage of error between experimental values and predicted values for the tool vibration is calculated. Mean error percentage is found as 3.454% for amplitude of tool vibration. From the results, it can be inferred that the proposed network model is capable of predicting the values of tool vibration. It also helps in the selection of cutting parameters to achieve good surface quality and also predicts the correct time for tool change.

5.1 Validation of ANN for prediction of tool vibration

The validation of ANN is done with rake angle 6° which is not a trained value. Table 7 shows the experimental and predicted values for rake angle 6°. Here the ANN gives a predicted tool vibration for particular rake angle (6°), cutting
speed and feed. The experimental value for this rake angle is close to the predicted values and which is shown in Figure 6. The mean error percentage was calculated.

Table 7. Predicted Value For 6°

Rake Angle (°)	Cutting Speed (m/min)	Feed (mm/rev)	Tool Vibration (mm/sec)	Predicted Tool Vibration (mm/sec)	Error
6	15.7	0.33	1.201	1.108	0.077
	23.5	0.31	1.633	1.741	0.066
	37.6	0.29	1.541	1.640	0.062
	56.5	0.22	1.871	1.781	0.048
				Error	0.064

IOP Conf. Series: Materials Science and Engineering 1013 (2021) 012003 doi:10.1088/1757-899X/1013/1/012003

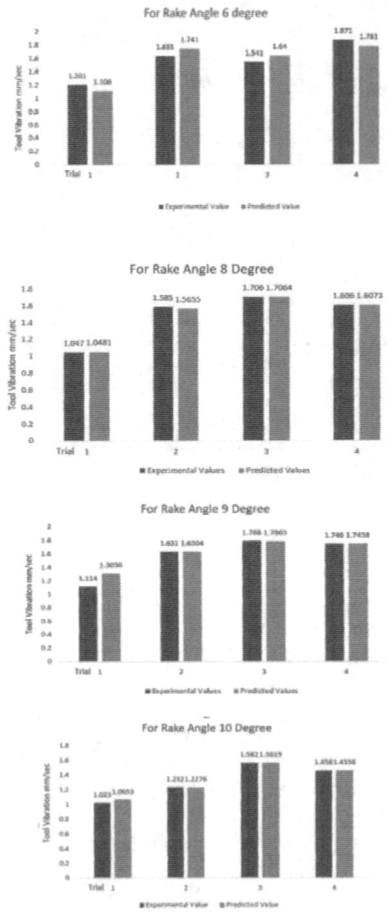

Figure 6. ANN Model for Tool Vibration

6. Conclusion

In this work, five experiments were conducted according to a proposed experiment with four levels of cutting parameters such as cutting speed, feed etc., A neural network (3-5-5-1) was used to learn the collected experimental data. The ANN is trained with 14 examples, validated with 3 examples and then tested with 3 examples

The trained ANN is used to predict the tool vibration. It was found that there is an agreement between experimental data and predicted values for tool vibration (3.454% of error). In that case, it is possible to change the cutting tool at the correct time in order to get a good quality of products. This neural network could help us to predict tool vibration and reduce surface roughness of the work piece.

References

[1] Anurag Naidu, and Sanjay Asati, Cutting Tool Condition Monitoring and Validation Using ANN- A Comprehensive Research, IJIRSET (2014) 80-98.

[2] Balla Srinivasa Prasad and Prakash Babu, Correlation between vibration amplitude and tool wear in turning: Numerical and experimental analysis, Engineering Science and Technology (2017)197-211.

[3] Chang Tseng, Influence of Tool Vibration on Machining Performance in of Ti-6Al-4V Alloy in Milling, Mechanical Engineering – Manufacturing, (2016) 38-78.

[4] Filippov, Nikonova, Rubtsova, Dmitrieva, and Tarasov Vibration and acoustic emission monitoring the stability of peakless tool turning: Experiment and modeling, Materials Process, (2017) 214-224.

[5] Gouarir, Martinez arellano, ratchev, Tool Wear prediction System Based on Machine Learning Techniques and Force Analysis, Advanced Manufacturing, (2018) 501-504.
[6] Krishnakumar, Rameshkumar, Ramachandran, Tool Wear Condition Prediction Using Vibration Signals in High Speed Machining (HSM) of Titanium (Ti-6Al-4V) Alloy, Machine Tools and Metal-Cutting Tools and Instruments, (2015) 270-275.
[7] Marcelo Mendes de Aguiar, Anselmo Eduardo Diniz, and Robson Pederiva, Correlating surface roughness tool wear and tool vibration in the milling process of hardened steel using long slender tools, Machine Tools & Manufacture, (2013) 1-10.
[8] Nexhat Qehaja, and AzemKyçyku, Tool Life Modeling Based On Cutting Parameters And Work Material Hardness In Turning Process, Machines Technologies Materials, (2017) 67- 79.
[9] Outeiro, Costes, and Kornmeier, Cyclic Variation of Residual Stress Induced by Tool Vibration in Machining Operations, Modeling of Machining Operations, (2013) 493-497.
[10] Sam Paul, Kolhar Sudeep Sunil, and Varadarajan, The Effect of Cutting Parameters on Tool Vibration During Magnetorheological Fluid Controlled Turning Bar, International Journal of Acoustics and Vibration, (2014) Vol. 22, No. 1, 27-33.
[11] Siamak Ghorbani, Vladimir Kopilov, Nikolay Polushin, and Vladimir Rogov, Experimental and analytical research on relationship between tool life and vibration in cutting process, Machine Tools and Metal-Cutting Tools and Instruments, (2018) 844-862.
[12] Venkata Rao Murthy and Mohan Rao, "Prediction of cutting tool wear, surface roughness and vibration of work piece in boring of AISI 316 steel with artificial neural network", Measurement, (2014) 63-70.
[13] Xu Chuangwen, Dou Jianming, Chai Yuzhen, Li Huaiyuan, Shi Zhicheng and Xu Jing, The relationships between cutting parameters, tool wear, cutting force and vibration, Advances in Mechanical Engineering, (2017) 1–14.

DESIGN AND FABRICATION OF A SCALED DOWN SELF LOAD PNEUMATIC MODERN TRAILER

Praveena B A[1*], Balachandra P Shetty[2], Arvind R[3], Gagan Deep J G[3], Gaurav Satish Kumar[3], Pavan Kalyan K[3], Vikram Kedambadi Vasu[1], Lokesh N[1]

1*, Assistant Professor, Department of Mechanical Engineering, Nitte Meenakshi Institute of Technology, Bangalore: 560064, Karnataka, India
2, Professor, Department of Mechanical Engineering, Nitte Meenakshi Institute of Technology, Bangalore, Karnataka, India
3, UG Student, Department of Mechanical Engineering, Nitte Meenakshi Institute of Technology, Bangalore, Karnataka, India

*E-mail: praveena404@gmail.com

Abstract: - This research work has mainly concentrated on the difficulty of the movement of 3 axes on a modern trailer and hence a suitable arrangement has been designed such that the materials can be unloaded from the trailer in three axes without application of any impact force. Our survey is in the regard of several automobile industries that revealed some facts which are most likely difficult methods that were adopted in weighing load especially in difficulties of unloading the materials from the trailer. It is difficult to unload the materials in small, compact streets and roads; therefore, in our project these issues are rectified to unload the trailer in all three sides with ease. The control valve of the trailer is activated manually and hence the compressed air passes to the pneumatic cylinder through the valve. The ram of the pneumatic cylinder acts as a lifting arm of the trailer cabin. This project also focuses on application of self-load weighing such that the material loading and unloading weight display is viewed to cross check while unloading. The automobile engine drive is coupled to the compressor engine, so that it stores the compressed air when the vehicle is running. This compressed air is used to activate the pneumatic cylinder when the valve is activated. Also, rotation of the trailer is done with the help of gear drive system operated by the motor. The whole mechanism is controlled by the operator or driver which is near the driving seat that helps them to operate safely and easily.

Keywords: 3 axes, compressed air, pneumatic cylinder, self-load weighing, gear drive system.

1. Introduction

Transport firms want to transfer goods quickly without being burdened by bureaucracy, paperwork and regular stoppages for authorities to test enforcement. In general, transport companies are also involved in fair competition, which involves applying and prosecuting regulatory standards fairly for all players in the industry Public authorities need to understand how to enhance the overall goals they want to accomplish, e.g. reduced maintenance costs for road infrastructure, reduced impact on the environment from road transport, increased road safety, more effective use of road transport [1]. Automation can be accomplished by computers, hydraulics, pneumatics, robots, electronics, etc. Public authorities need to understand how to enhance the overall goals they want to accomplish, e.g. reduced maintenance costs for road infrastructure, reduced impact on the environment from road transport, and increased road safety, more effective use of road transport. Automation can be accomplished by computers, hydraulics, pneumatics, robots, electronics [2] etc.

Figure 1. Pit mounted weighbridge

One aspect was surprising, that it was very difficult to unload material on complicated locations such as angular sides and directional sides (left and right) of dumper. Dumper truck remained suitable in situations such as these. It ate extra. Trailers are still the most common cause of construction site and factory accidents. A typical dump truck is fitted with an open box that is pneumatically controlled dead hinged at the rear [3]. The front of which can be raised to allow the contents to be deposited on the ground at the side of delivery behind the truck. Nowadays it is possible to rotate dumpers with swivel skips to sideways (3 directional trailer) which become common, particularly for working in narrow sites such as road works but even this technology is inadequate to fulfill our full unloading requirement. This technology is concerned only with solving the unloading problem on truck directional sides.

2. Literature review

Picking the "right" or "ideal" plan building has consistently been a significant worry of the creators. Decrease of frame weight is the most basic assignment of numerous sorts of boats in basic plan. Be that as it may, the estimation strategies accessible for this job truly limit the capacity of originators to make ideal plans of boat structures [4]. A cylindrical ram housing is pivotably connected to the chassis to be lifted in a self-stored chassis leveling hydraulic hammer. A cylindrical hydraulically actuated ram is carried telescopically inside the cylindrical ram frame. Within the ram is housed an electric motor. The engine drives a hydraulic pump which is also in the ram [5].

The pump output passes through a control valve to pipe the hydraulic fluid selectively between a ram extension chamber and a ram retracting chamber to selectively expand or retract the telescoping screw [6]. A three-point tractor hitch allowing an operator to connect or remove an implement without leaving the seat of the tractor. The upper hitch arm contains a linear actuator with hand grip and a manually controlled switch controlling upper arm extension and contraction. A latching device and a hand-controlled latch release are placed on the upper arm to allow connection and release of the upper arm to the upper connecting pin of a hitch mast of the implement [7].

3. Objectives

- To reduce the effort of man power and ensure sufficient accuracy level.
- To operate the trailers, lift at 360 degrees rotation so that the loading and unloading is done with ease at any direction and weight the load at shortest time.
- To incorporate load sensor for intimating the load or weight carried on to the trailer and to improve the loading and unloading system in a trailer lift of the truck or any load carrying trolleys.
- To ensure the availability of equipment at Low cost and to ensure less space is used to weigh large load.

4. Methodology

1) The first step was the selection of the project. After weighing in various factors, such as feasibility, cost, usefulness and the challenges involved, we decided on this project. This was done after extensive discussions with our guide.
2) After the selection of the project, we selected the various general elements necessary for the project. This involved the purchase of a welder and the receipt of a quotation on the steel pipes used.
3) Next, we designed an engine chassis and its accessories, receiver and transmitter unit. We based this on a guided planning structure and made the necessary changes.
4) The electrical components were selected to best suit the needs of the project. The circuit diagram was used as the basis for the selection of components.
5) The next step involved two almost simultaneous steps –the production of the model according to the design and the assembly of the electrical circuits, the calibration of the sensors and the microcontroller.
6) After the production of the model and the completion of the electrical circuits, the

mechanical and electronic components are integrated into one unit for further testing.

7) We test the integrated unit for accuracy and robustness.
8) If any changes or additions are required, we shall implement them and re-test them until satisfactory results are achieved.

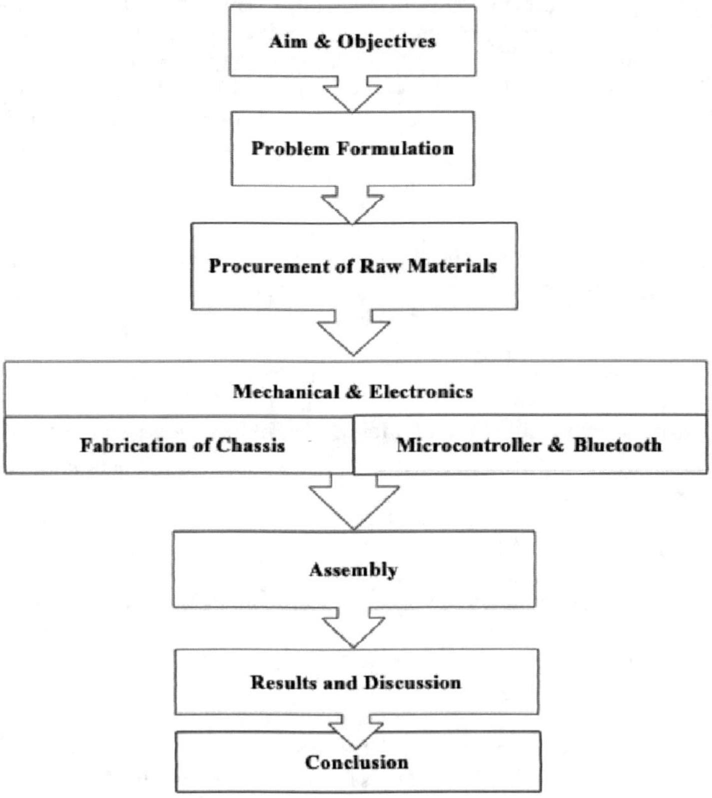

Figure 2. Flow Chart

5. Design calculations

5.1 Pneumatic Cylinder

Piston rod design:

Stacking because of gaseous tension,

Piston distance across, (d)	=	45 MM
Acting Weight (p)	=	8 KGF/CM^2
Materials used for the bar	=	c 45
Yield pressure (σ_y)	=	38 KGF/MM^2
Expecting FOS	=	2
Force following up on your bar (P)	=	Weight * area
	=	$p * \{\frac{\pi d^2}{4}\}$
	=	$8 * \{\frac{(\pi * 4.5^2)}{4}\}$
p	=	127.23 KGF

Stress design (σ_y) = $\sigma y / FOS$

= 38 / 2 = 19 KGF/MM²

= $px\{\frac{\pi d^2}{4}\}$

∴ d = $\sqrt{4p / \pi [\sigma y]}$

= $\sqrt{4 * 127.23 / \{\pi * 19\}}$

= $\sqrt{8.52}$ = 2.91 MM

∴ Least measurement of the bar required for stacking = 2.91 MM

We're accepting the distance across of the bar = 15 MM

Structure of the thickness of the chamber:

Material utilized = Cast iron

Accepting within measurement of the chamber = 45 MM

Extreme pressure on the tension = 255 N/MM² = 2550 GF/MM²

Working Stress = Extreme pressure on the tension/ FOS

Assuming, FOS = 4

Working stress (ft) = $\frac{2500}{4}$ = 625 KGF/CM²

According to 'LAMES EQUATION'

Minimum cylinder thickness (t) = $ri\{\sqrt{\frac{ft+p}{ft-p}} - 1\}$

Where,

R_i = Internal radius of the cylinder in CM.

F_t = Working stress (KGF/CM²)

p = Pressure to working KGF/CM²

∴ Substituting the values

t = $2.0 * \{\sqrt{\frac{625+8}{625-8}} - 1\}$

t = 0.322 CM = 3.2 MM

The thickness of the cylinder is assumed = 3.5 MM

The inside diameter of the barrel = 45 MM

External diameter of the barrel = 45 + 2

= 45 + (2 * 3.5) = 52 MM

Design of Piston rod:

Cylinder Rod Diameter:

Cylinder Rod Force (P) = Weight x area = $px\{\frac{\pi d^2}{4}\}$

= 8 x (π / 4) x (4.5) ²

= 127.23 KGF

Also, Cylinder Rod Force (P) = $(\pi/4) (d_p)^2 \times f_t$

P = $(\pi/4) (d_p)^2$ 625

127.23 = $(\pi/4) (d_p)^2$ 625

∴ d_p^2 = 75.39 * (4/π) * 1/625)

= 0.509

d_p = 0.509 CM

= 5.09 MM

By standardization, d_p = 20 MM

Piston rod length:

Approach stroke = 180 MM

Length of strings = 2 x 15 =30MM

Additional length due to front cover = 14 MM

Extra length of the head of accommodation = 25 MM

Total length of piston rod = 180 + 30 + 14 + 25

= 249 MM

Standardization of piston rod length = 250 MM

1. **BALL BEARING DESIGN**

No. 5202 Bearing

Outer Bearing Diameter (D)	=	40 MM
Thickness of bearings (B)	=	15 MM
Internal mounting diameter (d)	=	20 MM
r_1	=	Corner radii of the pole and lodging
r_1	=	1 (From DDB)
Maximum velocity	=	14,500 RPM (From DDB)
Diameter of the mean(d_m)	=	$\frac{D+d}{2}$
	=	$\frac{40+20}{2}$
d_m	=	30 MM

FACTOR WAHL STRESS

$$Ks = \frac{4c-1}{4C-4} + \frac{0.65}{C}$$

$$= \frac{(4*2.3)-1}{(4*2.3)-4} + \frac{0.65}{2.3}$$

$$Ks = 1.85$$

5.2 Specification

1. Double acting pneumatic cylinder

Technical Data

Stroke length	:	Length of the cylinder stroke180 mm = 0.18 m
Quantity	:	1
Seals	:	Nitride (Buna-N) Elastomer
End cones	:	Cast iron
Piston	:	EN – 8
Media	:	Air
Temperature	:	0-60 ° C
Pressure Range	:	8 N/m²

2. Solenoid Valve

Technical data

Max pressure range	:	0-12 x 10^5 N/m²
Quantity	:	3

3. Flow control Valve

Technical Data

Port size	:	0.653 x 10^{-2} m
Pressure	:	0-6 x 10^5 N/m²
Media	:	Air
Quantity	:	1

4. Connectors

Technical data

Max working pressure	:	12 x 10^5 N/m²
Temperature	:	0-120 ° C
Fluid media	:	Air
Material	:	Brass

5. Hoses

Technical date

Max pressure	:	12 x 10^5 N/m²
Outer diameter	:	8 mm = 8 x 10^{-3}m

Inner diameter : 4 mm = 4 x 10^{-3}m

6. Modeling

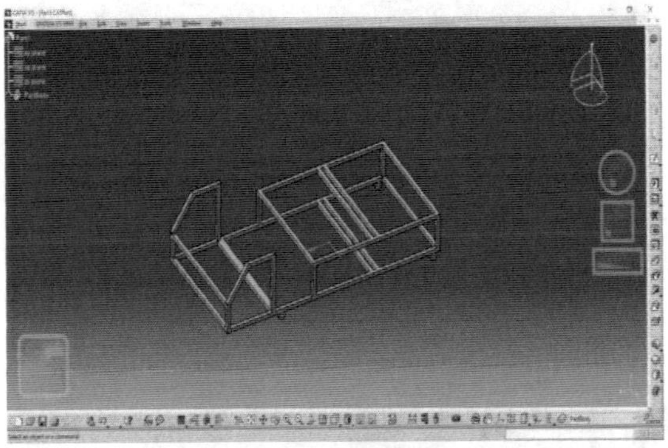

Figure 3. Frame Body

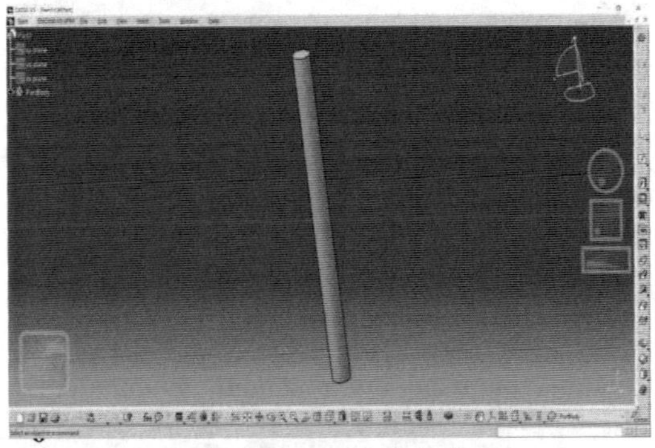

Figure 4. Spindle

Figure 5. Tray

Figure 6. Shaft

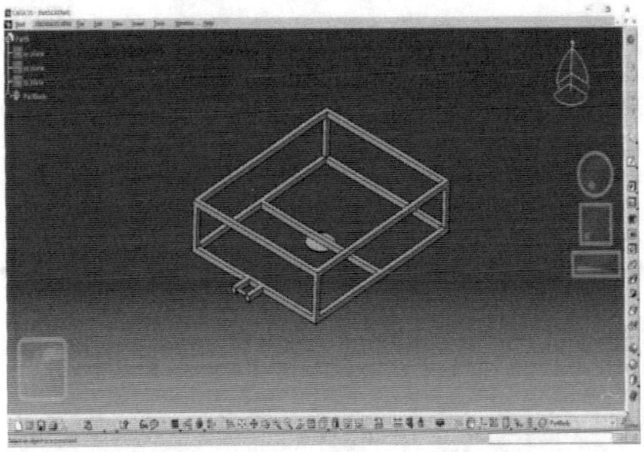

Figure 5. Middle Frame

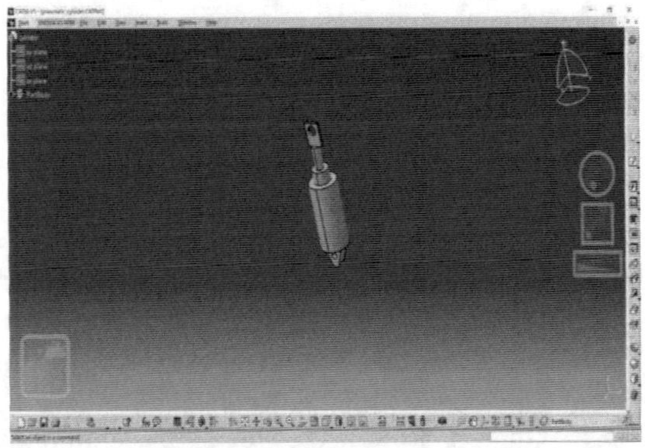

Figure 6. Pneumatic Cylinder

IOP Conf. Series: Materials Science and Engineering 1013 (2021) 012004 doi:10.1088/1757-899X/1013/1/012004

Figure 7. Wheel

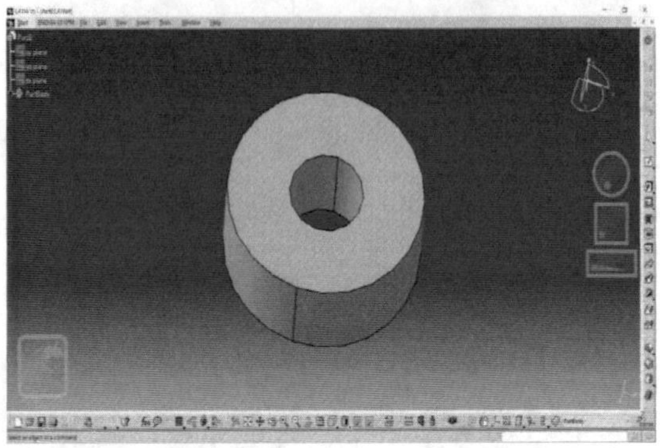

Figure 8. Collar

Figure 9. Full Assembly

7. Result and Discussions

A suitable arrangement has been designed such that the vehicles as LCD weighing display to show the exact materials loaded on the vehicle to cross check the loading materials. The trailer has multi direction axis can be easily rotated and tilted according to the comfort of the vehicle. There is the pneumatic system to tilt the trailer up to 30-40 degrees and rotation of the trailer with the help of gear drive system operated by motor. The whole mechanism is controlled by the operator or driver near the driving seat helps them to operate safely and easily.

8. Conclusion

This project work has given us an excellent opportunity and experience to use our manufacturing know-how. When doing this project research, we acquired a lot of practical information about preparation, ordering, assembly and machining. We believe the research of the project is a successful option for bridging the gaps between the institutions and business. We are happy to have effectively accomplished the research with the time restricted.

The "self-load weighing pneumatic modern trailer" is working as per required conditions. We are able to understand the difficulties in maintaining the tolerances and quality. We have used our ability and skill to make maximum use of available facilities. Thus, we have developed a "self-load weighing pneumatic modern trailer" which helps to know how to achieve low cost automation. This system's operating method is very basic and therefore every unqualified individual should be able to control the device. The prototype can be modified and developed according to the applications through the use of additional processes.

9. Scope of Future work

1. Mounting of shaft to the trailer base.
2. Motor with the electrical components to be connected to the spur gear.
3. Pneumatic cylinder to be connected to the air compressor.
4. Switches and levers are to be connected.
5. Load sensor to be placed.
6. Chassis for the electrical component to be placed are to be welded.

References

[1] Zbigniew Sekulski, Least-weight topology and size optimization of high-speed vehicle-passenger catamaran structure by genetic algorithm, Marine Structures, Elsevier, pages 691–711, 2009.

[2] Paterik Frank J JR, Self-stowing vehicle leveling hydraulic jack, US4235542A published on 1980, 11-25.

[3] Frank David Alexander "Hydraulic three-point tractor hitch" published on 2002-11-12.

[4] C H Neeraja, C.R.Sireesha and D. Jawaharlal, "Structural Analysis of Two Wheeler Suspension Frame", International Journal of Engineering Research &Technology, Vol.1 Issue 6, August 2012.

[5] Teo Han Fui, Roslan Abd. Rahman, Faculty of Mechanical Engineering, University Technology Malaysia, "Statics and Dynamics Structural Analysis of a 4.5 Ton Truck Chassis" December, 2007.

[6] S. Agostoni, A. Barbera, E. Leo, M. Pezzola, M. Vanali, "investigation on motor vehicle structural vibrations Caused by engine unbalances", Proceedings of the SEM Annual Conference June 1-4, 2009 Albuquerque New Mexico USA.

[7] Zbigniew Sekulski, "Least-weight topology and size optimization of high-speed vehicle-passenger catamaran structure by genetic algorithm.", Marine Structures, Elsevier, pages 691–711, 2009.

Optimization of WEDM Parameters for Machining of AZ31B Mg Alloy Using Taguchi Method

Karthik R[1], Viswanathan R[2], Balaji J[3], Sivashankar N[4], Arivazhagan R[5]

1, 4 and 5 Department of Mechanical Engineering, Kongunadu College of Engineering and Technology, Trichy, Tamilnadu
2 Department of Mechanical Engineering, Pallavi Engineering College, Hyderabad, India
3 Department of Mechanical Engineering, Jayalakshmi Institute of Technology, Dharmapuri, Tamilnadu
E-mail: vissona2005@yahoo.co.in

Abstract :- This work involves optimization of machining parameters of wire cut Electrical Discharge Machining (WEDM) used for maching of AZ31B magnesium (Mg) alloy with molybednum wire to attain the lowest surface roughness from the process. Mg alloys such as AZ31B are light-weight materials with high potential for industrial applications and used in the manufacture of airframes, aircraft engines, helicopter components, automotive parts and light trucks . L9 orthagonal array with three input parameters such as Voltage (V), Pulse on time (μs) and pulse off time (μs) at three levels are used to conduct the experiments. Taguchi method is utilized to study the effect of these parameters on the surface roughness (SR). From the investigation it was found that pulse on time exert the major influnce on surface roughness while Voltage and pulse off-time have lesser influnces on the process to reduce the surface roughness.

Keywords: WEDM, Molybednum Wire, Mg alloy, Surface Roughness, Taguchi Technique and optimization.

1. Introduction

Light alloys are getting great attention in mechanical industries such as automobile, aerospace and ship building constructions. Due to more industrial needs, Mg alloys are widely used because it is a light-weight metallic structural material with high specific strength and stiffness, good heat conductivity, good machinability and cast ability [1, 2]. Among the Mg alloys, AZ31B wrought Mg alloy has gained significant attention because of its lightness, good stiffness, heat conductivity, excellent weldability/machinability and higher corrosion/oxidation resistance [3]. WEDM is one of the unconventional machining processes, used to erode material from the workpiece to fabricate parts with intricate shape and profiles. In this process, the material is removed from the workpiece by a series of discrete spark discharges between a work and tool electrode immersed in a liquid dielectric medium. The spark developed by discharge of electricity melts and vaporizes minute amount of the work material. The removed material is ejected and flushed away by the dielectric [4]. After analysing the effects of the three input parameters used for the WEDM process that include pulse on time, pulse current and servo feed rate on the performance characteristics, namely; material removal rate (MRR), kerfwidth and surface roughness during machining of AZ91 Mg alloy, optimisation is carried out. From the experimental outcome the WEDM of Mg alloy it was observed that the material removal rate and the kerfwidth are higher than hard materials. It was noted during the processing of Mg alloys that at high cutting rate, the surface roughness is appropriate and also the surface topography and microstructural changes are studied using atomic force microscope, scanning electron microscope and EDX analysis [5]. The machining parameters of WEDM on Mg metal matrix composite reinforced with Reduced Graphene Oxide(r- GO) are optimised to reach maximum MRR with a lowest surface roughness (Ra) value. The optimal combination was found for a maximum MRR (18.38mm3/min) and minimum Ra (3.29 μm by using grey relation analysis [6]. To provoke the effect of machining operations over kerf and cutting speed in wire cut electrical discharge to machining Mg AZ91 the alloy material was analysed by MOORA method. The optimized results for kerf width (KW) and cutting speed (CS) were obtained with Taguchi L27 design matrix [7]. The turning work on AZ91Mg alloy with cemented carbide tools was carried out with an objective to minimize cutting zone temperature. Based on experimental outcome the regression model is developed by RSM technique. The model showed that cutting speed is the significant factor affecting cutting zone temperature. Apart from regression equation ANOVA was also carried out. It is found

that parameter combination of cutting speed at 40m/min, feed at 0.2mm/rev and depth of cut at 0.50mm for minimum cutting zone temperature using RSM [8]. The wire-cut EDM process using Cemented Tungsten Carbide tool has been investigated for obtaining better fine surface finish and faster material removal and the machining conditions influencing the process to optimise the parameters using Grey-Taguchi method. The pulse on time is found to be the dominant parameter for both, surface roughness and MRR. For the multiple performance characteristics the Grey-Taguchi method is recommended [9, 10].
In this study, using Taquchi technique, the process parameters of WEDM are optimised. ANOVA is also carried out to find the influence of each input parameter on the machining process. As the experimental design is taguchi based, the statistical analysis software, MINITAB 19 is used to perform ANOVA analysis.

2. Experimental method

2.1. Work material

In this work, AZ31B Mg alloy is used as test material and Table 1 provides the chemical properties of the base plate. The as-received Mg alloy optical and SEM microstructures are presented in Fig.1. The work material used in the experimental machining tests is a rectangular plate with a length, breadth and thickness of 10cm x 10cm x 12mm respectively.

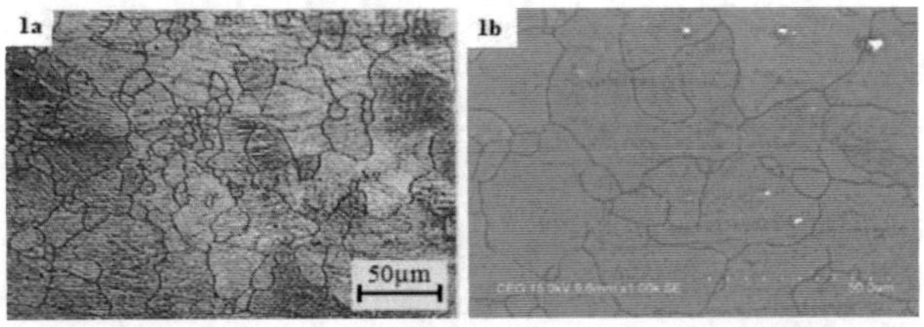

Figure 1. Microstructure of as received Mg alloy 1a) Optical Image 1b) SEM image

The machining was performed in WEDM, under different conditions of machining parameters using Concord DK-7732VC WEDM with a input capacity of 3 Phase, 415 Volts, 50 Hz and machine load of 1.5 KVA and the experimental setup is represented in Fig. 2. It is a 4-axis controlled CNC machine and the guide moves in X-Y plane. The deionised water is used as a dielectric fluid with the maximum tank capacity of 55 litres and a molybdenum wire of 0.2mm diameter is fed through the work piece for machining. The pictorial representation of machined out samples were presented in Figure 3.

Table 1. Chemical composition of work material (AZ31 Mg alloy) in wt%

Element	Al	Zn	Mn	Si	Cu	Fe	Mg
%	3.42	1.15	0.38	0.025	0.025	0.0041	Bal

The three levels of inputs values are specified for three process parameters in Table 2. The parameter values as recommended by the machine tool manufacturer were used in the experiments. Taguchi's L9 orthogonal array (OA) design was used to conduct the experiments to obtain smoothest surface by optimising EDM parameters and examining its effectiveness through confirmation test. After machining, the surface was measured using Surftest 211, Minutolo surface roughness tester for three times repeatedly to take the average.

Figure 2. WEDM of Mg alloy

Table 2. Selected parameters with their levels

Parameter	Units	Level		
		1	2	3
Voltage	Volts	15	25	35
Pulse on time	µs	10	20	30
Pulse off time	µs	12	22	32

3. Result and Discussion

From the experimental observations, the lower Ra and Higher Ra values measured and also the average Ra values calculated are shown in Table 3. ANOVA was carried out for the obtained results.

Table 3. L9 OA with Average Ra

Experiment	Voltage V	Pulse on time (µs)	Pulse off time (µs)	Lower Ra (µm)	Higher Ra (µm)	Average Ra (µm)
1	15	10	12	3.76	3.88	3.820
2	15	20	22	5.22	5.74	5.480
3	15	30	32	7.18	7.65	7.390
4	25	20	32	6.35	6.90	6.625

5	25	30	12	8.50	9.10	8.800
6	25	10	22	4.22	4.65	4.435
7	35	30	22	9.21	9.60	9.405
8	35	10	32	3.81	4.22	4.015
9	35	20	12	5.50	5.80	5.600

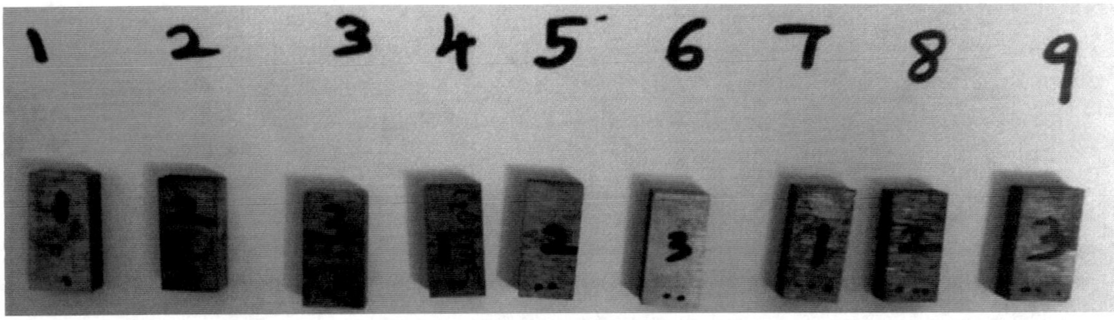

Figure 3. Specimens machined using different parameter combinations

Table 4 shows the response of means. From the responses, it could be seen that the most important parameter is pulse on-time, followed by Voltage. Among the three parameters, Pulse off time is less significant compared to others. Main effects plot of mean SR is presented in Fig. 4

Table 4. Response Table for Means

Level	Voltage V	Pulse on time (μs)	Pulse off time (μs)
1	5.563	4.090	6.073
2	6.620	5.902	6.440
3	6.340	8.532	6.010
Delta	1.057	4.442	0.430
Rank	2	1	3

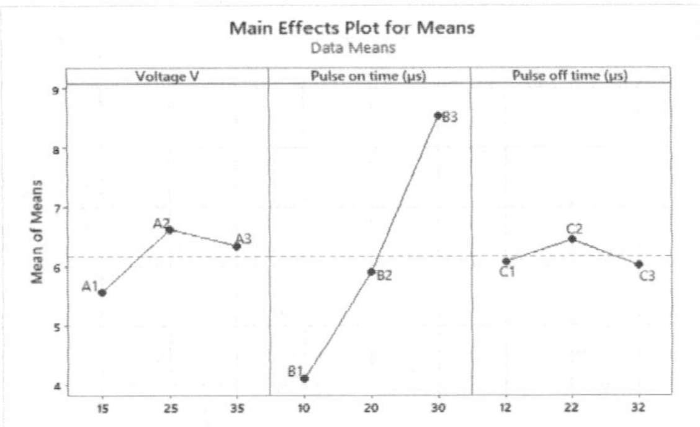

Figure 4. Main Effects plot of SR

From the Table 5 the optimum parameter combination for minimum SR is obtained which includes voltage at 15V, Pulse on time at 10 µs and Pulse off time at 12 µs. The Fig 5 shows the main effects plot for average surface roughness Ra. It reveals that Surface roughness increases as Pulse on time increases and pulse off time does not impact much on the surface roughness. The influence of each machining parameter is calculated and the difference in the S/N ratio is taken as the indicator to rank the dominance of the machining parameter on the performance parameter i.e surface roughness. In current work, smaller the better type S/N ratio is used to obtain optimal parameter combination for Ra and it is expressed in equation -1.

$$\frac{S}{N} = -10\log\frac{1}{n}\left(\sum_{i=1}^{n} y_i^2\right) \text{...........Eqn. (1)}$$

Table 5. Response Table for SN Ratios

Level	Voltage V	Pulse on time (µs)	Pulse off time (µs)
1	-14.60	-12.22	-15.16
2	-16.08	-15.39	-15.73
3	-15.50	-18.58	-15.29
Delta	1.49	6.36	0.56
Rank	2	1	3

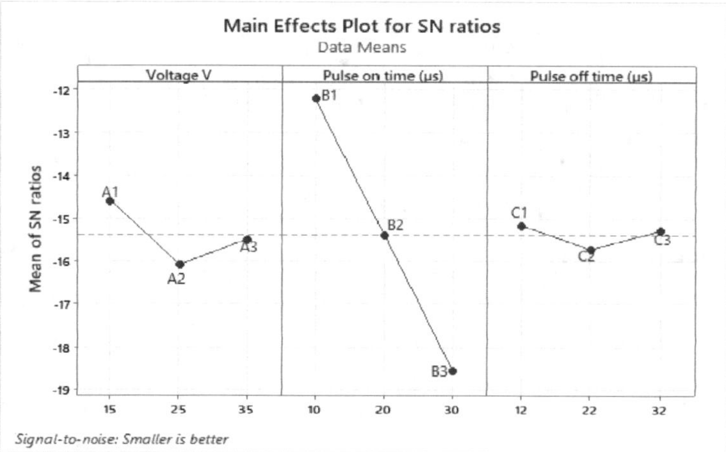

Figure 5. Main Effects Plot for SN ratios

According to the analysis of variance of S/N ratio, the Pulse on time is dominant parameter on response parameter and the results are presented in the Table 6. From the Table-6 it is seen that Pulse on time (µs) is the control factor that has the major contribution of 92.611% and is followed by Voltage (5.145%) whereas the contribution of Pulse off time is very small.

Table 6. Analysis of Variance for S/N ratios of SR

Source	DF	Seq SS	Adj SS	Adj MS	F	P	% of contribution
Voltage V	2	3.3695	3.3695	1.6848	3.56	0.219	5.145
Pulse on time (µs)	2	60.6547	60.6547	30.3273	64.01	0.015	92.611

Pulse off time (µs)	2	0.5223	0.5223	0.2611	0.55	0.645	0.797
Residual Error	2	0.9476	0.9476	0.4738			1.447
Total	8	65.4941					100
S=0.6883			R-Sq=98.55%			R-Sq(adj)= 94.21%	

Regression Analysis

The Table 7 gives the regression coefficients for Ra

Average Ra (µm) = 0.83 + 0.0388 Voltage V + 0.2221 Pulse on time (µs) - 0.0032 Pulse off time (µs)

Coefficients

Table 7. Regression coefficient of surface Roughness (Ra)

Term	Coef	SE Coef	T-Value	P-Value
Constant	0.83	1.16	0.72	0.505
Voltage (V)	0.0388	0.0292	1.33	0.240
Pulse on time (µs)	0.2221	0.0292	7.61	0.001
Pulse off time (µs)	-0.0032	0.0292	-0.11	0.918

From the Fig 6, it was found that the normal probability plot of the residuals follows a straight line and all the points are distributed evenly.

In Residual versus fitted value graph shows the random pattern of the residuals on both sides of zero. Histogram shows the pattern of distribution of residuals for the average surface roughness Ra.

Residual versus observation order shows all residuals in the order that the data were collected. Residual is higher for the observation order seven and for majority of the observation orders, the residual values are situated above and also near to the zero line.

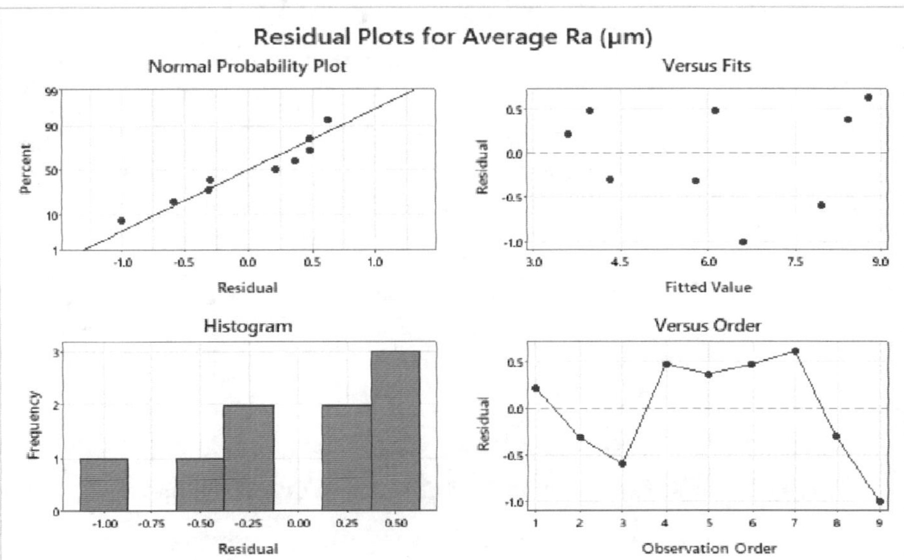

Figure 6. Residual plots for Average Ra from regression analysis

Figure 7 shows the graph for experimental and predicted values of Average surface roughness (Ra) plotted together. It is clear that the predicted values are in good agreement with the experiment values.

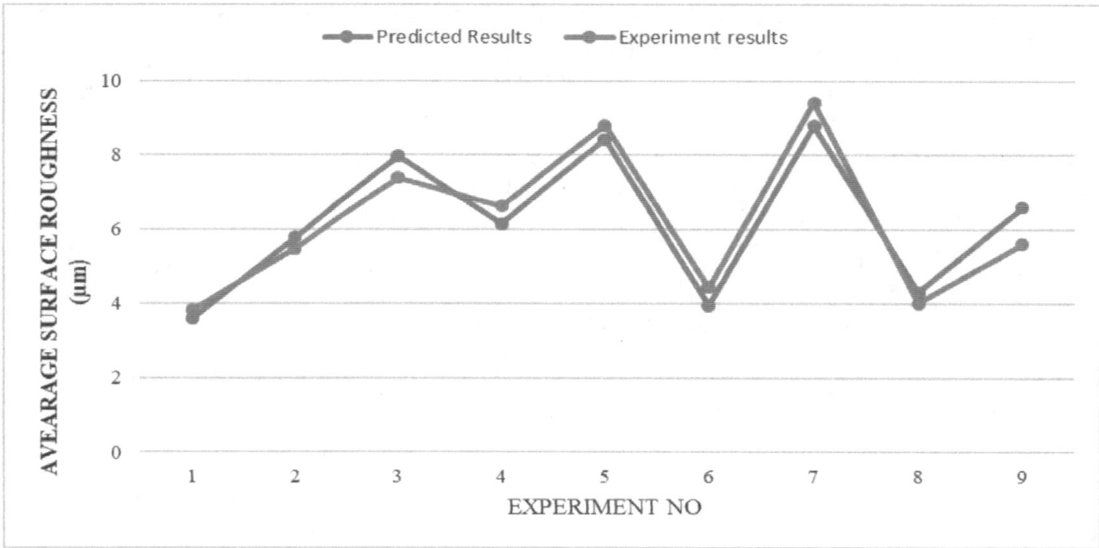

Figure 7. Comparison of actual and predicted values

The main objective of conducting the experiments for machining AZ31 through WEDM is to achieve minimum surface roughness and find the optimal level of control parameters. The RSM optimisation technique is used to determine the optimal machining parameters with an objective of minimizing the surface roughness (Ra). The predicted results of the RSM optimization for the surface roughness are shown in the Table-8 and Figure-8. The results reveal that surface roughness (Ra) mainly depends on Pulse on time (μs). The optimized cutting parameters are found to be cutting speed of 15 V, Pulse on time of 10 μs and Pulse off time of 32 μs. The optimised surface roughness predicted is Ra = 3.5306 μm.

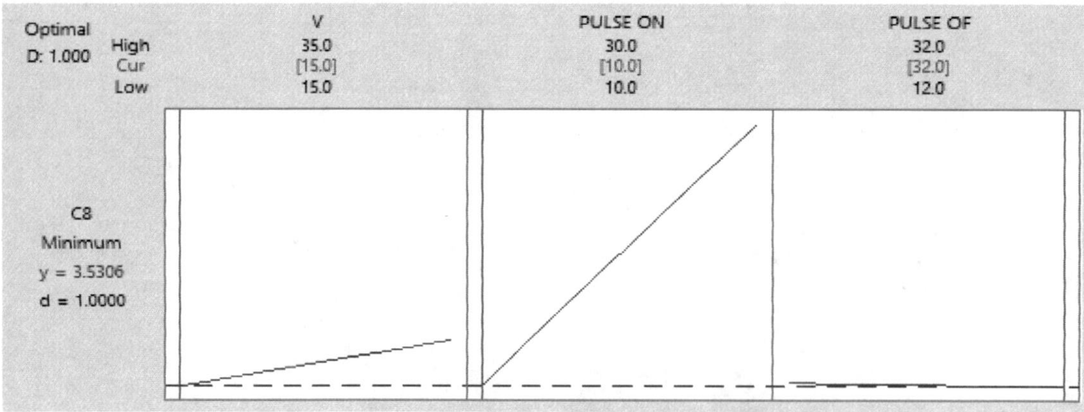

Figure 8. Parameters optimization for Surface Roughness

The verification test is carried out with optimal process parameter settings to confirm the optimal values. Table 9 shows the comparison of experimentally obtained Surface Roughness value with Taguchi predicted value. It was obtained with 95.41% similarity to the predicted value.

Table 8. Predicted results of the RSM optimization for the surface roughness

Response	Goal	Optimum condition			Lower	Target	Upper	Predicted Result	Desirability
		Voltage (V)	Pulse on time (μs)	Pulse off time (μs)					
Surface Roughness (Ra)	Minimum	15	10	32	3.82	3.82	9.41	3.53	1

Table 9 Conformation test values

Optimum Process parameters			Experimental Surface Roughness Ra (μm)	Predicted Surface Roughness (μm)
Voltage (V)	Pulse on time (μs)	Pulse off time (μs)		
15	**10**	**32**	**3.702**	**3.5306**

4. Conclusions

The conclusions drawn from the WEDM machining of AZ31B Mg alloy for achieving minimum surface roughness are as follows:

(a) The Pulse on time (μs) is the main dominant factor that influences Ra the most followed by voltage whose contribution is comparatively very less.

(b) The optimum EDM parameters have been obtained for machining AZ31B Mg alloy. Predicted Ra is in highly agreed with experimental responses.

(c) From RSM optimization, the set of input values obtained for machining AZ31B Mg alloy with minimum surface roughness are: Voltage at 15 V, Pulse on time at 10μs and Pulse off time at 32μs.

(d) The conformation test result proved a similarity of 95.4% between the developed regression model and the experimental value which shows the effectiveness of the analysis model employed.

References:

[1] S Ramesh, R Viswanathan, S Ambika, Measurement and optimization of surface roughness and tool wear via grey relational analysis, TOPSIS and RSA techniques, Measurement 78, 63-72, 2016.

[2] R Viswanathan, S Ramesh, V Subburam, Measurement and optimization of performance characteristics in turning of Mg alloy under dry and MQL conditions, Measurement 120, 107-113, 2018

[3] Vijayakumar Subramani, Balaji Jayavel, Ramesh Sengottuvelu & Prince JeyaLal Lazar 2019, 'Assessment of Microstructure and Mechanical Properties of Stir Zone Seam of Friction Stir Welded Magnesium AZ31B through Nano-SiC', Materials, vol.12, no.7, https://doi:10.3390/ma12071044.

[4] M. A. Razak, A. M. Abdul-Rani, T. V. V. L. N. Rao, S. R. Pedapati, and S. Kamal, "Electrical Discharge Machining on Biodegradable AZ31 Magnesium Alloy Using Taguchi Method," Procedia Eng., vol. 148, pp. 916–922, 2016, doi: 10.1016/j.proeng.2016.06.501.

[5] A. Mostafapor and H. Vahedi, "Wire electrical discharge machining of AZ91 magnesium alloy; investigation of effect of process input parameters on performance characteristics," Eng. Res. Express, vol. 1, no. 1, p. 015005, 2019, doi: 10.1088/2631-8695/ab26c8.
[6] V. Kavimani, K. S. Prakash, and T. Thankachan, "Influence of machining parameters on wire electrical discharge machining performance of reduced graphene oxide/magnesium composite and its surface integrity characteristics," Compos. Part B Eng., vol. 167, no. March, pp. 621–630, 2019, doi: 10.1016/j.compositesb.2019.03.031.
[7] A. Muniappan, M. Sriram, C. Thiagarajan, G. Bharathi Raja, and T. Shaafi, "Optimization of WEDM Process Parameters on Machining of AZ91 Magnesium alloy using MOORA method," IOP Conf. Ser. Mater. Sci. Eng., vol. 390, no. 1, 2018, doi: 10.1088/1757-899X/390/1/012107.
[8] R. Viswanathan, S. Ramesh, N. Elango, and D. Kamesh Kumar, "Temperature measurement and optimisation in machining magnesium alloy using RSM and ANOVA," Pertanika J. Sci. Technol., vol. 25, no. 1, pp. 255–262, 2017.
[9] S. Dhamotharan, B. Babu, "Optimisation of Machining Parameters in Wire Cut EDM for Cemented Tungsten Carbide using," International Journal for Research in Applied Science & Engineering Technology (IJRASET). ISSN: 2321-9653; IC Value: 45.98; Volume 6 Issue XI, Nov 2018
[10] Viswanathan R, Ramesh S, Maniraj S and Subburam V, "Measurement and Multi-response Optimization of Turning Parameters for Magnesium Alloy Using Hybrid Combination of Taguchi- GRA- PCA Technique", Measurement (ELSEVIER) Vol. 159 (2020).

IOP Conf. Series: Materials Science and Engineering 1013 (2021) 012006 doi:10.1088/1757-899X/1013/1/012006

Design and Development of Smart Manhole

Praveena B A[1*], Balachandra P Shetty[2], Murthy Samartha S Kalasi [3], Shivpratap Singh Yadav [1], Mahadeva Prasad [1], Sujith M N [3], Naresh Reddy [3], Mallikarjun L[3]

1, Assistant Professor, Department of Mechanical Engineering, Nitte Meenakshi Institute of Technology, Yelahanka, Bangalore: 560064, India.
2, Professor, Department of Mechanical Engineering, Nitte Meenakshi Institute of Technology, Yelahanka, Bangalore: 560064, India.
3, U G Student, Department of Mechanical Engineering, Nitte Meenakshi Institute of Technology, Yelahanka, Bangalore: 560064, India.

*E-mail: praveena404@gmail.com

Abstract: - In today's world, urban area house a majority of the population of any country. Analysis of both historical and real- time data results in an enhanced understanding of the urban infrastructures systems and consequently in an improvement in their management, By 2030 it is estimated to grow more. As urban areas are resorting to smart methods, either by IoT or by redesign of existing service sector, underground drainage plays a prominent role to provide a healthy environment, good hygiene and safety to a large population density. For this a good underground drainage system is necessary. We can bring about a change by simplifying and reducing the cost of maintenance of existing system. This can done by drastically reducing the time taken to pinpoint the area affected. By replacing existing manhole with sensor based manhole cover at certain key locations, real time data like flow rate of water, concentration of toxic gases can be determined. This data is sent to the authorized service station prior to its blockage thus preventing the flooding of streets through a speedy maintenance regime.

Keywords: sensor, power independent, manhole, flow rate, data center.

1. Introduction

Rapid development and the expansion of urban clusters are essential to our drainage network success in India. We do have a higher population level than many nations and so a reliable wastewater network is required. A research analysis lately has shown that a properly managed drainage network eliminates up to 40 percent and even up to 80 percent the chance of lethal diseases. Another research found that a healthy drainage network would stress pathogens without needing to encourage personal hygiene. Therefore, it is important for such a wellbeing of population to have a healthy drainage network. According to the context of 2011, at country level 48.9% of households neglect drainage facilities, although 33% of residences have public sewage facilities [1].

The National Family Health Survey 3 in 2006 indicates that in metropolitan settings, there was a 52.8% increase, which implies that the flush townhouses connecting with piped, septic sanitation systems were insufficient for 40% of residences, whereas public toilets were in need of about 23,7 percent and accessible protection was performed by about 17,9 percent. A second analysis indicates that 30% of metropolitan inhabitants remain in slum zones. At Bangalore, In Hyderabad, Andhra Pradesh 86 out of 124 cities had no sanitation or drinking water, but just half of the 52 towns were supplied by sanitation.

In metropolitan cities, there are most sewage treatment plants, 40 per cent of which generate sewage. 17 percent of all wastewater is generated in the cities of Delhi and Mumbai. Waste water in urban centers, as it can pollute ground water sources, is necessary to handle. A shift from unified supply to semi- centered supply is needed because it may reduce insecurity between quick urban growth and infrastructure supply for that growth [2].

IOP Conf. Series: Materials Science and Engineering 1013 (2021) 012006 doi:10.1088/1757-899X/1013/1/012006

2. Problem Statement

It is known that a major portion of the population move to cities, which means that by 2030 we would need more water to deliver up to 600 million people and that it would make India a sub urban country, with 2,86 treatment of wastewater, recycling and reuse every year. There is a huge pressure to provide clean water to the people of India and that could be a priority. Exported water may be supplied to urban areas for urban prerequisites from semi urban areas. Rough about 80% of all water is required in urban areas. In this report, we seek to understand the volume of waste produced in cities such as Ahmedabad, Hyderabad, New Delhi, Kolkata and Kanpur, as well as various livelihood application and agriculture. A combination of household and industrial waste is a major challenge in most of the cities which makes the system complex. Due to the lack of organized data on different types of water waste, estimation of quantity and quality are considered a problem [3].

Data from 900 cities of category I and of category II showed that the waste water from these sewages constitutes a mixture of urban and industrial wastewater. This makes it more difficult to process the generation of waste water than the treatment of water. In India the scheme used today is not that high technology, which leads to more obstructions in drains, and it is difficult to pin the position of the block when this occurs. This means that it takes a lot of time to repair the blockage. Because of such failures, people on the ground must suffer [4].

3. Methodology

- Old manholes are replaced by Smart Manholes. They're retro fit and can be designed to any shape and size to match existing manholes.
- They're designed to be power independent. So, solar panels with reinforced toughened glass is attached
- Smart Manholes are equipped with Flow Sensors, Microcontroller, Transmitter and Receiver.
- Once they start to collect data of flow rate from underground sewage system, it is stored and saved in data centres for further researches.
- If flow rate crosses threshold limit in any manhole, immediately message is sent to maintenance team to take necessary action.
- Service team will look after the issue with respective manhole with pinpoint accuracy and work can be easily done by advanced cleaning devices. (BandiKoot)

3.1 Redesign of Manhole Cover

An underground unit for the supply of accesses to utilities like a water system etc may be a manhole or inspection chamber. Underground facilities are then checked, updated, washed and preserved with the aid of a manhole. Typically manholes are 0.5 m from the roads' curb lines. Most of the house is too far away from the central traffic path. The top cover can be a plug to protect the manhole from any non-service access, which usually consists of rectangular, square or circular covers [5]. Canopy fabrics are usually precast, material or plastic materials enhanced by glass. Smart Manholes cover outdated manholes. These are retro designed and will conform to the predominant manholes of any form and scale. We took an existing manhole element of ward 18, Bengaluru for this phase.

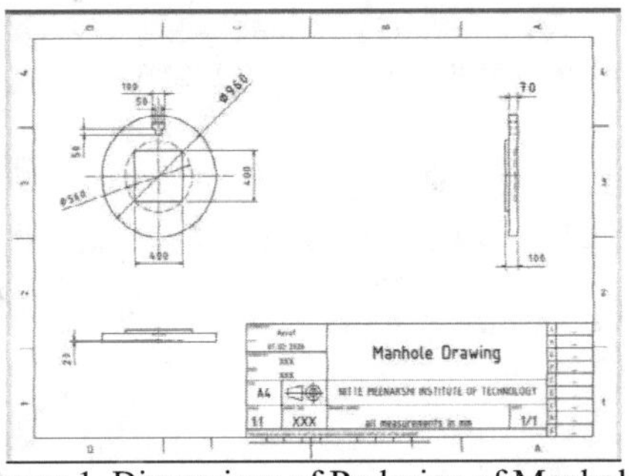

Figure 1. Dimensions of Redesign of Manhole

3.2 Solar Panel as power source

As global climate change comes, it is all the more important that we do our best to reduce greenhouse gas emissions pressure on our atmosphere. Solar panels do not have movable parts and require no upkeep. These are constructed robustly and when properly managed these last for many years. The storage is designed for charging small batteries of up to 7 Ah or 7000 mAh, with a capacity of 10 watts and a 12 volt solar rating that can be used to charge a 12V vehicle or a deep cycle battery [6].

They are designed to be autonomous. So, reinforced toughened glass solar panels are related. The 5V/1.25W manufacturing solar array is adequate to power the control panel. Now we have used the solar array with 12V/10W power. This is 24 * 14 cms in thickness.

Figure 2. Solar array retrofit Manhole

3.3 Installation of Controller

Arduino Mega, a programmable micro controller, is systematically configured according to the need and is linked to the information transfer module GSM 800A in an efficient manner. GSM serves as information relay to Data Centre. The Arduino network has only started with electronics and has been really fashionable men. The Arduino does not need a separate piece of equipment (called a programmer) as most other configurable circuit boards. You'll just load a fresh file onto the board with the USB cable. The IDE also uses a simplified C++ version, making searching for bent software easier. Arduino also offers a standard form factor that disconnects the roles of the microcontroller into a more functional bundle [7].

It's a microcontroller board promoting the at mega 2560 microcontroller Arduino Super 2560. With their simple to use interface, Arduino boards have revitalized the automation market, where some people with little to no technological background will begin studying basic programming and running skills on boards [8].

GSM can be a handheld modem for communication; it stands for a global mobile networking network (GSM). In 1970 Bell Laboratories developed the concept of GSM. This is commonly used worldwide mobile messaging system. GSM is an open and digital cellular technology that operates at the 850MHz, 900MHz, 1800MHz and 1900MHz frequency bands used to transmit mobile voice and data services [9].

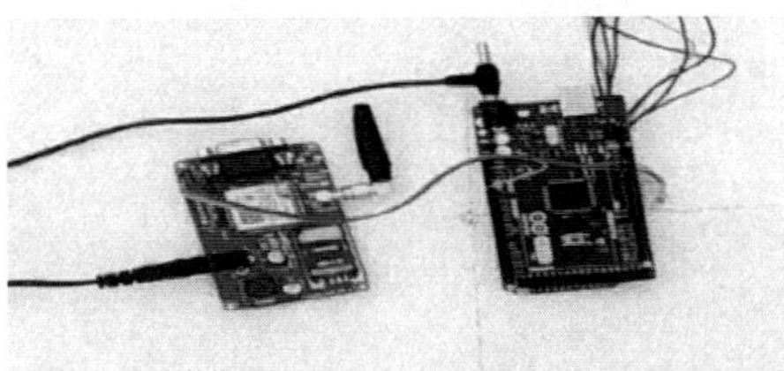

Figure 3. Controllers setup

IOP Conf. Series: Materials Science and Engineering 1013 (2021) 012006 doi:10.1088/1757-899X/1013/1/012006

SIM808 module SIM808 GSM / GPRS / GPS Bluetooth Compatible Wireless Board GPS Antenna (Arduino and Raspberry Pi Compatible) will be a computer board; SIM808 module allows you able to use GSM connectivity and GPS functionality in combination with the Arduino or Raspberry Pi device. You can send and receive SMS with this app; track a path and also create on-card functionality with your own telephone. SIM808 app as a GSM communicator and GPS receiver. The power supply module is 5 – 26V if the power supply is smaller than 2A, whichever is required for 9V. Next. Next. Street power supply port with an extra 3.5-4.2V battery power source lithium appropriate. USB device monitoring – TTL is also a lightweight [10].

3.4 Algorithm of the program

Step 1 : Start

Step 2: Initialize levels and controls Step 3: IF L>20 if NO back to step 2

Step 4: Warning message

Step 5: IF L> 26 if NO back to step 2 Step 6: CRITICAL ALERT MESSAGE

Step 7: Stop

3.5 Flow chart of Program

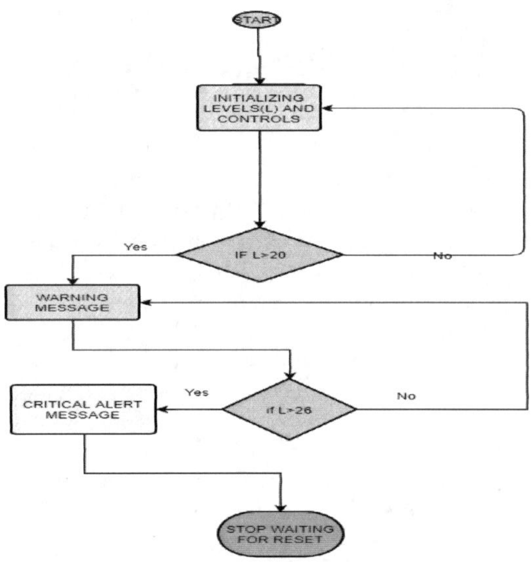

3.6 Sample code in C

```
void loop()
{
int s1 = sensorU1(); int s2 = sensorU1();

if (s1 == T1)
{
send_SMS ("S1 Threshold Level 1");
}
if (s1 = T2)
{
send_SMS ("S1 Threshold Level 2");
}
if (s2 == T1)
{
send_SMS ("S2 Threshold Level 1");
```

IOP Conf. Series: Materials Science and Engineering 1013 (2021) 012006 doi:10.1088/1757-899X/1013/1/012006

```
}
if (s2 = T2)
{
send_SMS ("S2 Threshold Level 2");
}
```

3.7 Assembly

Once all the parts are prepared with required codes and finishing it is assembled as shown below with all circuit connections.

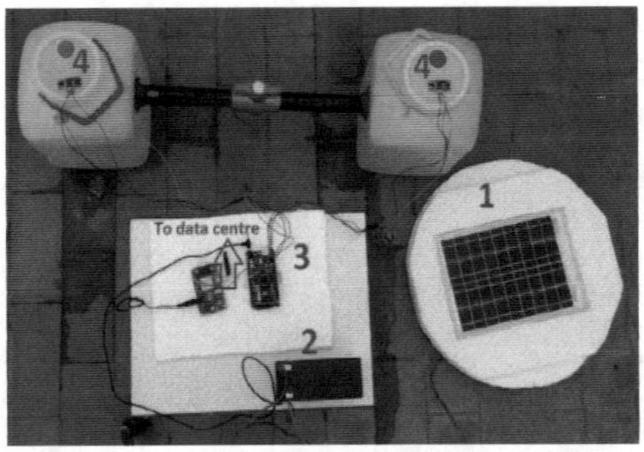

Figure 4. Complete setup of Experiment

- Redesigned Manhole fitted with Solar Panel
- Battery
- Programmable Microcontroller and GSM module
- Ultrasonic Sensor setup

3.8 Sample Output

Continuously tracking the water flow rate and automatically sending mails receive a text message when the water level is outside the normal range expected also feel and update.

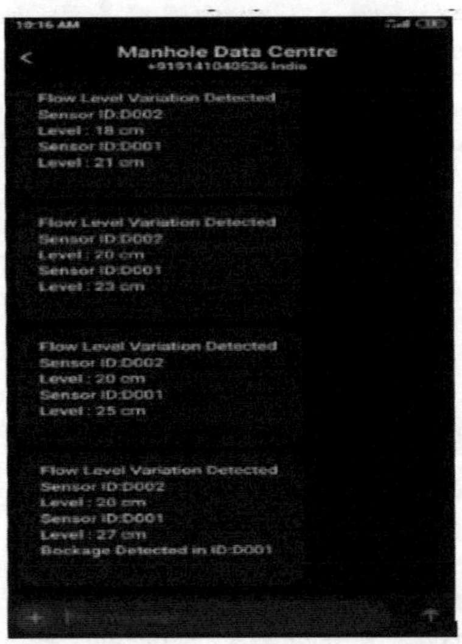

Figure 5. Sample output in mobile phone in text format

4. Results and Discussions

Once they start to collect data of flow rate from underground sewage system it is stored and saved in data centers for further research. Every season data is stored and referred timely to standard the threshold limits and critical limits.

If flow rate crosses threshold limit in any manhole, immediately message is sent to maintenance team to take necessary action.

- Alert before Clog
- Prevention of spillage and overflow at junctions
- Maintain hygiene
- Process and Cost efficient
- Checking and cleaning is easy, because the main parts are retro fit.
- Handling is easy
- No Manual power
- Easy to Repair.
- Replacement of parts is easy

5. Conclusions

A smart city is that the future goal to possess cleaner and better amenities for the society. Smart underground infrastructure is a crucial feature to be considered while implementing a sensible city. System monitoring plays an important role keep the town clean and healthy. Since manual monitoring is incompetent, this results in slow handling of problems in drainage and consumes longer to unravel.

To mitigate of these issues, the system employing a wireless sensor network, consisting of sensor nodes is meant. The proposed system is low cost, low maintenance, IoT based real time which alerts the managing station through an email when any manhole crosses its threshold values. This technique reduces the death risk of manual scavengers who clean the underground drainage and also benefits the general public.

Sensor unit automatically senses and updates the live values of the physical parameters like temperature, humidity, water level and flow, blockages, and manhole cap is open or closed through IoT. This makes the system smart and automatic.

The deployment of Wireless Sensor Networks (WSN), helps within the implementation of the Smart cities during a developing countries. This WSN also can be useful in designing of environmental monitoring systems, which helps in monitoring of volcanic activities, flood detectors and other system. by a little modification within the implementation, this project are often utilized in agriculture fields or other environmental fields to watch and control the systems. In future, Smart cities infrastructure might be modified for intelligent communication and management of traffic signals, street lights, transit applications, active lanes, and so on. With the mixing of smart devices during a city infrastructure can makes life during a city tons easier.

Also further by using PLC controller and SCADA systems, drainage water are often controlled, monitored and also this water are often wont to irrigate plants, clean toilets, etc. This PLC and SCADA systems are often used as a treatment system for drainage water. Primarily, PLC controls the method of sewage treatment plant and SCADA may be a link-attached terminal unit, which monitors and controls the whole area.

IOP Conf. Series: Materials Science and Engineering 1013 (2021) 012006 doi:10.1088/1757-899X/1013/1/012006

References

[1] Kumar SG, Kar SS, Jain A. Health and environmental sanitation in India: Issues for prioritizing control strategies. Indian J Occup Environ 2011; 15:93–6.

[2] Norman G, Pedley S, Takkouche B. Effects of sewerage on diarrhoea and enteric infections: A systematic review and meta-analysis. Lancet Infect Dis.2010; 10:536–44.

[3] Moraes LR, Cancio JA, Cairncross S, Huttly S. Impact of drainage and sewerage on diarrhoea in poor urban areas in Salvador, Brazil. Trans R Soc Trop Med Hyg. 2003; 97:1538.

[4] Report of the working group on urban and industrial water supply and sanitation for the twelfth five-year-plan (2012-2017)

[5] Kaushik A, Kumar K, Kanchan T, Sharma HR. Water quality index and suitability assessment of urban ground water of Hisar and Panipat in Haryana. J Environ Boil. 2002; 23:325–33.

[6] Weber B, Cornel P, Wagner M. Semi-centralised supply and treatment systems for (fast growing) urban areas. Water sci technol. 2007; 55:349–56.

[7] Saravanane, R., Ranade, V. V., Bhandari, V. M., & Seshagiri Rao, A. (2014). Urban Wastewater Treatment for Recycling and Reuse in Industrial Applications. Industrial Wastewater Treatment, Recycling and Reuse, 283–322. doi: 10.1016/b978-0-08-099968.

[8] Bieker S, Cornel P, Wagner M. Semi centralized supply and treatment systems: integrated infrastructure solutions for fast growing urban areas. Water Sci Technol. 2010; 61:2905–13.

[9] Sundaravadivel M, Vigneswaran S. Wastewater collection and treatment technologies for semi-urban areas of India: A case study. Water Sci Technol. 201; 43:329–36.

[10] Yash Narale, Apurva Jogal, Himani Choudhary, S. P Bhosale ,Underground Drainage Monitoring System Using IoT, ISSN: 2454-132X Impact factor: 4.295 (Volume 4, Issue 1).

MODELING AND CHARACTERIZATION OF AISI S-7 STEEL USING THE JOHNSON COOK MODEL – AN EXPLICIT ANALYSIS

Abhinav[1], Rahul Ribeiro[2],Ranjana Jain[3] ,Arunkumar.D[4]
1,2,4Alliance College of Engineering and Design, Alliance University,Bangaluru.
3Jain University, Bangaluru.

Email: abhinav@alliance.edu.in

Abstract : A two dimensional, uniformly distributed grains model has been developed for AISI S-7 grade Steel and the orthogonal machining process, in the pursuit of determining the various effects of temperature rise, residual stresses, chip morphology, and strain hardening. A Johnson-Cook material model, along with a numerical, technique was used to simulate the machining process using ANSYS 15 academic license. The machining has been carried out at different velocities: 60, 70, and 80 ms^{-1} and with depths of cut of 2, 3, and 5 mm. Results revealed that the accumulation of tandem grains offers a maximum resistance ahead of the tool-chip interface due to the strain hardening effect, during the metal removal process. This effect leads to a maximum rise in temperature up to 912.59 °C, which has been observed in the secondary shear zone. Serrated chip flow was observed mainly at a low speed of 60 ms^{-1}. The Strain hardening effect was more substantial at 60 ms^{-1} and 5 mm depth of cut compared to any other machining parameters. A great deal of discussion has been made on the above material machining process that may serve as a useful resource to the tool designer and manufacturing scientist.

Keywords: Finite Element Technique;Residual stress;Strain hardening;Chip morphology.

1.Introduction

Sam Coppieters, et al. [1] have proposed a theory on the identification of strain hardening phenomena in sheet metal at large plastic strains. The post necking strain hardening phenomena in ductile sheet metal are presented based on the minimization of external and internal work in the tensile test. In other studies R.K.Gupta, et al.[2] have explained strain hardening effect in the aerospace alloys. In their investigations, an attempt is made to explain the differential behavior of various alloys. In their investigations Austenitic steel identified to exhibit significant-high strain hardening effect compared to other alloys. A similar type of quality investigations carried out on the S30408 Austenitic Stainless-Steel pressure vessel and found that there is a significant effect on strain rate and yield strength at cryogenic temperature [3]. Earlier several investigations have been carried out on Aluminum grades, Stainless steel grades and Titanium-based alloys using a Johnson-Cook mathematical Model applying the FEA technique. One such work proposed and carried out on the determination of cutting forces a chip morphology by D. Yameogo et al. using the Lagrangian approach and found that segmentation and recrystallization of chips principally developed in the adiabatic shear band [4].

Ramy Hussein et al. [5] have proposed a theory on chip morphology and delamination characterization for vibration-assisted drilling of carbon Fiber-Reinforced Polymer. The results obtained showed, 31% reduction in cutting temperature and a significant augmentation in exit delamination [6]. J. Grum et al.[7] have explained about the residual stress in a Tungsten-chromium-vanadium (80WCrV8) steel grade after heat treatments and grinding. The result found that an annealed steel compressive residual stress is quite low. The relation between residual stress and temperature variation in a metal machining is of great technical significance especially in the study of metal machining and prediction of tool wear. Wen Chen et al. have carried out an experimental investigation on a lattice strain on the several families of grain in an additive manufactured (AM)316L stainless steel under the uniaxial condition and established a mechanistic connection between microscale residual stress and Mechanical behaviour of AM Stainless steel. An FEA study was carried out using Abaqus software and an effort is made to described a relationship between stress-strain and column behavior of stainless steel and found that residual stress leads to an increase of load-carrying capacity with varying Ramberg-Osgood strain hardening factor (n)[8].

From the literature survey it is understood that strain hardening is an important parameter that needs to be understood well and found to play a significant role in the study of machining characterization and wear control due to changing residual stress and temperature rise. An attempt is made to understand the thermomechanical behavior (strain hardening chip morphology and temperature change) of a specifically designed J-C model and implementing the same to AISI S-7 grade steel using the Finite Element Technique.

2. Mathematical Modeling and Analysis

The machining was carried out at different velocities: 60, 70, and 80 ms-1with varying depths of cut viz. 2, 3, and 5 mm. The thermomechanical properties of the materials have been taken from the material library database of the ANSYS 15 academic license software. The essential material properties and J-C material constants have been shown in Table 1. The mathematical expression for the J-C model has been shown in Equation 1[9]. The terminologies related to the equation are as follows: $\bar{\sigma}$, $\bar{\varepsilon}$, $\dot{\bar{\varepsilon}}$, $\dot{\varepsilon}_o$ T_{room}, and T_{melt} are flow stress, plastic strain, effective strain rate, reference strain rate ($1s^{-1}$), room temperature, and melting temperature, respectively. A, B (in MPa) and n represents the yield stress of the material at room temperature, strain hardening, and hardening modulus and work-hardening exponent, respectively. The constants C and m represent the strain rate hardening and thermal softening coefficient, respectively. For simplicity two-dimensional uniformly distributed grain model was developed. The quad-elements have been used to mesh the model. Approximately twenty thousand elements were generated for the numerical analysis. The essential dimensions and meshed models are shown in Fig.1 (a) & (b), respectively.

$$\bar{\sigma} = [A + B\bar{\varepsilon}^n]\left[1 + C\ \ln\left(\frac{\dot{\bar{\varepsilon}}}{\dot{\varepsilon}_o}\right)\right]\left[1 - \left(\frac{T - T_{room}}{T_{melt} - T_{room}}\right)^m\right] \quad \ldots\ldots \quad \text{Eq. 1}$$

Table 1. Johnson-Cook material model used in numerical technique

Parameters	AISI S-7
Density, Kg/m^3	7750
Specific Heat, J/kg/°C	477
Shear Modulus, Pa	8.18e10
Melting Temperature °C	1489.9
Johnson -Cook Material Constant	
A Pa	1.539e9
B Pa	4.77e8
n	0.18
c	0.012
m	1

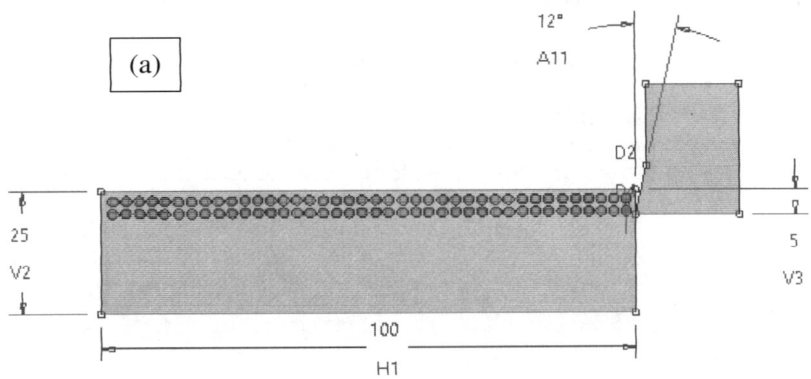

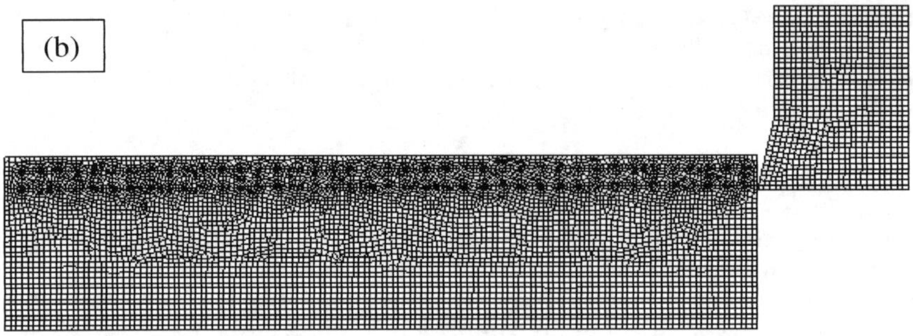

Figure 1. (a)Two-dimensional model with important dimensions in mm & (b) Meshed model

3. Results and Discussion

Sliding and shearing of grains observed to be a common phenomenon in all the machining processes irrespective of the depth of cut & speeds. Remarkably, the highest average temperature rises of 912.59 °C observed at 60 ms^{-1} at 5 mm depth of cut. The temperature versus time graph and numerical analysis results plot revealed that the accumulation of tandem grains found to offer a maximum resistance ahead of the tool-chip interface reason attributed to the collection of bulk grains leads to strain hardening. This strain hardening mainly observed near to the secondary shear zone and found to cross the value mentioned in the J-C Model, refer to Fig 2 (a) and Table 1. At 60 ms^{-1} speed, as the depth of cut decreased, temperature found to be lessening. At 2 mm depth of cut, the average temperature found to vary linearly and commenced early compared to other depth of cut. Non-linearity in the temperature observed more prominent at a lower speed, i.e., at 60 ms^{-1} compared to 70 and 80 ms^{-1}. As the cutting speed increased, the effect of grain orientation shadows the non-linear behavior in the temperature profile. The similar effect observed at 70 ms-1 refers to Fig 2 (b). However, at a higher speed of 80 ms-1, the significance of the depth of cut not found to have a significant role in the variation of temperature refer Fig 2. (c). The residual stress/von-mises stress found to vary in the range of 1.98e9 to 1.83e9 Pa. The residual stress found to accumulate in the primary shear zone irrespective of all the machining conditions and found to exceed the yield stress (A), i.e., at 1.539e9 considered in the J-C model parameter. In the present investigation, the flow stress was found to cross the assumed value, and it would be apt to address henceforth residual stress as flow stress.

The flow stress was found to be more crucial at higher depth of cut. Comparatively thick chips found at 5 mm and thin chips have formed at 2 mm depth of cut refer to Fig. 3, 4 & 5. Relatively longer curvilinear chips observed at 80 ms^{-1} also, at different depth of cut. Segmented short chips have seen at 70 ms^{-1}, and serrated chip flow followed at a low speed of 60 ms^{-1}. At this speed, the chips found to flow straight compared to other speeds. Also, the probability of breaking of chips found to be more consistent compared to different speeds, and depth of cut refers to Fig.4. A powerful relationship observed in the temperature plot with the flow stress plot. There is a change in the temperature with the breaking of grains. Early onset of breaking leads to more non-linearity in temperature versus time curve refer to Fig.2 & 3 (a). The continuous curvilinear chips were accountable in achieving constant steady-state conditions during the machining process refer to Fig.2 (a)(b), & (c). Depending on the machining requirements, the machinist may take prior precautions and guidelines to overcome the uncertainties during the machining process with the present set of simulation data.

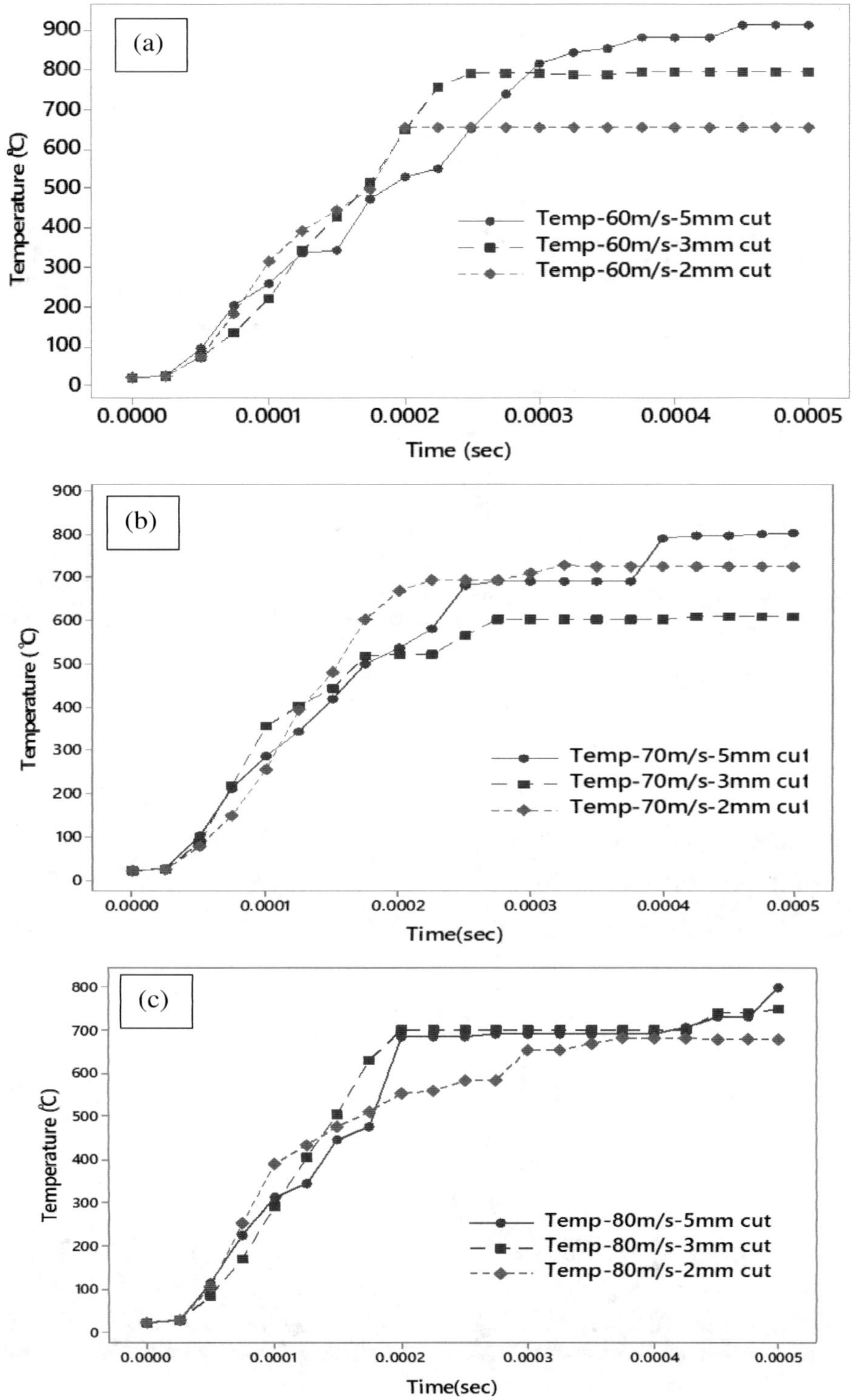

Figure 2. Temperature versus time plot at 5,3 & 2 mm depth of cut (a) 60 (b)70 & (c) 80 ms^{-1}

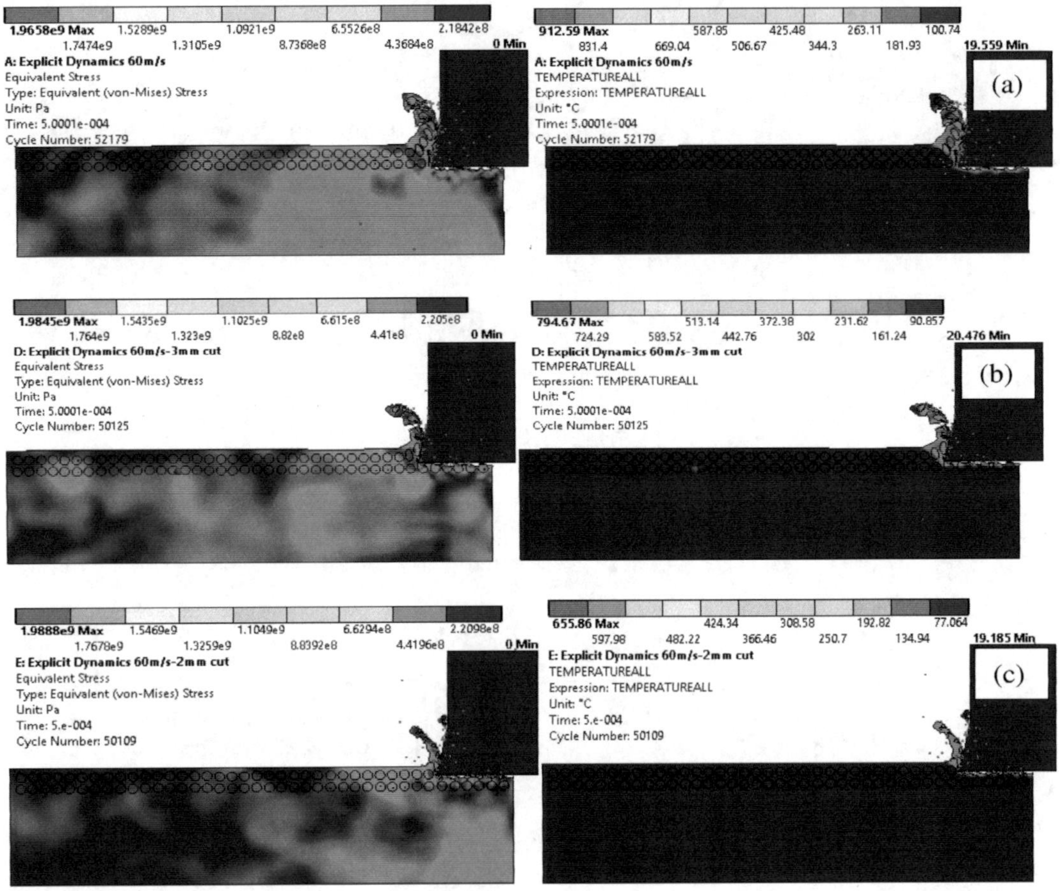

Figure 3. Von-Mises stress/ Residual stress and Temperature plot at 60 ms^{-1} (a) 5 (b) 3, & (c) 2 mm depth of cut.

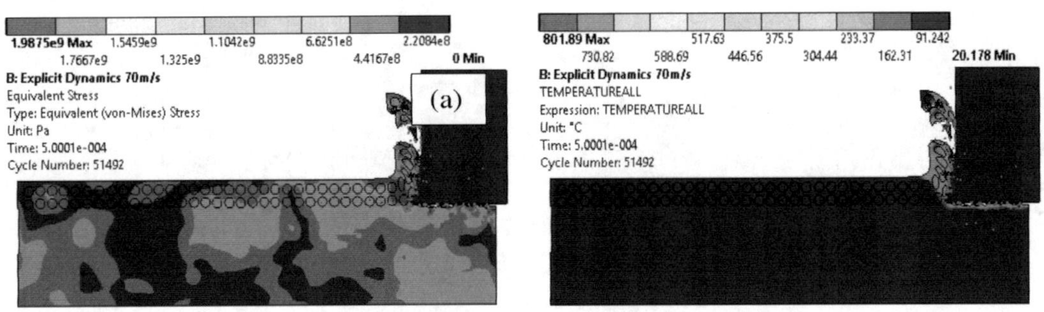

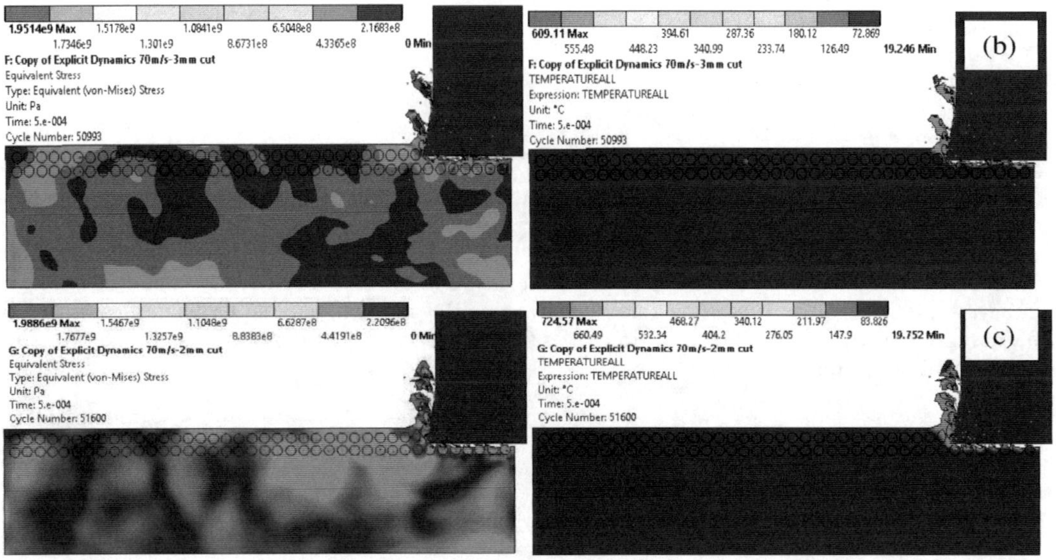

Figure 4. Von-Mises stress/ Residual stress and Temperature plot at 70 ms^{-1} (a) 5 (b) 3, & (c) 2 mm depth of cut.

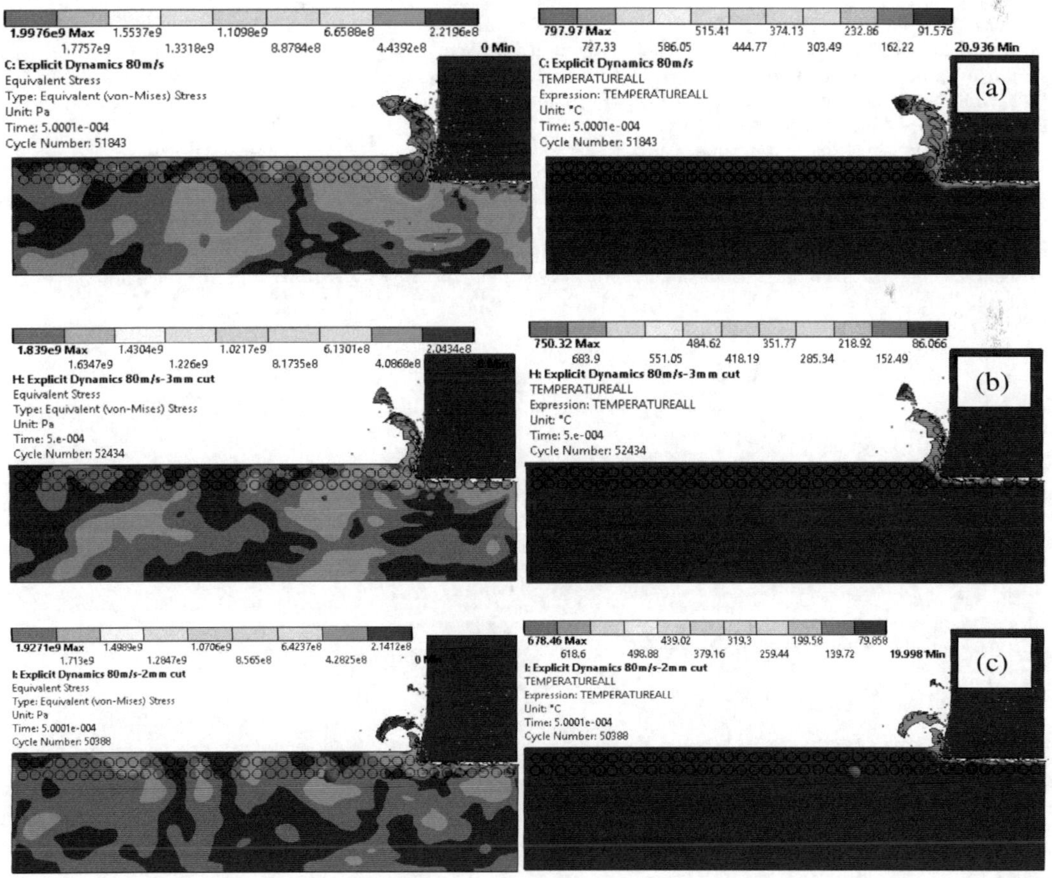

Figure 5. Von-Mises stress/Residual stress and Temperature plot at 80 ms^{-1} (a) 5 (b) 3, & (c) 2 mm depth of cut.

4. Conclusion

The machining was carried out at different velocities: 60, 70, and 80 ms^{-1} and with depths of cut of 2, 3, and 5 mm. Results revealed that the accumulation of tandem grains offers a maximum resistance ahead of the tool-chip interface due to the strain hardening effect, during the metal removal process. This leads to a maximum rise in temperature up to 912.59 °C which was observed in the secondary shear zone. Serrated chip flow was observed mainly at a low speed of 60 ms^{-1}. The Strain hardening effect was more substantial at 60 ms^{-1} and at 5 mm depth of cut compared to any other machining parameters.

References

[1] Sam Coppieters, Kazuhiro Ichikawa et al., Identification of strain hardening phenomena in sheet metal at large plastic strains. Science Direct Procedia engineering, Vol. 81, pp 1288 – 1293, (2014).
[2] R.K.Gupta, Cristy Mathew, P. Ramkumar. Strain hardening in aerospace alloys. Frontires Aerospace Engineering, DEStech Publications, Vol.4, pp 1-41, (2015).
[3] Bo Li, Fa Cai Ren, Xiao Ying et al., The effect of strain hardening on mechanical properties of S30408 Austenitic Stainless Steel: A Fundamental research for the quality evaluation of strain strengthened pressure vessel,IOP Conference Series: Material science and engineering,Vol. 382, (2018).
[4] D. Yameogo , B.Haddag, H.Makich and M.Nouari Predication of cutting forces and chip morphology when machining the Ti6A14V alloy using microstructural coupled model, Science direct Procedia CIRP,Vol . 58 ,pp.335-340, (2019).
[5] Ramy Hussein, Ahmad sadek et al.,Chip morphology and Delamination Characterization for vibration-assisted drilling of carbon Fiber-Reinforced Polymer, Journal of Manufacturing and Material processing ,Vol.3,pp 23 (2019).
[6] J. Grum and P. Zerovink, Residual stress in steel after heat treatments and grinding,Transactions on engineering sciences Vol. 17 (1997), ISSN 1743-3533.
[7] Chen, W., Voisin, T., Zhang, Y. et al., Microscale residual stresses in additively manufactured stainless steel. Nat Commun **10,** 4338 (2019). https://doi.org/10.1038/s41467-019-12265-8.
[8] M. Jandera and J Machacek, Residual stress influence on material properties and column behaviour of stainless steel SHS, M Jandera and J Machacek, Czech Technical University of Prague, Journal of Thin walled structures vol.83, pp.12-18 (2014).
[9] Feng Zhang et al. The modified temperature term on Johnson-Cook constitutive model of AZ31 magnesium alloy with {0002} texture, Journal of Magnesium and Alloys,Science Direct Vol.8, pp. 172-183 (2020).

Investigation of effect of mechanical properties and Influence of Heat treatment on AISI 4140 steel

Nagaraja T. K.[1]
1Department of Mechanical Engineering, JSS Academy of Technical Education, Bengaluru, India.

E-mail: tknmech@gmail.com

Abstract: - The optimization methods in metal heating processes, considered to be a vital tool for continual improvement of output quality in products. In order to optimize the gains from the raw material of AISI 4140 alloy steel, an accurate process must be constructed. The carbon content in AISI4140 steel ranges from 0.38 to 0.43%. In this approach the mechanical properties mainly like Hardness, Yield Strength, Wear Resistance are amenable to modification via heat treatment.In the present work the objective is to test the chosen steel with various heat treatment processes and also analyze their microstructure. A comparative study made across the properties vs. heat treat processes. Finally, to choose the best heat treatment process and optimal properties that can be achieved.

Keywords: heat treatment, hardness, alloy steel; microstructure; optimization; properties;

1. Introduction

In this study, heat insulation tests of AISI 4140 steel are carried out under different temperatures and holding times, and the grains sizes are measured. However,when the components made of this alloy steel are exposed to harsh operating conditions, such as surface rolling and sliding contact, during their service life, they are susceptible to serious surface damage, such as micro pitting, abrasive wear, and corrosion, which could accelerate premature failure and shorten the life cycle of these critical and expensive components.Thus, it is essential to restore the worn-out or damaged components so as to lengthen their service life.

2. Literature review

Literature survey is required to understand the specific objective of the research work. The previous research which gives clear contain and clear ideas about the work. Jon L.Dossett, P.E [1] et al, maximum hardness is a function of the carbon content of the steel being used and hardenability of steel and Determining the expected core hardness range can be determined by using the Jominy equivalent cooling rate (JEC) chart. Hamad K.Al-Khalid [2] et al, demonstrated that for a given microstructure, the apparent hardness of a ferrous P/M material is related to the density of the material,apparent hardness increases as density increases. A. Calika [3] et al, The hardness of borides, boride layer thickness and room temperature tensile properties were measured and it was observed that hardness and tensile properties strongly depend on chemical composition of steels.

3. Experimental details

IOP Conf. Series: Materials Science and Engineering 1013 (2021) 012008 doi:10.1088/1757-899X/1013/1/012008

3.1 Heat Treatment Process

The mechanical properties can be improved by heat treatment. These are stronger compared to low carbon steels. Medium carbon steels have low hardenability. Addition of Cr, Ni, Mo, improves the heat treating capacity [1]. Eutectoid steel is the stage in the iron carbide diagram, in which the steel contains completely pearlite in its microstructure. This forms at 727°C contains 0.76% carbon. At this temperature, ferrite and cementite is formed in combination called pearlite. Pearlite is a mixture of ferrite (0.022% C) and cementite (6.7% of C). Pearlite microstructure consists of alternate layers of α-ferrite and cementite. Hypo eutectoid steels contain less than 0.76% carbon where ferrite is in combination with cementite having thin lamellar called coarse pearlite. In which percentage of ferrite is decreasing. Eutectoid steel having 0.83% carbon where complete pearlite is formed.Hyper eutectoid steels having more than 0.83% percent carbon in which ferrite are completely decreased and cementite is increased in percentage where the cementite lamellar becomes thick called fine pearlite.

3.2 Experimental Procedure

Tensile test according to ASTM Standard's, Test Method: ASTM E8-15a:[15] These test methods cover the tension testing of metallic materials in any form at room temperature, specifically, the methods of determination of yield strength, yield point elongation, tensile strength, elongation, and reduction of area. In this, we made an attempt of testing 7 specimens. Which are, one specimen for Normal untreated, 3 specimens for Normalizing at 1100 $^{\circ}$C under holding time 30, 45 and 60 minutes. Three specimens are used for Annealing at 1000 $^{\circ}$C under holding time 30, 45 and 60 minutes.

4. Results and Discussions

Tensile test provides information on the strength and ductility of materials under uniaxial tension stresses. This evidence may be useful in comparisons of materials, alloy development, quality control, and design under certain conditions.

Table 1. Strength In Untreated V/S ASTM Standard

Sl No	Particular	ASTM Standard's	Untreated specimen
1	Tensile strength	655 MPa	997.87
2	Yield strength	415 MPa	827.47

Table 2. Obtained Young's Modulus results

Sl No	Particular	Holding Time (Min)	Young's Modulus(Mpa)
1	Normalizing at 1100 $^{\circ}$C	30	958.32
		45	1077.03
		60	**1082.53**
2	Annealing at 1000 $^{\circ}$C	**30**	761.90
		45	758
		60	**774.48**

Table 3. Obtained Yield Strength results

Sl No	Particular	Holding Time (Min)	Yield Strength (Mpa)
1	Normalizing at 1100°C	30	1017.54
		45	1127.86
		60	**1176.87**
2	Annealing at 1000°C	**30**	**787.58**
		45	767.54
		60	740.17

Table 4. Obtained Ultimate Strength results

Sl No	Particular	Holding Time (Min)	Ultimate Tensile Strength (Mpa)
1	Normalizing at 1100^{0}C	30	1147.54
		45	1274.58
		60	**1276.57**
2	Annealing at 1000^{0}C	**30**	917.57
		45	917.57
		60	**931.57**

As compared to the untreated specimen, In Normalizing process as the holding time increases the Young's modulus, Yield strength, and Ultimate strength are gradually increasing, where in case of Annealing process for different holding time the properties are varied which results annealing soften the specimen.

4.1 Brinell Hardness Test
Test Method: IS1500-2010, Ball: 5mm, Load: 3000Kg

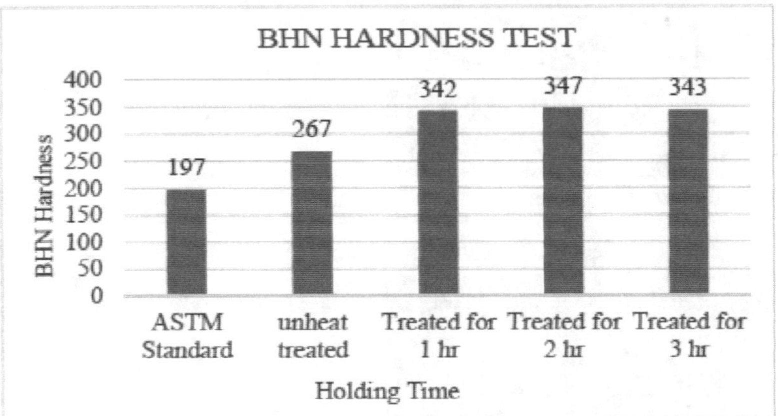

Figure 1. BHN v/s Holding time

4.2 Wear Test

Specimen Diameter = 7 mm, length = 30 mm.

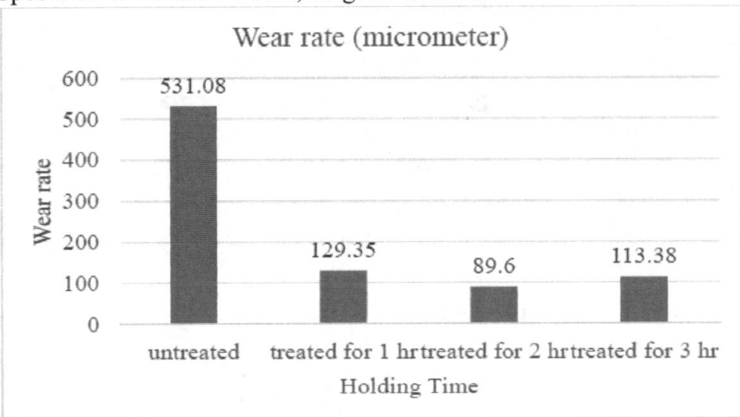

Figure 2. Wear rate v/s Holding time

4.3 Scanning Electron Microscopic(SEM)

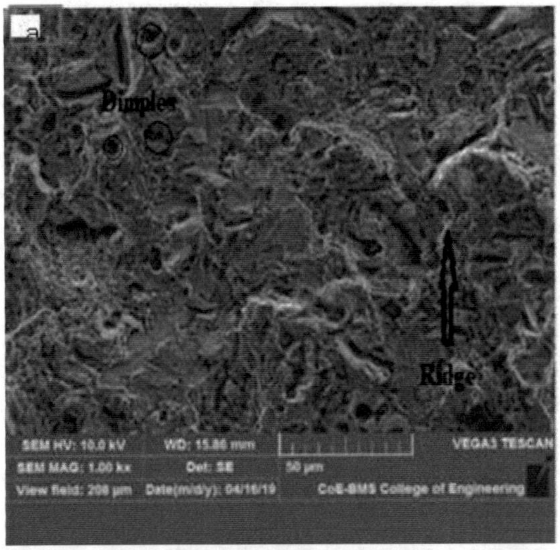

Figure 3. SEM on fractured specimen

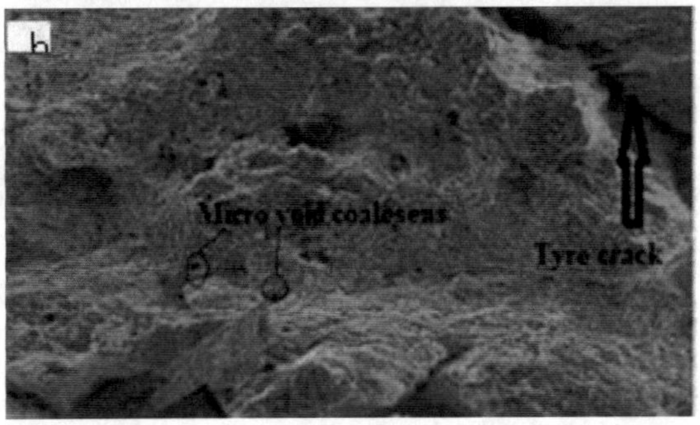

Figure 4. SEM on fractured specimen

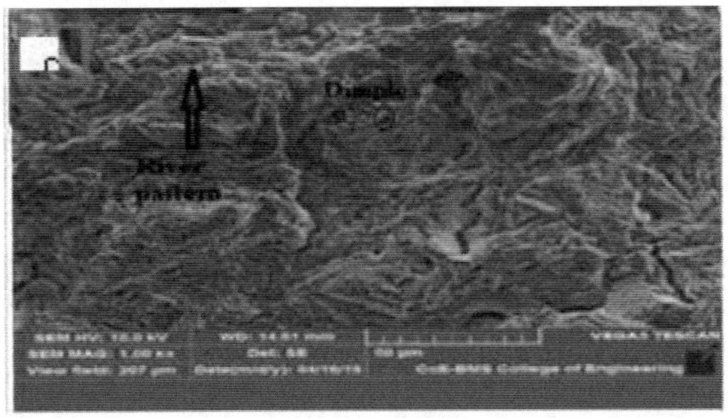

Figure 5. SEM on fractured specimen

From above Figure 3. magnification of untreated specimen there is Ridges and Dimples with a Interstitial free due to its carbon contain These steels have been designed to provide an excellent combination of drawability and mechanical strength based on their specific interstitial-free (IF) metallurgy. In Figure 4. microstructure there is a Tyretracks and Microvoid where this tyretracks which are perpendicular to crack propagation or crack growth. In Figure 5. the magnification image, there is a Dimples and river pattern with Quasi cleavage fracture, which is a fracture mode that combines the characteristics of cleavage fracture and dimpled rupture fracture.Cleavage is tendency of material to break along smooth plane parallel to the zones of weak bonding[14].

5. Conclusions

The AISI4140 material has undergone with different heat treatment process. The tensile test, hardness test and wear rate test were conducted for both treated and untreated specimens, from the results following points were noted.

- By comparing the values of yield strength for normalizing and annealing process,normalizing gives optmized results at 60 minutes of holding time during process.
- It has been observed that yeild strength increased gradually in normalizing process.
- Microstructure reveals globular structure with coarse carbides at grain boundaries in a matrix of ferrite.
- The optimized values of hardness and the wear rate is obtained for Two hours of holding time at 950°c during heat treatment process.

References

[1] Make Sure Your Specified Heat Treatment is Achievable- Heat treating process,Jon L. Dossett, P.E. March (2007).
[2] Hardness Variation as Affected by Bar- Hamad K. Al-Khalid, Ayman M. Alaskari and Samy E. Oraby, International Journal of Mechanical and Mechatronics Engineering Vol:5, No:3, (2011).
[3] Mechanical Properties of Boronized AISI 316, AISI 1040, AISI 1045 and AISI 4140 Steels, A. Calika , O. Sahinb and N. Ucarc, ACTA PHYSICA POLONICA A , Vol. 115 (2009)
[4]Assessment of mechanical properties of AISI4140 & AISI316 Dissimilar Weldments MP Reddy, A. Sam William, M. Prashant, S.N. Sabaresh Kumar, K. Devendranath Ram Kumar, N. Arivazhagan, S. Narayanan(2014).
[5]Normalizing effect on fatigue crack propagation at the heat-affected zone of AISI 4140 steel shielded metal arc welding - B. Vargas-Arista, J. Teran Guillen, J. Solis, G. Garcia- Cerecero, M. Martinez-Madrid(2013).
[6]Effects of gaseous nitriding AISI4140 alloy steel on corrosion and hardness properties Tamil Moli, N Wahab, M Gopinathan, K Karmegam and M Maniyarasi(2016).
[7]Effect of ion nitriding on fatigue behavior of AISI 4140 steel- Kenan Genel, Mehmet Demirkol, Mehmet Capa(1999)
[8]Statistical and graphical assessment of circumferential and radial hardness variation of AISI 4140- Hamad Al- Khalid, Ayman Alaskari and SamyOraby(2011).

IOP Conf. Series: Materials Science and Engineering 1013 (2021) 012008 doi:10.1088/1757-899X/1013/1/012008

[9]StudyonDual-PhaseStructureofAISI4140Steel-B.M.Gurumurthy,S.S.Sharma,M.C. Gowri Shankar, RomitWalia and Tanishq Agarwal (2016)
[10] R E Smallman, Modern Physical Metallurgy, Butterworths,1985.
[11] Aerospace Material Specification, AMS 6349A, SAE 4140 steel, revised1188.
[12] W F Smith, Principles of Materials Science and Engineering, McGraw Hill,1990.
[13] Investigation of the mechanical properties and microstructure of friction welded joints between AISI 4140 and AISI 1050 steels by SareCelik, IsmailErsozlu
[14] D J Davies and L Oelmann, The Structure, Properties and Heat Treatment of Metals.
[15] Standard Test Methods for Tension Testing of Metallic Materials(2010).

EFFECT OF HEAT TREATMENT ON FATIGUE CHARACTERISTICS OF EN8 STEEL

Manu Ravindran [1], Aswatha M[2], Santhosh N[3*], Ravichandran G[4], Madhusudhan M[5]
[1,2]Department of Mechanical Engineering, Bangalore Institute of Technology, Bengaluru, India.
[3]Department of Mechanical Engineering, MVJ College of Engineering, Bengaluru, India.
[4]Department of Mechanical Engineering and Automobile Engineering, School of Engineering and Technology, CHRIST (Deemed to be University), Bengaluru, India.
[5]Department of Mechanical Engineering, Presidency University, Bengaluru, India.

*E-mail: dr.santhosh@mvjce.edu.in

Abstract: - Fatigue failure is an important factor in most of the engineering applications, especially in steel materials, and among the steel materials, it is an important phenomena in medium carbon steels like EN8, which is very commonly used in components like shaft, gears etc., since it is prone to fatigue failure. Hence, without changing the composition, an attempt is made to enhance the fatigue strength by different heat treatment techniques. In this study, the investigation is carried out on heat treatment of EN8 steel material. Various kinds of heat treatment techniques like quench and temper, normalizing and annealing are performed on EN8 steel. After exposure to the heat treatment, the EN 8 steel material specimens are machined as per the ASTM standards and are subjected to RR MOORE test and SN-curves are plotted from the obtained results; the obtained results from the fatigue tests are further analyzed with the help of ANSYS software. Fatigue life and Factor of Safety (FOS) comparisons for EN 8 steel material is made with the structural steel material and it is found from the comparisons, that the heat treatment process enhances the fatigue strength and endurance limit.

Keywords: EN8 steel; Heat treatment; Fatigue strength; RR MOORE test; Basquin's equation, Endurance limit.

1. Introduction

The admixture of iron and carbon is steel. Steel is classified as an iron carbon in which carbon content is within 2.11% and all those iron that contain above 2.11% of carbon are called cast iron. The admixture elements collective effect and heat treatment can result in large varieties of microstructure and properties [1]. Medium carbon steel has carbon content between 0.25 and 0.60. These admixtures may be heat treated by heating it to austenizing temperature, quenched, and then tempered to advance mechanical properties. They are frequently made practical in tempered condition having microstructure of tempered martensite [2]. The medium carbon steel material is profusely used in railway wheel and tracks, gears, crankshaft, axles, screws, cylinders [2, 3]. Fatigue strength is modified by the factors like material composition, heat treatment, operating temperature, grain size and grain direction, stress concentration, surface condition, size, residual stresses, fretting, corrosion, operating speed, strength reliability etc. [4, 5, 6] Grain size and its dispersion have a significant influence on yielding specificities of steel.

In this research, influence on fatigue strength due to heat treatment of EN 8 steel is studied. The materials are subjected to different heat treatment conditions like quenching, tempering, normalizing, and annealing. After heat treatment, materials are machined as per ASTM standards and RR MOORE test is accomplished; values obtained from the fatigue tests are plotted in SN graph and endurance limit is found and the endurance limit obtained by testing different heat treatment specimens are compared; after comparison, the heat treatment condition which influences the fatigue strength the most is identified. The values found are entered in ANSYS and fatigue analysis of shaft is accomplished.

2. Literature review

Extensive literature review in the domain of heat treatment of structural steel material is carried out to formulate the research plan; hence sufficient research findings of several researchers are referred to study the effect of heat treatment on fatigue characteristics of EN 8 steel.

Mooljane et. al., [7] noticed that the grain size has very good impact on yield strength of steel. When the dispersion of grain is not uniform, the plastic deformation occurs more in weaker phase and stresses will be higher in stronger phase. The temperature and soaking time in annealing process has considerable effect on the grain size and phase dispersion. Among the metals, low cost metals are versatile in use because of its less expensive, convenient and economical production. Kareem et al., [8] has investigated consequence of heat treatment on fatigue strength of low carbon steel. The fatigue specimen was exposed to various heat treatments like annealing, normalizing and quenching before tests. The study concluded that annealed specimen has higher fatigue strength succeeded by specimen that is quenched in silica sand and normalized specimen. Phoumiphon et al., [9] had concluded that cold rolled specimen; when subjected to inter-critical annealing at different temperatures, there is increase in micro hardness and tensile strength; the research concluded that even though low carbon steel consists less pearlite, the inter-critical annealed specimen has better mechanical properties than as received specimen and cold rolled specimen. Li et al., [10] has demonstrated computational analysis of fatigue life of 0.44% carbon steel at various tempering temperatures. Two different methods estimating mean fatigue life are provided. One based on dislocation dipole accumulation and Paris law, another is Kringing model. The study concluded that the fatigue life decreased with increasing temperature. The dipole accumulation model was found to be more superior to the Kringing model. The fatigue crack growth behavior of hot rolled medium carbon steel specimens are studied by Mossaab et al., [11] after subjecting the specimen to heat treatment, they have concluded that refined microstructure is obtained by subjecting the specimens to different heat treatment conditions. The quenched specimen has higher strength compared to other heat treated samples. The normalized specimen exhibited higher malleability than other heat treated specimen. With increase in tempering temperatures there is decrease in hardenability. Heat treatment influences fatigue life, higher fatigue life is obtained after normalization treatment. Biswas et al., [12] has studied hardness and tensile strength of EN 8 steel that is heat treated at different temperatures, the study concludes that there is declination in hardness after tempering. For particular application, hardness is not preferred, because it will lead to failure of component, so it is recommended to temper after normalizing. Sanusi et al., [13] has studied the modification of mechanical properties due to heat treatment. The result suggests that annealing and normalizing can result in drop of material's ductility. The soaking time in furnace has an effect on surface finish. The soaking time also has effect on corrosion rate.

From the literature review, it is reported that the heat treatment has an influence on mechanical properties and microstructure, which has provided sufficient base for studying the effect of heat treatment on heat treatment characteristics of EN 8 steel.

3. Experimental details

3.1 Workpiece Materials

The type of steel used in this research is EN8 steel, and the chemical composition of material is shown in Table 1.

Table 1. Chemical Composition of EN8 Steel

Composition	Mn	P	C	S	Si	Fe
Amount in %	0.77	0.02	0.42	0.02	0.22	98.55

The dimension of EN8 steel rod is as follows, diameter of the specimen of the steel rod is 15 mm, while the length of specimen of the steel rod is 250 mm; the total number of steel rods considered for studying the effect of heat treatment are 60 no.'s, with 10 steel rods each for the different heat treatment processes.

3.2 Heat Treatment Conditions

The effect of heat treatment on fatigue characteristics of the EN8 steel material is of prime concern, thus there is a need for subjecting the EN8 steel rods to different heat treatment conditions.

Figure 1., shows the procedure for heat treatment, the EN8 steel materials are exposed to process annealing, normalizing and quenching. First 10 numbers of the specimens are considered as it is, for unheat treated condiditions, while the next 10 numbers of the specimens are heated to 650°C, soaked for 1 hour and allowed to cool in furnace and remaining 40 numbers are heated to 900°C soaked for 1 hour; out of these 40 numbers, 10 numbers of the specimens are allowed to cool in air and remaining 30 numbers of specimens are quenched in water.

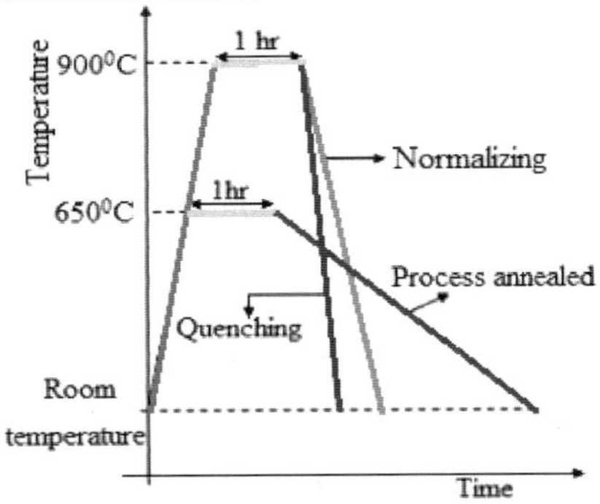

Figure 1. Temperature v/s time diagram for process annealing, normalizing and quenching [14, 15]

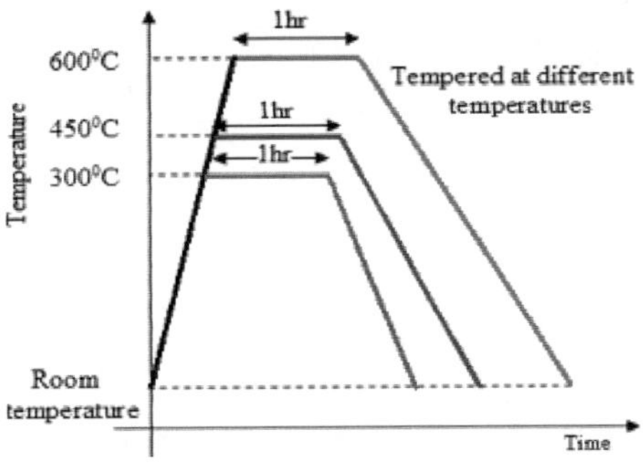

Figure 2. Temperature v/s time diagrams for tempering [16, 17]

The procedures of tempering as shown in **Figure 2,** are followed. From 30 quenched EN8 steel, 10 numbers are heated to 300⁰C soaked for 1 hour, 10 numbers are heated to 450⁰C soaked for 1 hour, remaining 10 numbers are heated to 600⁰C soaked for 1 hour, and all the tempered specimens are cooled in air after soaking time.

3.3 Specimen Preparation for Testing

Materials are machined according to ASTM specifications, viz., ASTM F1160, ISO 1143. After machining, the materials are subjected to RR-MOORE test. **Figure 3**, shows the draft of the sketch of the specimen for fatigue tests.

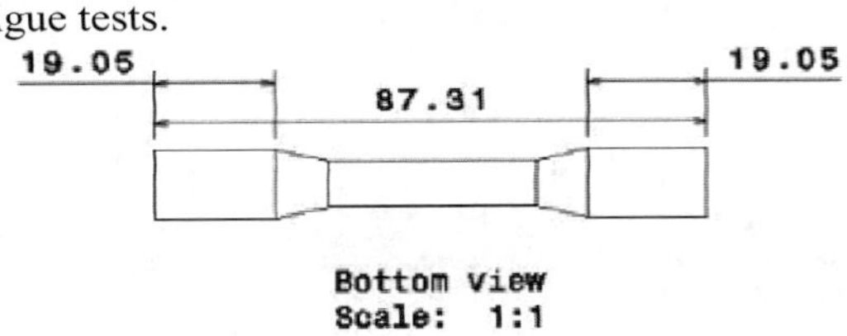

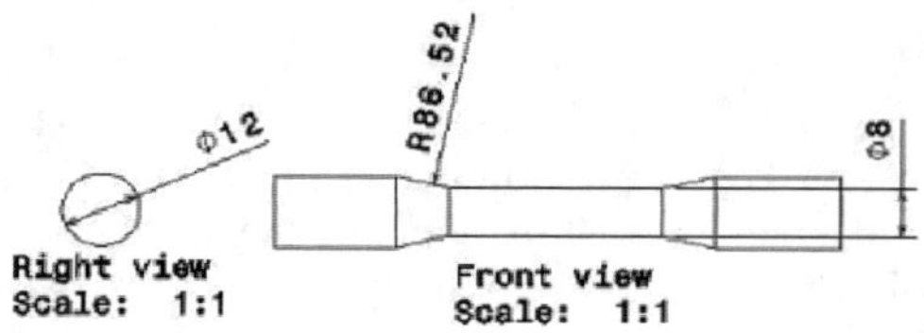

Figure 3. Specimen dimensions for fatigue test [18]

3.4 Design Calculations

The set of the equations used for design calculations for the fatigue strength estimations of the un- heat treated and heat treated specimens are as follows,

Bending stresses, [19]

$$\frac{\sigma_b}{c} = \frac{M_B}{I} \quad (1)$$

Where moment of inertia *I* for a circular cross section is $I = \frac{\pi \times d^4}{64}$ (2)

Positive maximum bending moment, [19, 20]

$$M_B = \frac{1}{24} \times Wl \quad \text{At } x = \frac{l}{2} \quad (3)$$

Where W=w×l (4)

Negative maximum bending moment

$$M_B = \frac{1}{12} \times Wl \quad (5)$$

c = distance from neutral fiber, for circular shaft $c = \frac{d}{2}$

Displacement at $x = \frac{l}{2}$

$$y_{max} = -\frac{1}{384} \times \frac{W \times l^3}{EI} \quad (6)$$

Goodman criteria [19, 20]

$$\frac{\sigma_a}{S_e} + \frac{\sigma_m}{S_u} = \frac{1}{n} \quad (7)$$

Basquin's equation [19, 20]

$$\sigma_f = \sigma_f' (2N_f)^b \quad (8)$$

3.5 Analysis of Both Side Fixed Beam Using ANSYS Software

Condition 1

L=2400 mm, d =50 mm

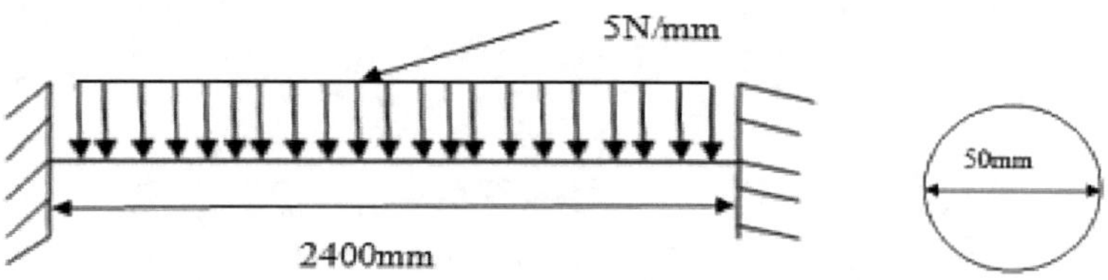

Figure 4. Beam under load and constraints

Material: structural steel
Loading and support conditions

- Uniformly distributed load = 5 N/mm
- Both ends fixed support condition

Condition 2

The constraints and loading conditions are same as in condition 1; only material condition is considered to be tempered EN 8 steel at 300°C after hardening. For both the conditions, shear force, bending moment diagram, stresses, displacements and fatigue life, safety factors are estimated and the relative change in fatigue life after changing the material and constraints are determined.

4. Results and Discussions

The experiments were conducted for the unheat-treated and heat-treated specimens and the SN curves for fatigue strength were plotted to study the fatigue behavior of EN 8 steel specimens subjected to different heat treatment conditions.

4.1 SN curves from experiments for different heat treatment conditions

The SN Curves from experiment for different heat treatment conditions on the EN 8 specimens are depicted in the **Figure 5**.

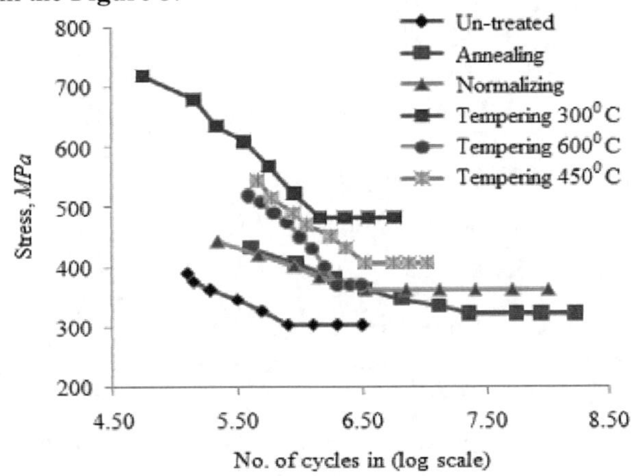

Figure 5. SN curves for different heat treated specimens

Figure 5, shows the SN graphs for stress in MPa v/s Number of cycles of non-heat treated specimen and heat treated specimens; further, for the unheat-treated specimens, it is observed that the endurance limit is 306 MPa at 1.26 E+06 cycles and the fatigue stress, which is calculated from the basquin's equation is 311.46 MPa.

σ_f=306 MPa, N_f=1.26E+06 cycles (9)

$b = \frac{\Delta\sigma}{\Delta N_f}$= -0.112033 (10)

From Basquin's equation

$$\sigma_f = 328\left(\frac{7.94E+05}{5.01E+05}\right)^{-0.112033} = 311.46 \text{ MPa} \quad (11)$$

Similarly, the fatigue stress is calculated from basquins equation for other heat treated specimens, and the results are tabulated. The table 2 gives the experimental and numerical fatigue limit of specimens for different heat treatment conditions.

Table 2. Fatigue limit of specimens for different heat treatment conditions

Heat Treatment Condition	EXPERIMENT Fatigue limit (MPa)	Fatigue limit from BASQUIN equation MPa	Error in %
Non heat treated	306	311.46	-1.78
Process annealed	323	321.70	0.4
Normalization	364	357.13	1.88
Temper 600^{0}C	370	376.92	-1.87
Temper 450^{0}C	407	411.97	-1.22
Tempering, 300^{0}C	483	489.4	-1.32

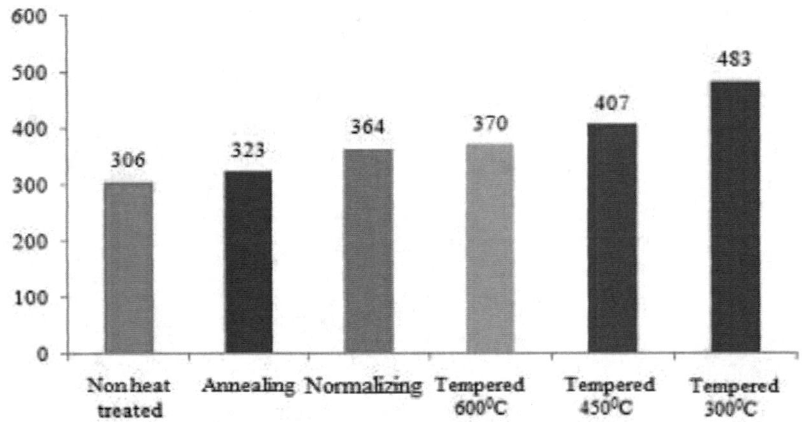

Figure 6. Comparison of fatigue limit of specimens subjected to different heat treatment conditions.

Figure 6, shows the comparison of fatigue limit for specimens subjected to different heat treatment conditions, the result shows that fatigue strength increases after heat treatment. It is clear that EN 8 steel tempered at 300°C has a fatigue limit of 483 MPa which is high when compared to non-heat treated specimens and the specimens subjected to other heat treatment conditions.

4.2 Fracture Studies

Figure 7, shows the optical fractographs of fatigue cracked specimens. One of the main features associated with fatigue crack growth is the appearance of fatigue striations. Striations are identified by distinct ridges, grooves and channels [21]. While such marks are due to crack extension probably by fatigue growth, the mechanism of formation requires the hard inclusions, combined mode 1(tension) and 2(shear direction of crack growth) or mode 1(tension) and 3(shear direction perpendicular to crack) [22, 23]. These marks are called the tyre tracks which are visible in untreated steel specimens [24, 25]; however, these are not distinctly visible in heat treated specimens due to micro-coring and segregation among the residual atoms in the steel solutionized phases [26].

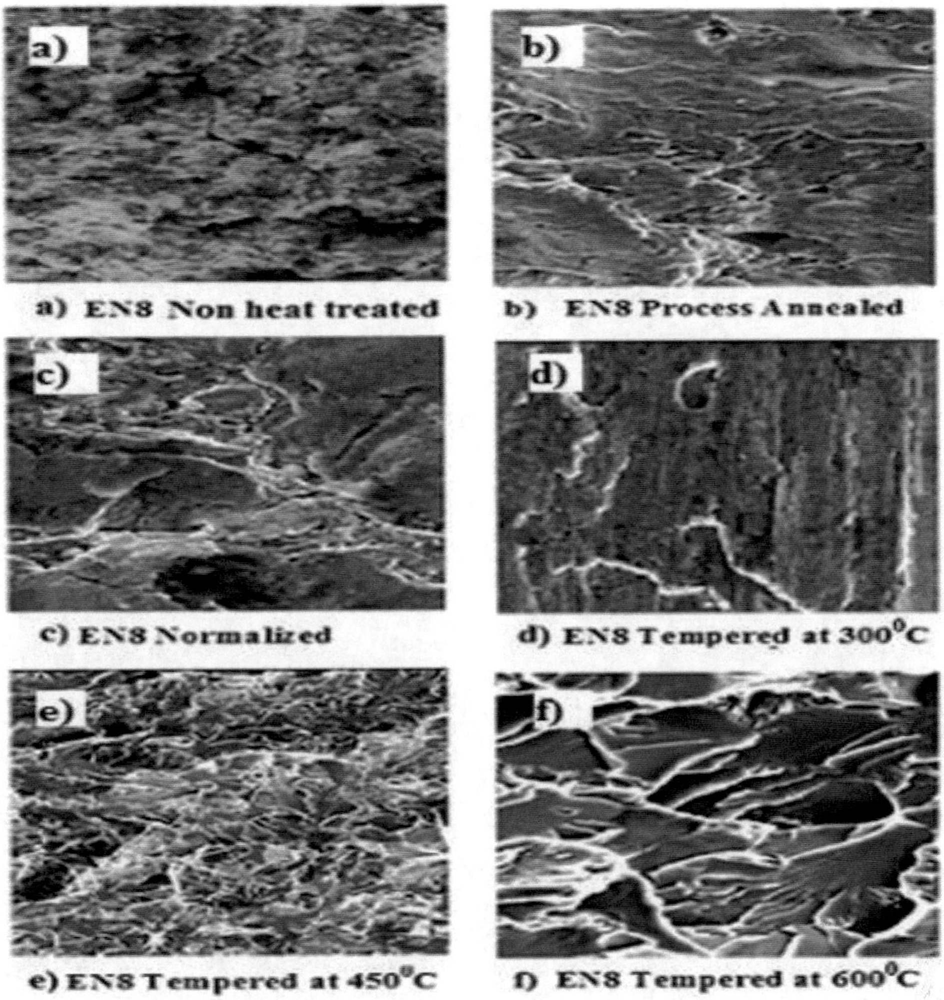

Figure 7. Optical fractographs of fatigue cracked specimens for different heat treatment conditions.

4.3 Analysis of Beam Using ANSYS

The analysis of beam is carried out using ANSYS for different end conditions and the comparative evaluation of the results are accomplished for structural steel and EN 8 steel.

The analysis of beam is carried out for 2 ends fixed condition for structural steel and EN 8 steel material tempered at 300°C conditions.

Analysis of beam fixed at 2 ends (Structural Steel)

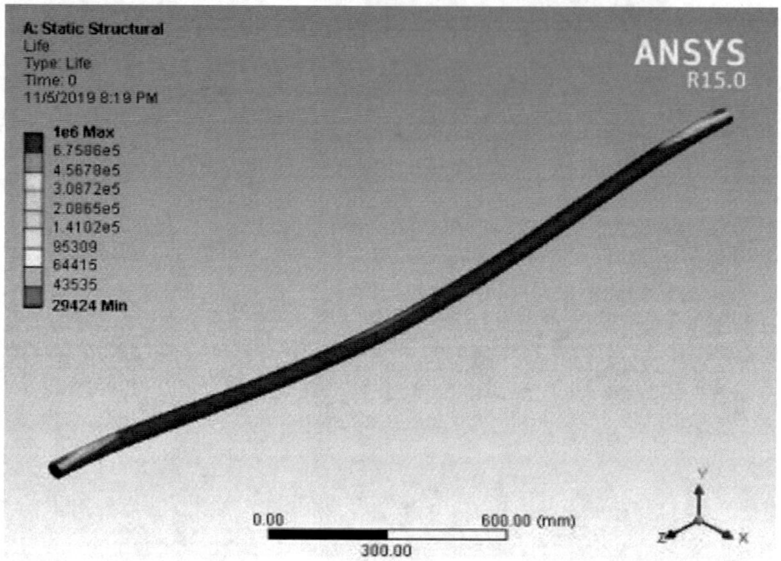

Figure 8. Fatigue life of beam fixed at 2 ends

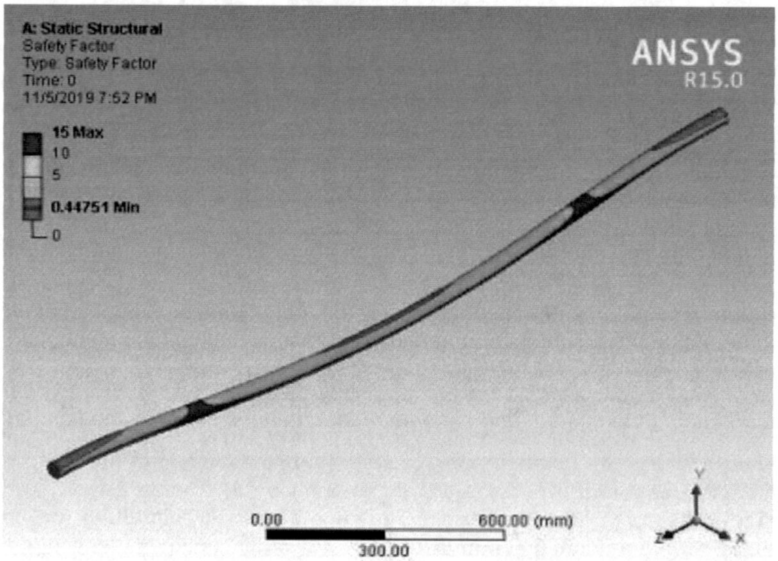

Figure 9. FOS of beam fixed at 2 ends

Table 3. Comparison of the results of a beam fixed at 2 ends

Material	Structural steel	
	Calculations	**ANSYS**
Total load	12000 N	12000 N
Positive maximum bending moment	1.2×10^6 Nmm	1.2×10^6 Nmm
Negative maximum bending moment	2.4×10^6 Nmm	2.4×10^6 Nmm
Maximum Bending Stresses	195.56 MPa	195.97 MPa
Displacement	7.04 mm	7.08 mm
Life (no. of cycles)	-	29424
F.O.S(min)	0.44	0.44

Analysis of beam fixed at 2 ends (material conditions EN8 steel tempered at 300^{0}C)

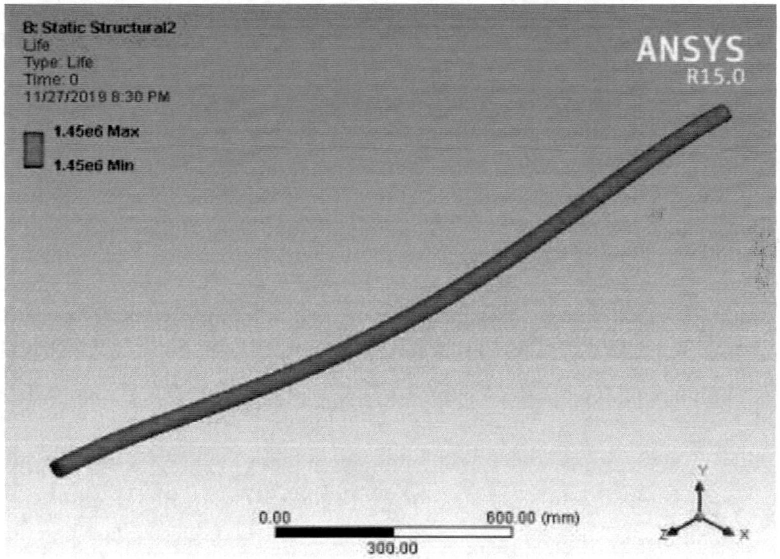

Figure 10. Fatigue life of beam fixed at 2 ends

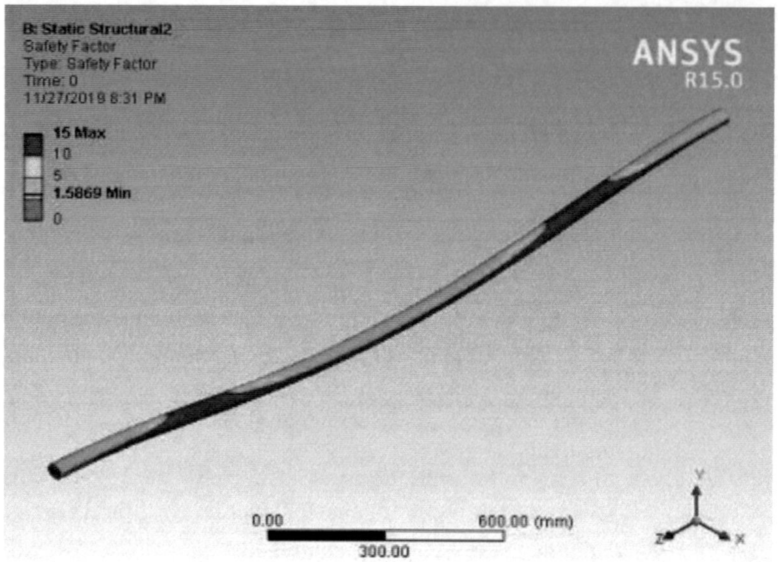

Figure 11. FOS of beam fixed at 2 ends

Table 4. Comparison of the results of a beam (material EN 8 steel tempered at 300^{0}C) fixed at 2 ends

Material	**EN 8 300^{0}C tempered**	
	Calculations	**ANSYS**
Total load	12000 N	12000 N
Positive maximum bending moment	1.2×10^6 Nmm	1.2×10^6 Nmm
Negative maximum bending moment	2.4×10^6 Nmm	2.4×10^6 Nmm
Maximum Bending Stresses	195.56 MPa	195.97 MPa
Displacement	6.70 mm	6.75 mm
Life (no. of cycles)	-	1.45×10^6
F.O.S(min)	1.58	1.58

Table 5. Comparison of fatigue life and FOS for different materials and conditions

Fatigue parameters	**Structural steel fixed at 2 points**	**EN 8 steel tempered at 300^{0}C for 1 hour, fixed at 2 points**
Life (no. of cycles)	29424	1.45×10^6
F.O.S(min)	0.4475	1.58

Life estimation of the specimens was carried out using the RR Moore fatigue test, for "as brought" and "heat treated" conditions like annealing, normalizing, and tempering at 300°C, 450°C and 600°C. After plotting the obtained values in a graph, it is observed that fatigue strength of annealed specimen is higher than untreated specimen, i.e., 323 MPa. Normalized specimen has higher fatigue limit when compared to annealed specimen i.e., 364 MPa. When compared to normalized specimen, the tempered specimen at 600°C, 1 hour, has fatigue limit of 370 MPa, while the tempered specimen at 450°C, 1 hour has fatigue limit of 407 MPa; the tempered specimen at 300°C, 1 hour, has a fatigue limit of 483 MPa. Thus, the EN8 Steel tempered at 300°C is considered for analytical evaluation and subsequent comparison with the structural steel material.

Analysis of beam at uniformly distributed load condition for structural steel gives a fatigue life of 29424 cycles and FOS of 0.4475.

For the improvement of fatigue life of the beam, two end support condition is considered from the initial trial and error in ANSYS software, which shows improvement in fatigue life and FOS i.e., the application of two end fixed boundary condition gives a fatigue life of 1×10^6 cycles and FOS of 1.7.

Improvement of fatigue life and factor of safety is observed, when material is changed, i.e., for EN 8 tempered steel fixed at 2 ends; the fatigue life is 1.45×10^6 cycles and FOS is 1.58, the bending stress 195.56 MPa is within safety limit, since the yield limit for EN8 tempered steel is 255 MPa.

5. Conclusions

The conclusions for the effect of heat treatment on fatigue strength of EN 8 steel is discussed with respect to the fatigue strength and subsequently, the improvement of life and factor of safety of EN 8 steel beam fixed at two points is validated with computational results obtained from ANSYS software.

Conclusions for the effect of heat treatment on Fatigue Strength

- It is observed from the experimental results that the fatigue strength increases after heat treatment.
- It is clearly seen from the results, that the fatigue strength is high for tempering as compared to other heat treated specimens.
- In tempering, one which is tempered at lower temperature i.e., 300°C has high fatigue strength compared to other tempered specimens.
- It is found that heat treatment can increase fatigue strength and life of the EN8 steel material along with other properties viz., FOS for the material.
- It is concluded that, Basquin's equation can be used to predict endurance limit from the values of test.

Analysis of Beam fixed at two ends

- If supports are used, fatigue life and FOS increases but usually supports can add weight and will modify the design because additional brackets and bearing is needed, that will sometimes increase maintenance.
- By change of material from structural steel to EN8 steel, fatigue strength and FOS are increased, while the bending stresses are within the allowable stress limit.

Thus, the EN8 steel with effective heat treatment viz., tempering at 300°C can be used for structural applications requiring better fatigue life, when subjected to cyclical loading in their routine operations.

References

[1] George E Totten Steel Heat Treatment Metallurgy and Technology Handbook 2nd Edition, Taylor and Francis, pp. 91-163, 2006.
[2] William D Callister Jr.; David G. Retwisch. Fundamentals of Material Science and Engineering., An Integrated Approach 3rd Edition John Wiley and Sons inc., pp380-390, 2007.
[3] Dr Dmitri Kopeliovich, Materials engineering, https://www.substech.com/dokuwiki/doku.php?id=dmitri_kopeliovich accessed on 10 Oct 2020.
[4] Jack A Collins, Henry R Busby, Gearge H Staab. Mechanical Design of Machine Elements and Machines 2nd edition John Wiley and Sons Co. pp 248-253, 2009.
[5] Ralph I Stephens, Ali Fatemi, Robert R Stephens, Henry O Fuchs. Metal Fatigue in Engineering 2nd Edition John Wiley and Sons Co, pp 59-90, 2000.
[6] Julie A Bannantine, Jess J Comer, James L Handrock. Fundamentals of Metal Fatigue Analysis, Prentice hall publications, New Jersey, pp 1-29, 1989.

[7] Cullen M. Moleejane, Kazeem O. Sanus, Olukayode L. Ayodele, Graeme J. Oliver. Microstructural Features and Mechanical behaviour of Unalloyed Medium Carbon Steel (EN8 Steel) after Subsequent Heat Treatment. World Congress on Engineering and Computer Science, vol. II, pp 22-24, Oct 2014.
[8] S. Abdul Kareem, S. I. Talabi; J. A. Adebis, T. Yahaya, O. H. Amuda. Fatigue Behaviour of Medium Carbon Steel of Different Grain Structures. IISTE, vol.5, pp 10-14, 2015.
[9] Phoumiphon. N, Radzali Othman, Ahmad Badri Ismail. Improvement in Mechanical Properties Plain Low Carbon Steel Via Cold Rolling and Intercritical Annealing Elsevier B.V. Procedia Chemistry 19, pp 822 – 827,2016.
[10]Changyou Li, Shuangfeng Li, Fei Duan , Yuewu Wang, Yimin Zhang, David He, Zhenyuan Li, WeiWang. Statistical Analysis and Fatigue Life Estimations for Quenched and Tempered Steel at Different Tempering Temperatures met7080312 MDP, August 2017.
[11]Blaoui Mohammed Mossaab, Mokhtar Zemri, Mustapha Arab. Effect of Medium Carbon Steel Microstructure on Tensile Strength and Fatigue Crack Growth 1757-9864 International Journal of Structural Integrity, vol.10 issue1, pp. 67-75, 2018.
[12]Palash Biswas, Arnab Kundul, Dhiraj Mondal, Prasanta Kumar Bardhan. Effect of heat treatment on microstructure behavior and hardness of EN 8 steel Materials Science and Engineering 377 IOP 012065 International Conference on Mechanical, Materials and Renewable Energy, 2018.
[13]Kazeem O. Sanusi, Esther T. Akinlabi. Experiment on Effect of heat treatment on mechanical and microstructure properties of AISI steel Elsevier B.V. Materials Today: Proceedings 5, pp17996–18001, 2018.
[14]Ramesh Singh Applied Welding Engineering, 2nd edition Elsevier pp-58-64, 2016.
[15]George E dieter Mechanical Metallurgy SI Metric edition, Mc-Graw-Hill Book ISBN: 0-07-100406-8, 1988.
[16]Oberg E, Machinery's Handbook 29th edition industrial press Inc, 2012.
[17]Sweta Rani Biswas, "Fatigue Behaviour Analysis of Differently Heat Treated Medium Carbon Steel", M tech (Res.) Thesis, Dept. of metallurgical and material engineering, National Institute of Technology, Rourkela, India, Jan 2013.
[18]Santhosh N, U N Kempaiah, Ganesh Sajjan, Ashwin C Gowda, Fatigue behaviour of silicon carbide and fly ash dispersion strengthened high performance hybrid Al 5083 metal matrix composites, Journal of Minerals and Materials Characterization and Engineering, Vol. 5, no. 5, pp 274 – 287, 2017.
[19]K.Lingaiah Machine Design Databook 2nd Edition, Mc-Graw-Hill Book ISBN: 0-07-136707-1, 2001.
[20]K.H Prabhudev Handbook of Heat Treatment of Steels, Mc-Graw-Hill education ISBN: 978007518311, 1988.
[21]Bohler shop service, A Practical Guide to Successful Processing of Special Steel Bohler Edelstahl GMBH, 1999.
[22]Tubal cain, Hardening And Tempering Heat Treatment Workshop Practice Series 1, ISBN: 978085248375, 1984.
[23]Heat treatment, ASM handbook, vol.4 ISBN: 0871703793, 2000.
[24]Avner, Sydney.H . Introduction to Physical Metallurgy 2ndedition, ISBN: 9780070024991, 1994.
[25]Robert.E.Reed-Hill Physical Metallurgy Principles 2nd Edition, ISBN: 0442268688/ ISBN-13: 9780442268688, 1979.
[26]Raymond A Higgins BSc (Brim.), C.Eng.F.I.M., Engineering Metallurgy Part Applied Physical Metallurgy, 6th edition ISBN:9780340568309 , 1993.

Fracture Toughness Measurement of Aluminium and Steel Inserted Aluminium Using Numerical Method

Mohan Kumar[1], Dr. Neelakantha V Londhe[2]
1,2Department of Mechanical Engineering, MITE, Mangaluru, VTU, Belagavi, India
E-mail: haimohanji@gmail.com

Abstract: - The aim of the current work is to test the possibility of improvement of fracture properties of a shaft fabricated by press fitting a rod inside a tube and explore the effect of interface on the crack propagation. Measurement of Mode II fracture toughness is considered for this purpose. In this study numerical method developed by Yong Li and Viola based on Kienzler and Herrmann's method by conducting a simple shear test on round bar specimen is adopted. Round bar specimens with a circumferential notch are subjected to shear test. Three specimens are prepared using Al6061T6 alloy tube inserted with rod of three different materials A6061T6 alloy, commercial pure Copper, and AISI1045 steel. Also a plain rod of Al6061T6 alloy is tested under shear to compare the results. Shear tests are conducted on Universal Testing Machine and fracture toughness values are calculated using modified Kienzler and Herrmann's equations. The results indicated an improvement of 18% in the fracture toughness values in the case of A6061T6 alloy rod inserted A6061T6 alloy tube compared to the fracture toughness of plain A6061T6 alloy rod. And when steel rod is inserted inside the A6061T6 alloy tube an improvement of 33.33% in the fracture toughness.

Keywords: Fracture toughness; Mode II; layered shaft; shear test; Al6061 alloy

1. Introduction

Fracture is the major failure mode in mechanical structures which lead to breaking down of components without any prior indication before the estimated safe working life. Many researchers have conducted extensive work on fracture mechanics to understand the phenomenon and developed many theories. But still many areas are left unexplored. Determination of fracture toughness is a significant approach to estimate the fracture strength of the materials. Researchers have established many ways to measure the fracture toughness of the materials. But these methods suffer a major problem that, they require very expensive equipment and accurate preparation of the specimens and require lot of time and cost. To overcome these problems many researchers have developed nonstandard methods to measure fracture toughness which are economical in terms of time and money, but achieved good approximation of the results. In most of the cases this approximate values are well sufficient [1, 2].

2. Earlier Work

2.1 Stress Intensity Factor – KI

The analytical procedure used for Linear elastic fracture mechanics relates the distribution of stress field and its magnitude near the crack tip, the stress applied to a specimen, orientation, size, and shape of a crack. In fracture mechanics, the stress intensity factor K means the stress field ahead of the crack tip in a structural member. The unit of parameter K is $MPa\sqrt{m}$, which indicates that it depends on the nominal stress (σ) in the member and the size of the crack. Hence different members with cracks or defects can be loaded to different values of K. Fracture of the materials have been classified into three major modes, called opening mode or Mode I, shearing mode or Mode II, and tearing mode or Mode III. The Fig. 1 illustrates the three modes of fracture [3].

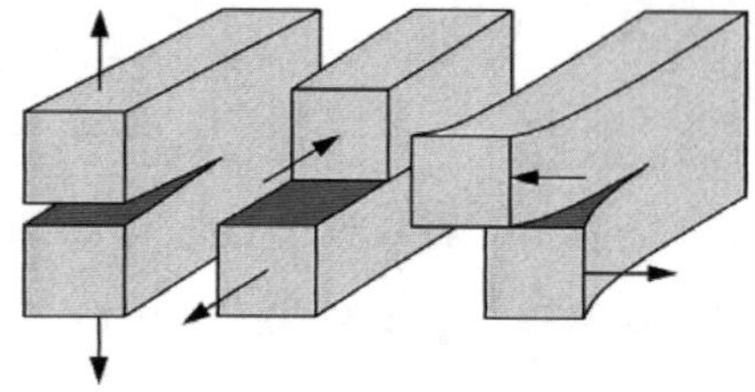

Figure 1. Three basic modes of crack surface displacements

2.2 The Critical Stress Intensity Factor

A structural member with crack can be loaded up to the complete fracture when the stress intensity factor reaches a limiting value. This limiting value before fracture is called fracture toughness, KC, which is a material property. The crack propagates when $K_I \geq K_{IC}$ is the fracture criterion by K_{IC}. [2–4]. ASTM E-399 is the standard test methodology for determination of plane-strain fracture toughness (K_{IC}) of metallic materials which fail predominantly in a brittle manner. The standard is designed to ensure that linear-elastic conditions prevail throughout the test. This is achieved by requiring a sufficiently large specimen for the particular toughness. Due to the possibility of brittle fracture in the presence of the notch, for the validation of fracture toughness values the plane-strain condition is required. The measured critical stress intensity factor (SIF) is then called the plane-strain fracture toughness for mode I (designated as K_{IC}) and is a true material property. It is important to note that the K_{IC} value to be obtained must be known or estimated before a K_{IC} test specimen is machined [5]. This test method (ASTM E-399) is not practical for materials showing significant plasticity before the beginning of stable crack growth because of the size limitations. Industry requires a simple straightaway approach to investigate the failed members.

Plane-strain fracture toughness testing of aluminum alloys is performed essentially in accordance with ASTM E 399 method [6]. However there are certain areas in which experience has shown a need for tighter requirements for aluminum alloys than specified in test method E 399, and others where some relaxation of validity requirements is justified on the basis of controlled experimental studies. The application of this practice will be of particular importance for quality assurance and material release in cases where valid plane-strain fracture toughness data cannot be obtained because of limitations of specimen size or some aspect of the fatigue cracking or resultant fatigue crack front, or both. To obtain valid K_{IC} value requires large size specimens as specified in ASTM E399. But in many cases it is practically difficult to obtain the required specimen size of the material to be tested. This necessitates the investigation of small sized specimen in fracture toughness test. Circumferentially notched round bar specimen (CNRB) is promising specimen geometry in this context [7].

The traditional standard fracture mechanics test procedures are very expensive, consuming longer time and require bulky specimens. Also, sometimes required size of sample may not be produced. The above said limitations lead to a development of a new approach for the determination of stress intensity factor (KI) using circumferential notch tensile (CNT) test specimens. This (CNT) is the smallest possible specimen which can give a valid plane-strain crack loading conditions within ± 3% of the values obtained using ASTM standard compact tension (CT) test specimens [8].

It is sometimes impractical and difficult to meet the dimensional requirements of specimens according to standard test procedures; may be due to insufficient volume of material available or due to geometric unsuitability of the of the sample material such as in pressure vessels and piping (PVP) systems materials. Adequate toughness is an essential characteristic of PVP system design. From the available material if it is not possible to prepare a specimen that can meet the specimen size criteria as per ASTM E399 specifications, then it is impossible to obtain a valid K_{IC} according to E 399. To overcome the inherent limitations of standard plane-strain fracture toughness test procedures, various new test approaches are proposed by investigators using either spiral notch torsion test or circumferentially cracked round bar subjected to torsion and tensile load or single edge rod specimen under mode I loading [9]. Standard procedures for fracture tests require elaborate specimen preparation, pre-fatigue cracking of the notch, and fatigue crack growth rate measurement, thorough analysis and interpretation of the data; which are all very expensive and time consuming. Furthermore, to obtain effective material data for all cases of specific batch of material, considerable numbers of tests required to be conducted. Therefore any method that can decrease the number of experimental tests and improve and optimize this process, and reduce the overall cost without compromising safety issues, is an advantage [10].

Neelakantha et al. conducted experiments on circumferentially notched round rod specimens of high strength aluminium alloy Al2014T651 and calculated K_{IC} and fatigue crack growth rate. The test procedure followed was same as explained earlier [11]. The round rod specimens with 'V' notch were pre-cracked on fatigue testing machine and then loaded till failure on universal testing machine. The Mode I fracture toughness was calculated using available equations. The values thus obtained were very close to the standard K_{IC} recorded by many researchers using standard compact tension (CT) and single edge notched bend (SENB) specimen configurations. The study reveals that use of circumferentially cracked round bar specimens to determine fracture toughness and fatigue crack growth rate is a modest, reliable, quicker and cost-effective method. This methodology was later used by many researchers and was recommended for fracture toughness as well as crack growth measurement as a standard test method [12].

2.3 Advantages of Using CNRB for Fracture Toughness Test

Common practice is to use a circumferential 60^0 V-notch in round specimens with cross sectional area under the notch equal to half that of the un-notched section. The surface effects and residual stresses at the notch root are very influential in this geometry because of small volume of the material at the notch root. The root radius is generally 0.13 mm or less. The diameter of notched section and the diameter at the root of the notch should be given a close tolerance within ± 25 microns [13].

The analytical fracture behavior of a CCRB specimen in tension is studied and the maximum fracture toughness as a function of specimen size is evaluated [14]. It is a practical test specimen used by various researchers for determining the fracture toughness [15]. Christopher et al. [16] lists following advantages;

- The CCRB specimen under tension is an axisymmetric geometry which develops uniform stress triaxilty along the circumferential notch front.
- The geometry allows equal access to the notch tip, making it ideal for stress corrosion cracking tests.
- Geometry is ideal for fracture toughness testing in mixed mode I and III.
- The geometry is attractive for dynamic or impact fracture toughness testing.

In the linear elastic fracture mechanics study of cracked surfaces determination of the stress intensity factor (SIF) plays a vital role. Three dimensional analysis would be required for the study of crack problems in round bar specimens having surface or part-through cracks. However, three dimensional analysis are complex in nature and exact solutions for surface cracks in round bars are not available. It has been proved by many researchers that methods involving minimum amount of computational work are suitable for the determination of strength of cracked bodies, even if the accuracy is lower [18]. E. Viola et. al., used two simple methods to determine quick and approximate values of stress intensity factor in cracked beams shown in Fig.2. The methods are, a) Kienzler and Herrmann's Method; b) Section Method.

The schematic of the specimen used in their study is as shown in the Fig.2.

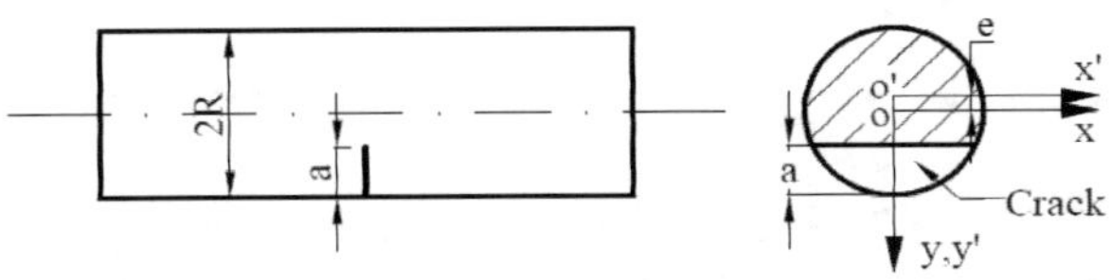

Figure 2. Cracked round bar and cross-section at the crack

In the current study the fracture toughness of aluminium rod and steel rod inserted aluminium beam is determined by Kienzler and Herrmann's Method.

2.4 Kienzler and Herrmann's Method

It's a numerical method of fracture toughness evaluation which gives approximate values which are sufficient for safe design of structures. This method enables to evaluate K_I, K_{II}, and K_{III} fracture toughness from simple tests of bending, shear and torsion respectively. The results obtained by this method for the beams with symmetrical cracks about the neutral axis are matching well with other known numerical results, and can thus serve as a good estimation tool in engineering design. [17].

When a beam of circular cross section with an edge crack is subjected to the bending moment M, the mode I SIF K_{IM} is given by Equation (1).

$$K_{I_M} = \frac{M}{R^{5/2}} F_M(\eta) \tag{1}$$

Where, $F_M(\eta)$ is the geometric function under bending moment [17].

If a cracked beam of circular section is subjected to a torque T, the expression of the SIF is given by Equation (2),

$$K_{III_T} = \frac{T}{R^{5/2}} F_T(\eta) \tag{2}$$

Where, $F_T(\eta)$ is the geometric function under torsion [17].

Similarly in the case of a cracked beam with circular section is acted upon by a shear force S, the expression for K_{II} is given by Equation (3),

$$K_{II_S} = \frac{S}{R^{3/2}} F_S(\eta) \tag{3}$$

Where, $F_S(\eta)$ is the geometric function under shear [17].

3. The Current Work

In the present work, fracture toughness of Aluminium 6061T6 is determined using Kienzler and Herrmann's Method and is compared with the fracture toughness of AISI 1045 medium carbon steel rod inserted inside aluminium tube. The aim is to enhance the fracture toughness of aluminium by reinforcing steel rod. The Mode I stress intensity factor K_I is measured by conducting bending test on UTM. Also the Mode II fracture toughness K_{II} is determined by conducting double shear test on UTM. Two types of specimens are prepared, one plain aluminium rod and second type is steel rod inserted inside aluminium tube by interference fitting. Both the specimens are of outside diameter 12 mm. The steel rod is 6 mm diameter solid. A circumferential crack is machined at the center of bending specimens. On the shear test specimen a very shallow crack is made using a hacksaw blade. The minimum crack depth required as available in the literature is assumed for shear test. Five specimens are tested in both the bending and shear.

After conduction of tests the fracture toughness is determined using Kienzler and Herrmann's Method which allows the quicker calculation of fracture toughness. Though it gives approximate value, in terms of time consumed it is beneficial, as the value obtained is sufficient for most of the applications.

4. Results and Discussions

The calculated values of fracture toughness (K_I) due to bending moment are illustrated in the following Table 1. The Table 2 lists the K_{II} fracture toughness values determined by shear test.

Table 1. K_I Fracture Toughness for Two Types of Specimens

Specimen No.	K_I Fracture toughness of solid aluminium rod (MPa$\sqrt{m}$)	K_I Fracture toughness of steel rod inserted aluminium (MPa$\sqrt{m}$)
1	35.8	41.3
2	35.6	41.5
3	35.8	41.0
4	36.1	41.2
5	35.7	41.2

Table 2. K_{II} Fracture Toughness for Two Types of Specimens

Specimen No.	K_{II} Fracture toughness of solid aluminium rod (MPa$\sqrt{m}$)	K_{II} Fracture toughness of steel rod inserted aluminium (MPa$\sqrt{m}$)
1	21.3	24.0
2	21.4	23.8
3	21.1	23.9
4	21.2	24.2
5	21.1	24.4

The K_I fracture toughness calculated for bending tests indicate that the rod inserted tube is having higher fracture toughness compared to solid aluminium rod. There is an improvement of 14% in the fracture strength, whereas the weight of steel rod inserted specimen increases by 44.5% but at the same time there is a saving in the costlier material like aluminium. Usually the strength and fracture toughness contradictory parameters, by combining high strength steel and tougher material like aluminium both toughness and strength are combined. This is also evident from the K_{II} fracture toughness values obtained by shear test. There is an improvement of 14.5% in K_{II} values in the case of steel inserted aluminium.

5. Conclusion

Enhancement of fracture toughness is a major challenge to the design engineer and always the strength and toughness does not exist together. The current work is an attempt to improve the fracture properties of aluminium. Bending and shear tests are carried out to determine the failure load and using Kienzler and Herrmann's Method the fracture toughness is measured. An improvement of around 14% is obtained for both K_I and K_{II} fracture toughness.

References

[1] Duga, J.J., Fisher, W.H., Buxbaum, R.W., Rosenfield, A.R., Burh, A.R., Honton, E.J., and McMillan, S.C., ''The Economic Effects of Fracture in the United States." NBS Special Publication 647-2, U.S. Department of Commerce, Washington, DC, March 1983.

[2] T. L. Anderson, *Fracture Mechanics: Fundamentals and Applications*, 3 rd ed, CRC Press, Taylor & Francis Group, N.W.,2005, ch. 1, pp 3 – 15.

[3] J. M. Barsom, and S. T. Rolfe, 1999, Fracture and Fatigue Control in Structures: Applications of Fracture Mechanics, ASTM, West Conshohocken, PA.

[4] N. Perez, 2005, Fracture Mechanics, Springer London Net Library, Inc. distributor, Guildford; Boulder.

[5] *ASTM E 399-90, Test method for plane-strain fracture toughness of metallic materials*, Philadelphia, PA, American Society for Testing and Materials, ASTM International, 2004.

[6] *ASTM-B645,Standard practice for plane-strain fracture toughness testing of aluminum alloys*, ASTM International, Philadelphia, PA., American Society for Testing and Materials,1995

[7] Steven. R.Lampman, Alloy Design for Fatigue and Fracture, Fatigue and Fracture; ASM Handbook, Handbook of Fatigue And Fracture, ASM International, Vol 19, 1997

[8] R.N. Ibrahim and H.L. Stark, Validity requirements for fracture toughness measurements obtained from small circumferentially notched cylindrical specimens, Engineering Fracture Mechanics, Vol 28(4),1987, pp 465-460

[9] J.A. Wang and K.C. Liu, A new approach to evaluate fracture toughness of structural materials, Transactions of the ASME, Vol 26, 2004, pp 534-540

[10] B. Farahmand, K. Nickbin, Predicting Fracture and Fatigue crack growth properties using tensile properties, Engineering Fracture Mechanics, Vol 75, 2008, pp 2144-2155

[11] N. V. Londe, T. Jayaraju, C. Rajashekar, and T. K. Chandrashekar, 2015, "Determination of Fracture Toughness in Tensile Fracture and Fatigue Fracture of Steels," pp. 92–97.

[12] V.L. Neelakantha, T. Jayaraju, P. Naik, D. K. K, C.R. Rajashekar, and Mohankumar, 2015, "Determination of Fracture Toughness and Fatigue Crack Growth Rate Using Circumferentially Cracked Round Bar Specimens of Al2014T651," Aerosp. Sci. Technol., 47, pp. 92–97.

[13] Ashok Saxena, Georgia Institute of Technology; Christopher L. Muhlstein, University of California, Berkeley, Fatigue Crack Growth Testing, ASM Handbook, Mechanical Testing and Evaluation,ASM international, Vol 8, 2000, pp 741-755

[14] B.K.Neale, The analytical fracture behaviour of a bar in tension containing a circumferential edge crack, UK. International Journal of Pressure Vessels and Piping, Vol 73, 1997, pp 191-198

[15] B.I.G.Barr and E.B.D.Hasso, Fracture toughness testing by means of the SECRBB test specimen, International Journal of Cement Composites and Lightweight Concrete, Vol 8(1), 1986, pp 81-87

[16] D.Christopher, Wilson and John D. Landes, P. C. Paris and K. L. Jerine, Fracture toughness testing with notched round bars, Fatigue and Fracture Mechanics, ASTM STP 1360, Eds., American Society for Testing and Materials, West Conshohocken, Vol 30, 2000, pp 69-82

[17] Erasmo Violaa, Yong Lib, Nicholas Fantuzzi On the Stress Intensity Factors of Cracked Beams for Structural Analysis, Key Engineering Materials, Trans Tech Publications, Switzerland Vols 488-489 (2012) pp 379-382

[18] Kienzler R, Herrmann G. An elementary theory of defective beams. Acta Mech 1986; 62:37-46.

[19] Viola E, Li Y, Fantuzzi N. The stiffness matrix of cracked beams for structural analysis. Technical report –DICAM Department (2011), University of Bologna, Italy.

Solid State Diffusion Bonding Process-A Review

Aravinda T [1]*, Dr. Niranjan H B[2], Dr. Satish Babu B[1], Dr. Udaya Ravi M[1]
1 School of Engineering, Presidency University/Department of Mechanical Engineering, Bengaluru, India
2 Sambhram Institute of Technology/Department of Mechanical Engineering, Bengaluru, Karnataka, India

*Email: sagar.aravind@gmail.com

Abstract— The present paper reports on the various diffusion bonding techniques using atomic diffusion of elements at the interface. The bonding process is used mainly for joining dissimilar metals. However, two similar metals can also be bonded with or without using an interlayer. Dissimilar materials include metal-ceramic; interlayer can also be used if necessary, in between the interfaces. The other methods of joining would probably interfere with the microstructure of the base metal at the point of interface. Various methods of joining like welding create reaction between the materials in the molten pool. Residual stresses would not appear at the bonding site, neither has it created thermal gradients using diffusion technique. Distortion observed in diffusion bonding is quite less compared to other joining process. It is also observed that the interface retains similar mechanical and physical properties as that of the base material. The strength at the interface depends on parameters like temperature, pressure and time taken for contact while forming the solid state bond.

Keywords: — Diffusion Bonding; Interlayer; Interface;

1. Introduction

Aluminium matrix composites (AMCs) are prospective materials for innumerable applications owing to their good physical properties. Several reinforcements can be added to AMCs in order to increase their mechanical properties [1-3]. Nowadays, structural materials in vehicles are being replaced by composite materials [4]. The idea of diffusion bonding to join MMCs is extensively used since there are complications of using conventional methods during fusion process [5]. These were problems related to formation of brittle phases during the interfacial reactions between the matrix alloy and reinforcements. Processes like fusion welding are not feasible for joining of composites. Metal Matrix Composites (MMCs) may not mix with metals owing to decreased flowability of the welding pool and viscosity [6]. There is also an undesirable chemical reaction between reinforcement and matrix which is undesirable. Flaws like porosity are commonly observed in solidified form. A review is made on the recent developments involving diffusion bonding processes.

2. Diffusion bonding involving powders:

Ji-hua Huang et al. [7] used Al-Cu and Al-Si powders during the reactive bonding of SiCp/6063 MMCs. Slurries were prepared by mixing suitable powders of Cu, Si, SiC and Al powders with ethanol in required proportions. SiCp/6063 MMC were then coated by the slurries prepared. Finally, reactive bonding process was performed under a definite temperature and pressure for a specified time under a vacuum. The interlayers formed by the powders were found to have a dense joining layer. A hypoeutectic microstructure was observed in the bonding. Upon close observation, they could find porous zones in large numbers in the site of joining layer. This could bring down the shear strength of the joints.

Urena et al. [8] used AA2024 alloy, reinforced with SiC particles with a volume % of 13. The composite was prepared by using compo casting process. A 13 mm thick plate was extruded to perform the diffusion process. The specimen surfaces were thoroughly mechanical grinded with SiC particles embedded in the emery sheets. The material was then degreased using ultrasonic waves in an acetone bath. An aluminium-lithium alloy in the form of a foil was used as an interlayer. A temperature of the range of 480 to 520° C, pressure between 3-6MPa and time of 60-120 min was maintained during the entire process. Vacuum was used to stop oxidation if any. Microstructural characterization was carried out using SEM and mechanical tests were conducted on the bonded specimen using Universal Testing Machine (UTM). Urena et al. concluded that if higher deformations were used with thickness reductions up to 25%, the Aluminium-Lithium layer would plastically flow thereby reducing voids from bonding interfaces. The maximum shear strength of 50 MPa was recorded under normal conditions of applying deformation at 12%, however no much increase was observed even though the bonding deformation was increased to 25%.

Janghorbhan [9] successfully produced multilayered ceramic-metal composites using polycrystalline alumina and different Al alloys through solid state diffusion bonding. The composites were fabricated at a temperature of 610°C. A 2-hour duration time was used under a compressive stress of 3 MPa. A spinel structure was observed for one of the composites containing Mg and Si in the alloy. However, another phase that is rich in Si and Mg was also noticed.

Several researchers [10-11] have reported that the bonding interface between matrix/matrix is mainly responsible for obtaining better joint during composite diffusion welding compared to reinforcement/reinforcement and matrix/reinforcement. The welding joint produced by diffusion, hinders to have a strong joint at the interfaces mentioned. Liu Liming et al. [12] worked on the welded joint for the composite Al2O3p/Al 6061 and predicted the existence of a critical temperature at the matrix/reinforcement interface. The composite exhibited better diffusion bonding at the critical temperature. The bonding at the critical temperature involved all the three interfaces namely reinforcement/matrix/reinforcement.

3. Diffusion bonding involving metal alloys in a solid form:

Mojtaba Jafariana et al. [13] worked on the interfaces of joints involving AZ31-O Mg and Al 6061-T6 diffusion bonding. Mechanical properties like hardness were measured using a standard hardness tester. A thermally O and T6 tempered aluminium alloy was used for the aluminium and magnesium joints. Diffusion of maximum number of magnesium atoms occurred to aluminium alloy while using AZ31-Mg/Al-6061-O. A micro hardness of 140 HV was observed in the diffusion zone. However, failure was observed while evaluating fracture surfaces from brittle intermetallic phases.

The difficulty in joining dissimilar metals like magnesium-aluminium is due to the formation of brittle intermetallic and oxide films in the joint region. Mahendran et al. [14] discusses about joining magnesium-aluminium through diffusion bonding technique. However, three main variables are responsible for the integrity of the bond namely, reaction time, pressure and temperature. Hardness and strength of the dissimilar joint was found out using response surface method and concluded that reaction temperature is mainly responsible for the strength of the joint.

Seyyed Salman Seyyed Afghani et al. [15] studied on the diffusion bonding method involving dissimilar alloys of Al 7075 and Mg AZ31. A pressure of around10-35 MPa, temperature of 430-450 °C with a reaction time of 60 min were selected as process parameters under a vacuum of 13.3MPa.The interfacial transition zone was characterized using energy-dispersive X-ray Spectroscopy (EDS), X-ray Diffraction (XRD) and Scanning Electron Microscopy (SEM), and found the presence of interfacial compounds, such as Al3Mg, Al12Mg17, and α(Al) solid solution. With increasing temperature, there was an increase in diffusion of aluminium atoms in to magnesium alloy. Fractographic analysis indicated predominantly of brittle and cleavage type, with cracks propagating from the brittle phase of Al_3Mg_2.

4. Diffusion bonding involving interlayers:

Jian Zhang [16] Used Ni interlayer while joining magnesium and aluminium through diffusion bonding technique. A vacuum furnace was used under a pressure of 1 X 10-3 Pa to hold the Mg-Ni-Al assembly. A reaction temperature of 673, 683, 703, 713, 723 and 733 K was used for 90 min. The results suggested that the interlayer was responsible to eliminate the formation of Al-Mg interfacial compounds. The micro hardness was recorded at 255 HV due to the presence of Mg2Ni phase. With increasing reaction temperature, the shear strength of the diffusion joints increased and then gradually decreased.

A copper foil was used as an interlayer by Nami et al. [17] in between two samples of insitu processed Al/Mg2Si composite prepared through gravity metal mold casting. The composite with the interlayer was placed in a vacuum chamber (30Pa) and the bonding was carried out at 540 °C for a time duration of 90, 120 and 150 min at a pressure of 6 MPa. The shear strength of the prepared samples increased with increasing bonding duration.

5. Diffusion bonding involving glass as substrate:

Field-assisted diffusion bonding is used as a major process during the preparation and assembly of pressure sensors, microfluidic devices and accelerometers [18- 19]. Liu CR et al. [20] worked on the field assisted diffusion bonding process of glass wafers with aluminium foils. A three layer sample of glass/aluminium/glass was placed in an oven so that the two glass wafers were connected to the negative electrode and aluminium film was electrically to the positive electrode. A temperature of 350-450°C and voltage of 400-700 V was used under a pressure of 0.05MPa.The bonded material showed an increase in tensile strength with increasing temperature and applied voltage.

6. Importance of process parameters in diffusion bonding:

Reference maps were prepared by Balasubramanian et al. [21] for aiding welding and design engineers while selecting proper diffusion bonding parameters. Temperature–time and pressure–time diagrams were treated as diffusion bonding windows while working on diffusion bonding of magnesium and aluminium alloys. To predict lap shear strength of the bonded joints, empirical relationships were developed. Further, to optimize the parameters involved in diffusion process, response surface methodology was used.

7. Conclusion

Diffusion bonding could be used efficiently to join substrates in powder form/solid form with metal alloys. Diffusion bonding can be used to overcome the problems associated with fusion welding.
Process parameters like temperature, pressure and holding time play a vital role in the diffusion bonding phenomenon. A critical temperature was recorded for the bonding process for getting an ideal bond. Shear strength and tensile strength of the joints was directly dependent on the reaction temperature used during the bonding process. Diffusion bonding technique at a critical temperature aids in having an efficient joint since all the interfaces like reinforcement/matrix/reinforcement are involved in the joining process unlike in conventional diffusion welding joining process.

8. Conflict of Interest

"The authors declare no conflict of interest".

9. Author Contributions

Aravinda collected information from various research related papers; Niranjan analyzed the paper; Aravinda, Udaya Ravi and Satish wrote the paper; all authors had approved the final version.

10.References

[1] Boppana, S. (2020) In Situ Synthesis of Titanium Carbide in Pure Aluminium. Journal of Materials Science and Chemical Engineering, 8, 1-10. doi: 10.4236/msce.2020.81001.
[2] Boppana, S.B., Dayanand, S. Impact of Heat Treatment on Mechanical, Wear and Corrosion Behaviour of In Situ AlB2 Reinforced Metal Matrix Composites Produced by Liquid Metallurgy Route. J Bio Tribo Corros 6, 33 (2020). https://doi.org/10.1007/s40735-020-0324-7.
[3] Boppana, Satish Babu, Samuel Dayanand, MR Anil Kumar, Vijee Kumar, and T. Aravinda. "Synthesis and characterization of nano graphene and ZrO2 reinforced Al 6061 metal matrix composites." Journal of Materials Research and Technology 9, no. 4 (2020): 7354-7362.
[4] Guruprasad T, Satish B, Maruhti B, Pramod K, Manjunatha H. Bus body structural strength analysis through FEA. Int. J. Technol. Res. Eng, vol. 2, pp. 2494-8, July 2015.
[5] Partridge PG, Dunford DV (1991) The role of interlayers in diffusion bonded joints in metal-matrix composites. J Mater Sci 26: 2255-2258.

[6] Zhang XP, Ye L, Mai YW, Wei GF (1999) Investigation on diffusion bonding characteristics of SiC particulate reinforced Al metal matrix composites. Compos A 30: 1415-1421.
[7] Jihua Huang, Yueling Dong, Jiangang Zhang, Yun Wan, Guoan Zhou (2005) Reactive Diffusion Bonding of SiCp/Al Composites by Insert Powder Layers with Eutectic Composition. J Mater Sci Technol 21(5): 779-781.
[8] Urena A, Gomez de Salazar JM, Escalera MD (1996) Diffusion bonding of an aluminium-copper alloy reinforced with silicon carbide particles using metallic interlayers. Scr Mater 35(11):1285-1293.
[9] Janghorbhan K (2009) Multilayer diffusion bonding of polycrystalline Al2O3 to Al alloys containing Sr, Mg and Si. Iran J Sci Technol B, 33 (B3): 241-251.
[10] Ellis MBD, Joining of aluminium based metal matrix composites (2013) Int Mater Rev 42 (2): 41-54.
[11] Liu YL (1996) A study on flow stress of Al-SiC composites deformed in a large range of strain. Scr Mater 35: 253-259.
[12] Liu Liming, Zhu Meili, Pan Longxiu, Wu Lin (2001) Studying of micro-bonding in diffusion welding joint for composite. Mater Sci and Eng A 315: 103–107.
[13] Mojtaba Jafariana, Mohsen Saboktakin Rizib, Morteza Jafariand, Mehrdad Honarmande, Hamid Reza Javadinejadb, Ali Ghaherib, Mohammad Taghi (2016) Effect of thermal tempering on microstructure and mechanical properties of Mg AZ31/Al-6061 diffusion bonding. Mater Sci and Eng A 666:372-379. http://dx.doi.org/10.1016/j.msea.2016.04.011.
[14] Mahendran G, Babu S, and Balasubramanian V (2009) Analyzing the effect of diffusion bonding process parameters on bond characteristics of Mg-Al dissimilar joints. J Mater Eng Perform 19(5):657-665.
[15] Seyyed Salman Seyyed Afghani, Mojtaba Jafrian, Moslem Paidar, Morteza Jafarian (2016) Diffusion bonding of Al 7075 and Mg AZ31 alloys: Process parameters, microstructural analysis and mechanical properties. Trans. Nonferrous Met Soc. China 26:1843-1851.
[16] Jian Zhang, Guo-Qiang Luo, Qiang Shen, Lian-Meng Zhang, Zhi-Jun Huang (2016) Characterization of diffusion-bonded joint between Al and Mg using a Ni interlayer. Rare Met 35(7): 537-542.
[17] Nami A, Halraee A, Adgi H, Hadian A (2010) Investigation on microstructure and mechanical properties of diffusion bonded Al/Mg2Si metal matrix composite using copper interlayer. J Mater Process Technol 210: 1282-1289.
[18] Wallis G, Pomerantz DI (1969) Field assisted glass-metal sealing. J Appl Phys 40 (10):3946-3949. https://doi: 10.1063/1.1657121.[19] Shoji S, Kikuchi H and Torigoe H (1998) Low temperature anodic bonding using lithium aluminosilicate-p-quartz glass ceramic sensors. Sens Actuators A 64:95-100.
[20] Liu CR, Zhao EJF, Lu XY, Meng QS, Zhao YP, Munir ZA (2008) Field-assisted diffusion bonding and bond characterization of glass to aluminum. J Mater Sci 43:5076-5082.
[21] Balasubramanian V, Joseph Fernandus M, Senthilkumar T (2013) Development of processing windows for diffusion bonding of aluminium/magnesium dissimilar materials. Weld World 57:523-539. https://doi.org/10.1007/s40194-013-0048-0.

LASER SURFACE TEXTURING OF EUTECTIC Al-Si ALLOY

Nandakumar M B[1], Lokesh G N[2], Harshad Natu[3] Mahesh B R[1], and Venkatesh M K[4]
1Department of Automobile Engineering, Dayananda Sagar College of Engineering. Bengaluru -560078, INDIA
2School of Engineering, Presidency University, Itgalpura, Rajanukunte, Bengaluru -560064, INDIA
3Director, Technical, Magod Fusion Technologies Pvt. Ltd., Bhosari, Pune - 411026, INDIA
4Department of Mechanical Engineering, Dayananda Sagar College of Engineering. Bengaluru -560078, INDIA

E-mail: nandakumarmb@gmail.com

Abstract: - Relative movement between piston and cylinder of an IC engine contributes to more than 60% of the friction experienced by the engine - in turn contributing to the wear of the piston and hence loss in its operational performance and efficiency. It is pertinent to explore technological solutions to overcome this situation. Texturing of surfaces appears to be an useful technique to enhance the tribological performance of the piston and cylinder assembly and information about the same are available in published literature. Laser surface texturing of a typical piston material such as an eutectic Al-Si alloy employing high power CO2-CW laser, firstly to examine the magnitudes of optimum laser processing conditions to realize surfaces which can retain oil and secondly to carry out characteristic analysis of the surface quality of the test material subjected to laser texturing, constitute the main thrust of this work. Measurements and analysis of the areal roughness parameters on the laser textured surfaces of the samples on a con-focal microscope reveal their capacity for enhanced oil retention, in turn ensuring reduced chances of scuffing of the sliding surfaces in comparison with non-textured samples, examined under similar test conditions.
Keywords: Laser Surface Texturing (LST); Tribology; Areal Surface Roughness Parameters.

1. Introduction

Automobile Industry continues to remain competitive both in terms of technology and cost. Each opportunity to improve Efficiency, Performance and service life of the components with cost effective methods is highly appreciated. Research endeavours have paved the way to develop components and methods with improved efficiency, performance and service life. Honing the inside surface of the cylinder has been found to create pockets for retaining oil, ensuring reduced levels of friction and wear. The honing of cylinders bores has been traditionally used to reduce friction and wear by creating oil pockets to retain oil for lubrication. This oil is at a risk of being burnt by the combustion gases leading to oil consumption.

2. Literature review

The oil retained in these tiny pockets are also found to be at a risk of getting burnt by the combustion gases leading to enhanced oil consumption. [1]. It is also evidenced that the lubrication oil can be best held in the piston skirt in comparison with the cylinder bore since it will not get exposed to the combustion gases directly[2]. The piston skirt, however, suffers from the side thrust due to the obliquity of the connecting rod which squeezes the lubricating oil present in the gap between the piston and cylinder thus making it to starve from lubrication. Modification of the topology of thrust side surface of the piston skirt by high power lasers appears to be a novel exercise to overcome the cited situation. The benefits derived from such an approach are also available in published information. It has also been found that the laser surface texturing of the piston skirt are stable and lasts longer in comparison with surfaces textured by honing process. Surface texturing has also been employed to produce micro dimples of varying geometries, resembling mini reservoirs and acting as micro hydrodynamic bearings[3,4,5] Laser surface texturing have been found to contribute to frictional reduction in lubricated tribo pairs which are operating under boundary lubrication regime[6,7]. Investigations on laser surface micro texturing of diesel engine cylinders have shown reduction in fuel consumption in comparison with a non textured liner surface under similar loading conditions[8]. Several methods are in practice to assess the oil retention capacity of the surfaces. Calculated magnitudes of oil capacity employing different methods appears to be in agreement with the results obtained by employing the family of Sk parameters[9]. Skewness(Rsk) establishes the symmetry of the profile of the surface about the mean line. A surface with equal peaks and valleys will have skewness value of zero whereas the one with more peaks than valleys will have magnitude of skewness which is positive and vice-versa. Surfaces with negative skewness has more bearing surface for sliding which is illustrated by the figure 1. The surface with zero skewness when plateaued can give negative skewness due to decrease in peaks but with no change in valleys. The plateaued surfaces have been known for lower friction having the capacity to hold oil in deep pockets[10]. Earlier results related to assessment of quality of the textured surfaces have also shown that the plateaued surfaces ensures lower friction due to its higher capacity to hold oil in deep pockets[11,12,13]. It is an established fact that the thickness of the oil film is a direct function of the surface roughness of the piston skirt and liners. The thickness of the oil film increases with an increase in the roughness characteristics of the surface[14]. It has also been conclusively evidenced that it is not only possible to selectively texture the surface by a laser but also can achieve consistency of the process[15]. In contrast to the belief that the smoother surfaces posses better scuffing resistances, research investigations have also established that the laser treated lapped surfaces posses better scuffing resistance in comparison with the ground and polished surface subjected to similar conditions of friction and wear[16,17]. Parameters influencing the quality of surfaces are defined schematically in figures 2 and 3.

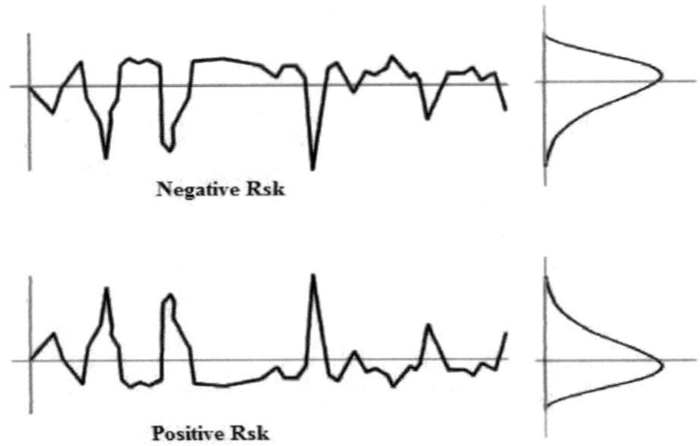

Fig. 1. Surface profile for Negative and Positive skewness.

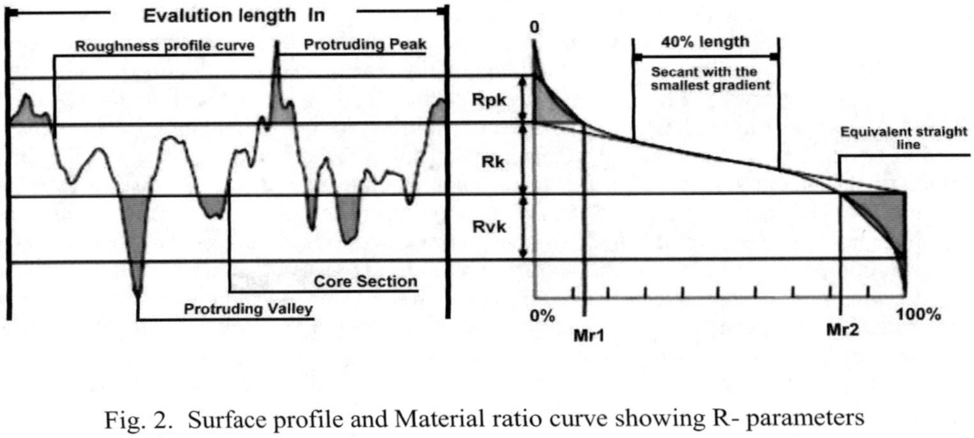

Fig. 2. Surface profile and Material ratio curve showing R- parameters

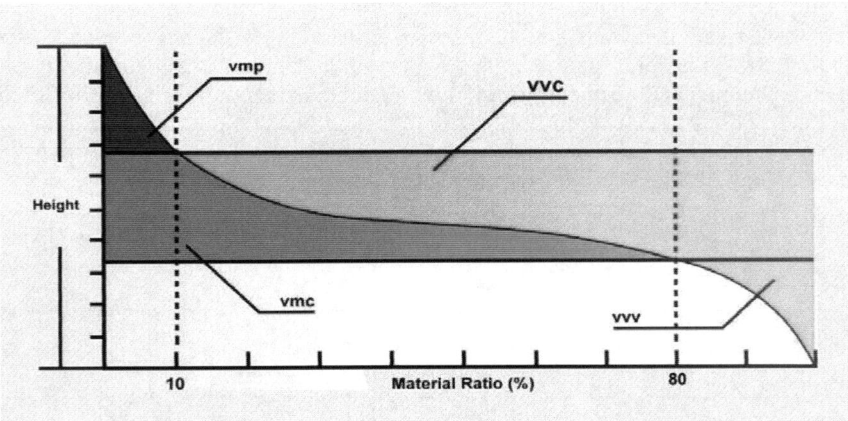

Fig. 3. Material ratio curve showing V- parameters[12]

3. Experimental details

3.1 Test Material

Eutectic Al–Si alloy test samples, 15x15x5mm geometry were directly machined out of old pistons of a single 6 cylinder IC engine of a commercial transport vehicle. The geometry of the test sample is shown in fig 4 and the composition in table 1.

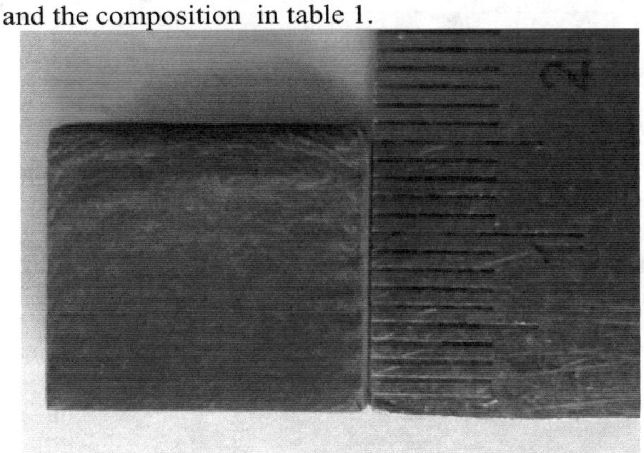

Fig 4: Test Sample

Table 1: Chemical Composition

Al	Si	Cu	Mg	Zn	Ni	Fe	Mn	Na
Remainder	11.85	1.5	1.3	1.5	1.5	0.8	0.3	0.25

The test samples were then ground to ensure an average surface roughness, Ra of 6μm and chemically treated with NaOH of 1N to increase laser absorptivity before being surface textured with a high power CO2 – CW laser.

3.2 Laser Parameters for Surface Texturing

Ground surface of the test samples were first textured over a range of laser processing parameters viz., power(0.4 to 0.8KW), beam diameter (0.4mm to 0.8mm) and scan speed (0.3m/min to 2.7m/min) to establish the basic and critical information about the depth and the quality of the laser textured surfaces. Energy density employed during the texturing process for each combination of the processing parameters in the cited range were computed employing scientific information available in published literature[3], i.e.,Energy Density (J/mm2) is calculated as
Laser Power (W)/[Beam Diameter(mm)x Scan Speed(mm/s)] and their magnitudes are shown in table 2.

Table 2: Energy Density

Specimen	Laser Power(kW)	Beam Diameter(mm)	Scan Speed(m/min)	Energy Density (J/mm2)
1	0.4	0.4	0.3	200.00
2	0.4	0.6	2.7	14.81
3	0.4	0.8	1.5	20.00
4	0.6	0.4	2.7	33.33
5	0.6	0.6	1.5	40.00
6	0.6	0.8	0.3	150.00
7	0.8	0.4	1.5	80.00
8	0.8	0.6	0.3	266.67
9	0.8	0.8	2.7	22.22

3.3 Optimization of Laser Parameters

In the first trials, nine (9) test samples were laser textured and the results are tabulated in table 3, which includes the actual photographs of the laser surface textured test samples over a range of laser processing conditions

Table 3. Lasing Parameters and Surface Photographs

Specimen	Laser Parameters	Photographs showing the distortion and surface texture of the sample
1	Power = 0.4kW Beam diameter = 0.4mm Scan speed =0.3 m/min Energy density = 200J/mm2	
2	Power = 0.4kW Beam diameter = 0.6mm Scan speed = 2.7 m/min Energy density = 14.81 J/mm2	

3	Power = 0.4kW Beam diameter = 0.8mm Scan speed =1.5 m/min Energy density = 20J/mm2	
4	Power = 0.6kW Beam diameter = 0.4mm Scan speed =2.7 m/min Energy density = 33.33 J/mm2	
5	Power = 0.6kW Beam diameter = 0.6mm Scan speed =1.5m/min Energy density = 40J/mm2	
6	Power = 0.6kW Beam diameter = 0.8mm Scan speed =0.3m/min Energy density = 150J/mm2	
7	Power = 0.8kW Beam diameter = 0.4mm Scan speed = 1.5m/min Energy density = 80J/mm2	
8	Power = 0.8kW Beam diameter = 0.6mm Scan speed =0.3m/min Energy density =266.67J/mm2	
9	Power = 0.8kW Beam diameter = 0.8mm Scan speed = 2.7m/min Energy density = 22.22J/mm2	

Table 3 clearly reveals that the surface of the test samples 1, 6, 7 and 8 are over melted or distorted, possibly due to higher magnitudes of input energy densities whereas the surfaces of the test samples 2 and 3 appear to have not been discernibly affected by laser texturing process. However in the case of samples 4, 5 and 9, the quality of their surfaces appears to be moderately affected by laser texturing

process. The magnitudes of the energy densities employed to laser texture the surfaces of these test samples are in the range of 20 to 40 J/mm2, considered as optimum in this work.
Accordingly, three new samples were laser textured at powers varying from 0.4 to 0.8kW and scan speeds varying from 2 to 4 m/min at a constant beam diameter of 0.4mm. Energy densities in all the cases were kept constant at 30 J/mm2 which is the mid value between 20 to 40 J/mm2 considered as optimum range in this work. The laser treatment scheme is tabulated in table 4.

Table 4. Laser Texturing with Optimum Process Parameters

1. Sample	2. Laser Power(kW)	3. Scan Speed(m/min)
4. 1	5. 0.4	6. 2
7. 2	8. 0.6	9. 3
10. 3	11. 0.8	12. 4

3.4 Surface Topography Analysis

Surface topology of the laser textured test samples shown in table 2 were captured on a Olympus LEXT OLS 4000 laser con focal microscope. The 3D surface topology of the laser textured test samples and the parameters for assessing the characteristic nature of the surfaces are shown in figs. 5,6 and 7.

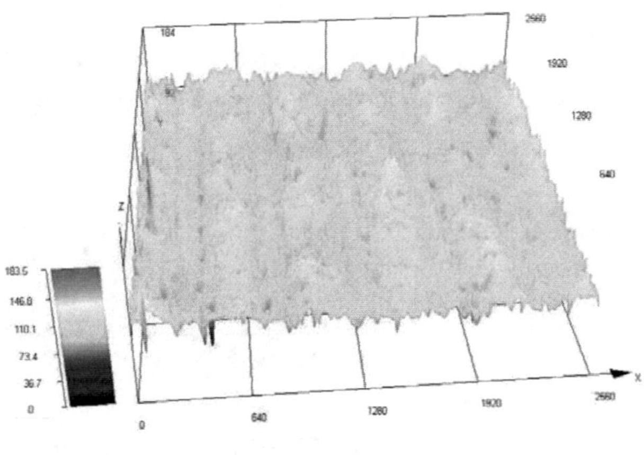

Figure 5. Surface Topology (P - 0.4kW, U - 2m/min)

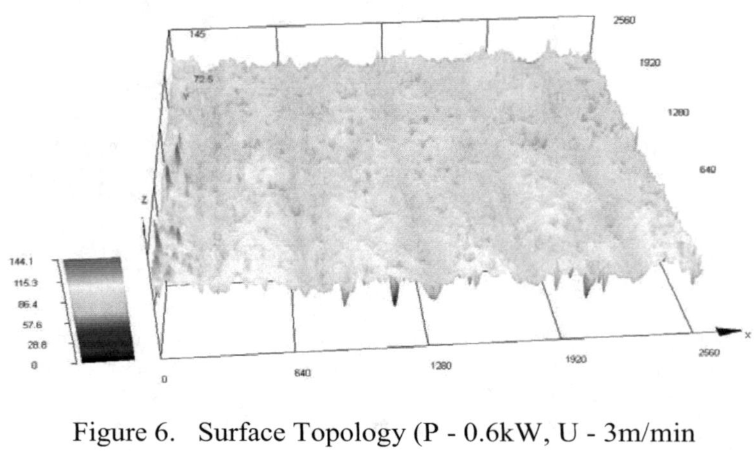

Figure 6. Surface Topology (P - 0.6kW, U - 3m/min

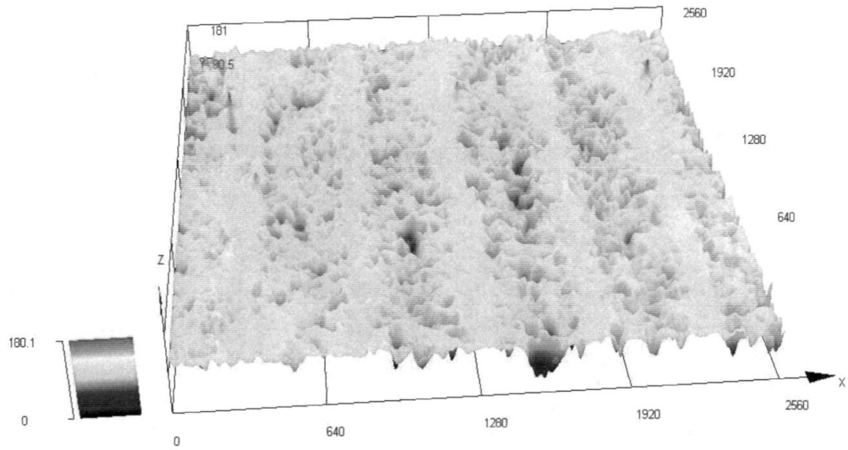

Figure 7. Surface Topology (P - 0.8kW, U - 4m/min)

4. Results and Discussions

Analysis of the topology of the laser textured surfaces of the test samples clearly reveal that the magnitudes of the volume of the voids are 1.171 $\mu m^3/\mu m^2$, 1.319 $\mu m^3/\mu m^2$ and 3.254 $\mu m^3/\mu m^2$ for samples 1(P – 0.4 KW, U – 2 mm/s), 2 (P – 0.6 KW, U – 3 mm/s) and 3 (P – 0.8 KW, U – 4 mm/s) respectively clearly indicating that at identical magnitudes of energy density, input laser power determines the magnitude of the volume of voids with higher laser power facilitating the larger volume of voids which in turn can hold higher quantity of oil during the lubricated tribo process. Further, the volume of peaks per unit area are 0.425$\mu m^3/\mu m^2$, 0.428$\mu m^3/\mu m^2$ and 0.326 $\mu m^3/\mu m^2$ for samples 1, 2 and 3 respectively, probably influenced by the cooling rates encountered during the texturing process, with higher scan speeds of the beam ensuring higher cooling rates- characteristic of the process. Further, the magnitudes of S_{sk}, representing the number of valleys on the textured surface are all negative indicating the presence of more number of valleys per unit area compared to number of peaks per unit area clearly suggesting that the textured surfaces have more number of valleys which can hold more oil during the lubricated tribo process and hence contributing reduction in friction and wear. The results of this investigation highlights the novelty of employing high power laser for texturing the surfaces of materials besides suggesting the importance of this technique to realise the much needed techno economic benefits in automobile environments.

5. Conclusions

The results of this investigation reveal the following

Laser texturing of the surfaces of pistons of an IC engine can significantly contribute to their reduced friction and wear. Optimum energy densities, between 20 to 40 J/mm^2 appears to be appropriate to realise surfaces which can retain oil during the lubricated tribo process for the test materials used in this investigation. The results of this investigation has considerable relevance and significance in the contexts of employing test samples which are directly machined out from an in service vehicle

References

[1] Furuhama, S., Hiruma, M., and Yoshida, H.: "An Increase of Engine Oil Consumption at High Temperature of Piston and Cylinder", SAE paper 810976, 1981.

[2] David P. Hoult & Byron T. Shaw II, 'The Puddle Theory of Oil Consumption©', Tribology Transactions, Volume 37, 1994 - Issue 1, Pages 75-82
[3] J. A. Wahab, M. J. Ghazali, W. M. W. Yusoff & Z. Sajuri (2016) Enhancing material performance through laser surface texturing: a review, Transactions of the IMF, 94:4, 193-198, DOI: 10.1080/00202967.2016.1191141
[4] M. Afzal Malik, S Adnan Qasim, Mumtaz Ali Khan, Riaz Mufti, Shear-Heating Effects on Piston Skirts Lubrication in the Initial Engine Start Up, Proceedings of the World Congress on Engineering 2010 Vol II WCE 2010, June 30 - July 2, 2010, London, U.K.
[5] Izhak Etsion, "State of the Art in Laser Surface Texturing" Vol. 127, JANUARY 2005, Transactions of the ASME
[6] AndriyKovalchenkoa, OyelayoAjayia, Ali Erdemira, George Fenskea, IzhakEtsionb. "The effect of laser surface texturing on transitions in lubrication regimes during unidirectional sliding contact", Tribology International 38 (2005) 219–225.
[7] G. Ryk , Y. Kligerman & I. Etsion (2002) Experimental Investigation of Laser
Surface Texturing for Reciprocating Automotive Components, Tribology Transactions, 45:4, 444-449, DOI: 10.1080/10402000208982572
[8] XijunHua, JianguoSun, PeiyunZhang, HanqingGe, YonghongFu, JinghuJi, Bifeng Yin, "Research on discriminating partition laser surface micro-texturing technology of engine cylinder" TribologyInternational 98(2016)190–196
[9] Wieslaw Grabon, Pawel Pawlus, Waldemar Koszela, Rafal Reizer, "Proposals of methods of oil capacity calculation", Tribology International 75 (2014) 117–122
[10]E. S. Gadelmawla, M. M. Koura, T. M. A. Maksoud, I. M. Elewa, H. H. Soliman, "Roughness Paraemeters", The Journal of MaterialProcessing Technology, 123(2002), 133-145
[11] Jeng, Y. R., 1996, ''Impact of Plateaued Surfaces on Tribological Performance,'' Tribol. Trans., 39, pp. 354–361.
[12] Marko Sedlaček, Peter Gregorčič & Bojan Podgornik (2017) Use of the Roughness Parameters Ssk and Sku to Control Friction—A Method for Designing Surface Texturing, Tribology Transactions, 60:2, 260-266, DOI:10.1080/10402004.2016.1159358
[13] Tianchang Hu & Litian Hu (2011) Tribological Properties of Lubricating Films on the Al-Si Alloy Surface via Laser Surface Texturing, Tribology Transactions, 54:5, 800-805, DOI:10.1080/10402004.2011.604761
[14] Staffan Johansson, Per H. Nilsson, Robert Ohlsson, Cecilia Anderberg, Bengt-Goran Rose´n, "New cylinder liner surfaces for low oil consumption", Tribology International, Volume 41, Issues 9–10, September–October 2008, Pages 854–859.
[15] M. Shuster , F. Mahler & D. Crysler, "Metallurgical and Metrological Examinations of the Cylinder Liner-Piston Ring Surfaces AfterHeavy Duty Diesel Engine Testing" Tribology Transactions, 42:1, 116-125, ISSN: 1040-2004 (Print)
[16] Lugt, P. M., Severt, R. W. M., Fogelström, J., and Tripp, J. H., 2001, "Influence of surface ... on friction, film breakdown and running-in in the mixed lubrication regime," Proc. Inst. Mech. Eng., Part J: J. Eng. Tribol., 215, pp. 519–533
[17] Jon Sundh , Ulf Skytte af Satra, "Influence of surface topography and surface modifications on seizure initiation in lean lubricated sliding contacts" Wear 262 (2007) 986–995.

Mechanical Properties of Aluminum Metal Matrix Composite Reinforced with Silicon Carbide using FEM

Mohammed Faraz [1], Haseebuddin M R [2], Harishraj [1], Bhaskar Pal [3]

1 UG Scholar, Department of Mechanical Engineering, Dayananda Sagar College of Engineering, Bengaluru.

2 Associate Professor, Department of Mechanical Engineering, Dayananda Sagar College of Engineering, Bengaluru.

3 Professor, Department of Mechanical Engineering, Presidency University, Bengaluru.

Abstract

In this work the SiC reinforced Alumunium composites were studied and Finite Element Method (FEM) analysis carried out and the results of FEM analysis were compared with the actual experimental results. As the Finite Element Analysis has become a major tool for an engineer, the Finite Element Analysis is done to solve advanced analysis issues. In this paper, Finite Element Method based ANSYS software is used to determine the mechanical properties in the aluminium metal matrix composite reinforced with Silicon Carbide by changing the volume fraction of the Silicon Carbide in aluminium. The FEM analysis results were compared with the actual experimental result. The testing if the mechanical properties such as tensile testing and compression testing we carried out in the FEM analysis.

Keywords: FEM, Mechanical Properties, SiC, Aluminium, MMCs.

1. Introduction

Aluminium is a widely used metal in most engineering applications. This is mainly because of its beneficial properties such as its low weight, low density and good formability. Aluminium finds its use in various high-performance applications such as automotive, military, aerospace and electrical industries. In the automobile industry, the Aluminium metal is used in the fabrication of suspension parts, cylinder blocks inside the engine, radiators and transmission bodies. The disadvantage of pure aluminium is its low strength and the fact that it is prone to corrosion. Due to this, Aluminium is reinforced with other metals or non-metals to form suitable alloys and composites to further improve its properties and overcome its disadvantages. One such reinforcement of nano-particles which is used is Silicon Carbide (SiC). Silicon Carbide has been the most broadly utilized material for the utilization of ceramics in structures. Attributes, for example, low coefficient of thermal expansion, high power to-weight span, high thermal conductivity, hardness, corrosion and abrasion resistant, and in particular, the support to elasticity at temperatures up to 1650 ° C, have prompted a wide scope of employments. A Metal Matrix Composite (MMC) is a composite material constituting two parts, one being a metal mostly and the other part can be a metal or a different kind of material such as an organic compound. The creation of an MMC of a metal improves its properties with respect to its base metal. MMCs combine metallic properties with ceramic properties of the added reinforcement and hence possess greater strength in shear and compression and also ensure high service temperature capabilities. They induce in the Aluminium significant increase in strength and the strength of the composites were found to be directly proportional to the percentage contents and fineness of the reinforced particles up to a certain point.

2. Experimental Procedure

Finite Element Analysis or FEM is a computerized, computational method for predicting force, damping, and other effects. Analysis of FEA, the material breakage, strength, wear and tear can be identified. To solve the exact solution, we will figure out the need to obtain the exact solution. A cylinder was manufactured in this work using the produced aluminium matrix composite and the value of the properties.

2.1 Structural Analysis by Workbench:

The ANSYS workbench program is ideal for estimating the engineering problem complex and providing effective product design. FEA is a tool that can optimize the simulation and produce it, and get resolution. This ANSYS workbench program is designed to solve the physical work quite easily. It was pointed out about the stress distribution of pushrod components. The flow of solving a problem by ANSYS FEA is shown below.

IOP Conf. Series: Materials Science and Engineering 1013 (2021) 012013 doi:10.1088/1757-899X/1013/1/012013

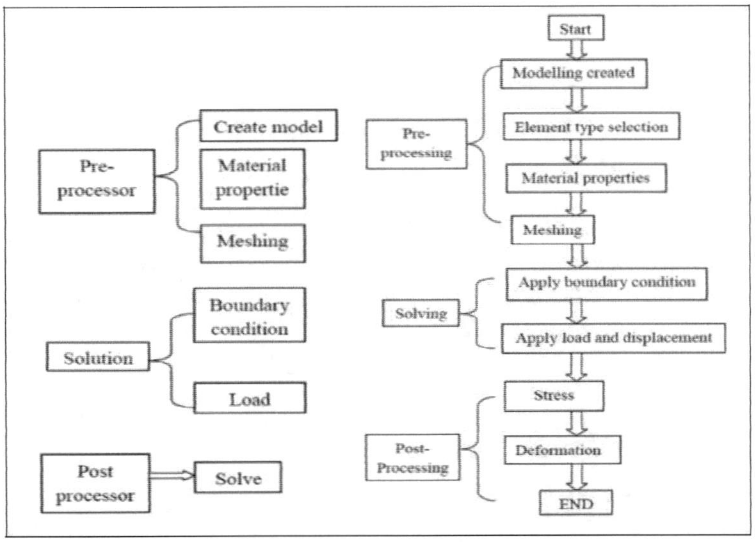

Figure 2.1: Flow in ANSYS

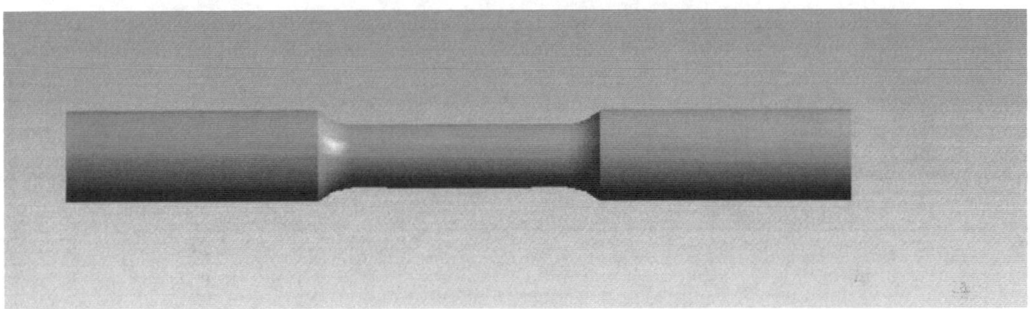

Figure 2.2: Cylindrical Model

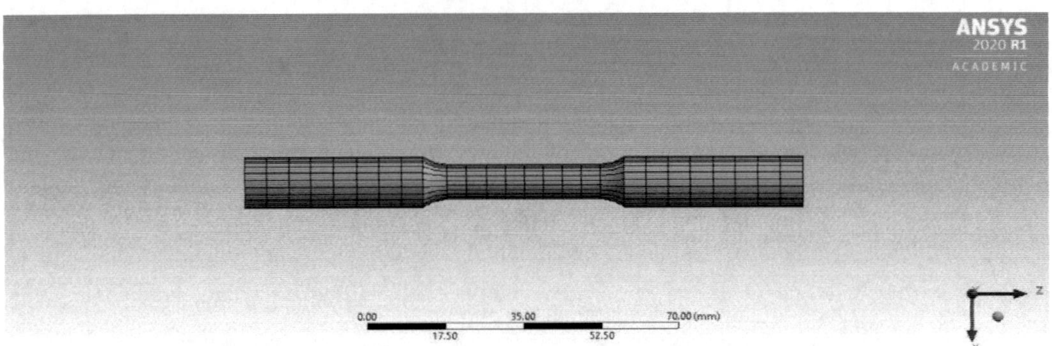

Figure 2.3: Meshed Model

2.2 Tensile Test

Object Name	*Fixed Support*	*Force*
State	Fully Defined	
Scope		
Scoping Method	Geometry Selection	
Geometry	1 Face	
Definition		
Type	Fixed Support	Force
Suppressed	No	
Define By		Components
Applied By		Surface Effect
Coordinate System		Global Coordinate System
X Component		0. N (ramped)
Y Component		0. N (ramped)
Z Component		-11000 N (ramped)

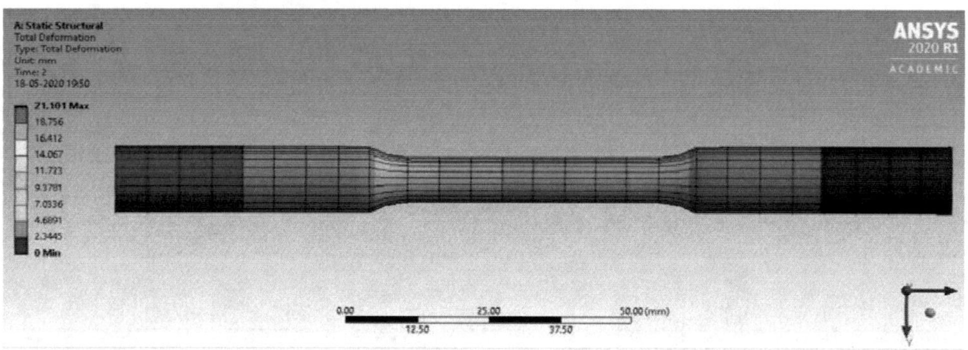

Figure 2.4: Von Mises Stress for specimen with 5% SiC Reinforcement

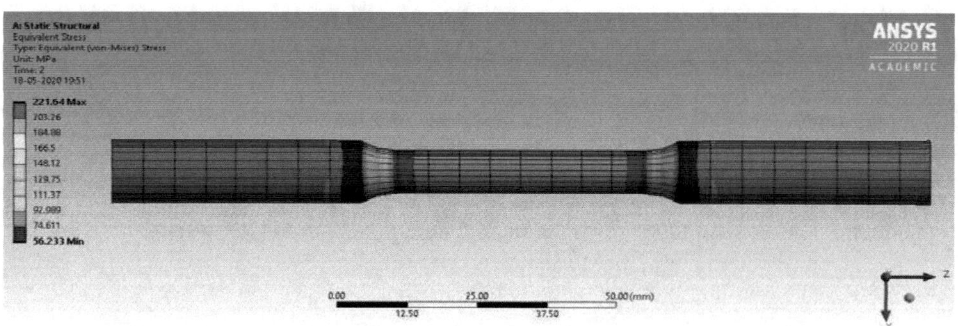

Figure 2.5: Total Deformation for specimen with 5% SiC Reinforcement

IOP Conf. Series: Materials Science and Engineering 1013 (2021) 012013 doi:10.1088/1757-899X/1013/1/012013

Figure 2.6: Stress for specimen with 10% SiC Reinforcement

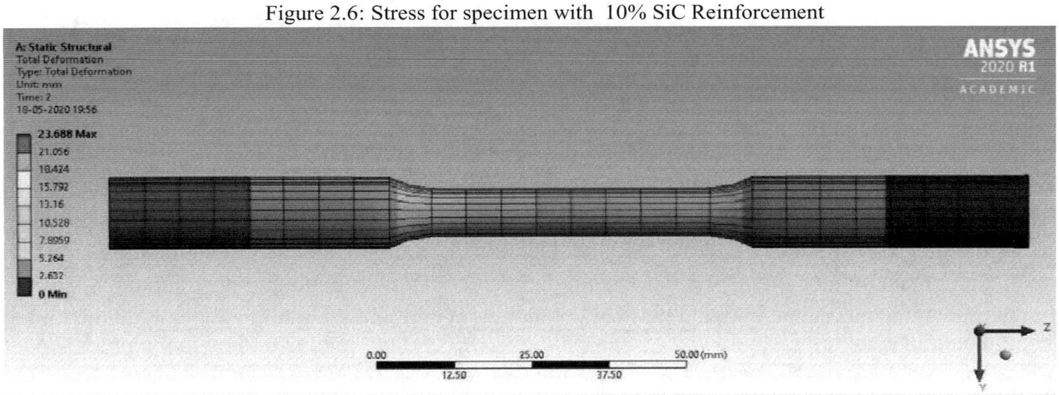

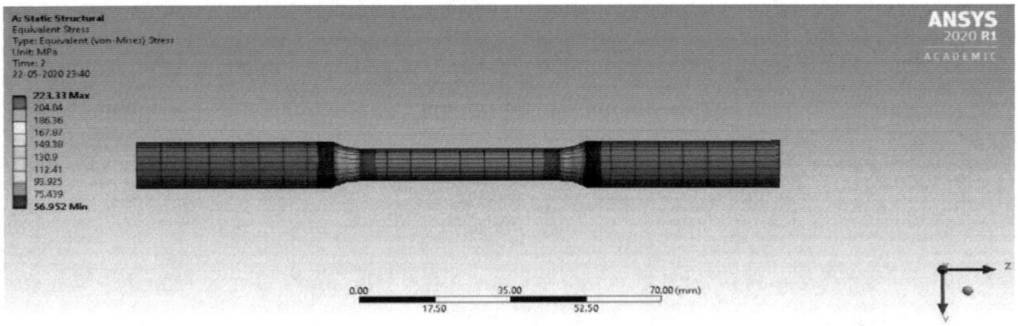

Figure 2.7: Total Deformation for specimen with 10% SiC Reinforcement

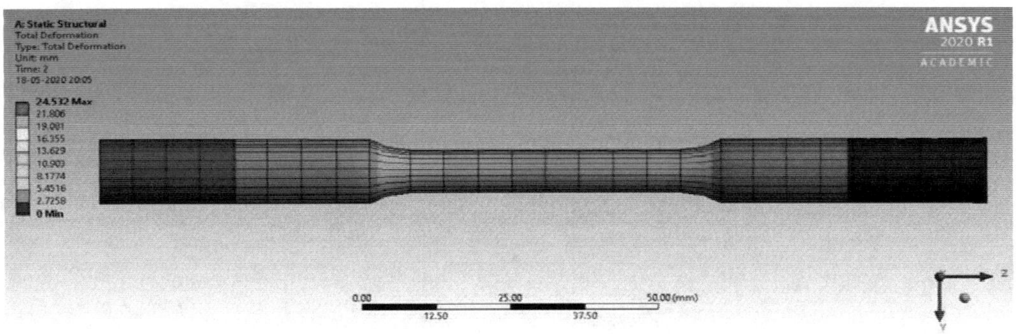

Figure 2.8: Stress for specimen with 15% SiC Reinforcement

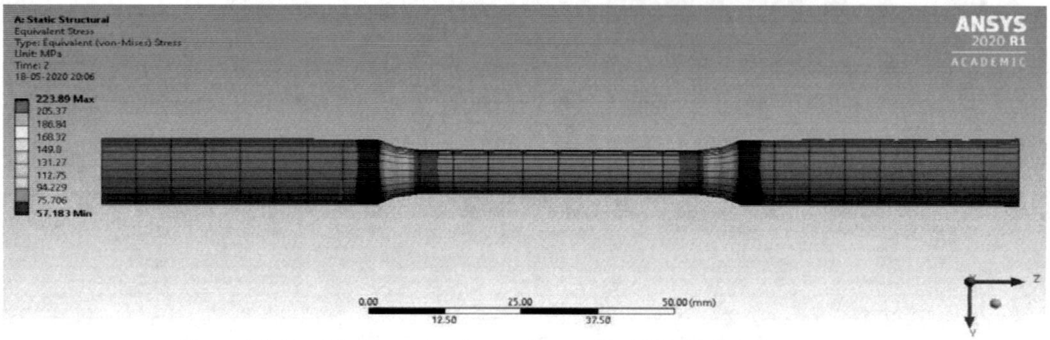

Figure 2.9: Total Deformation for specimen with 15% SiC Reinforcement

Table 2.2: Stress and Total deformation results

Weight Percentage Volume	Equivalent Stress (MPa)		Total Deformation (mm)	
	Maximum	Minimum	Maximum	Minimum
5%	221.64	56.233	21.101	0
10%	222.46	56.756	23.688	0
15%	223.89	57.183	24.532	0

2.3 Compression test

Figure 2.3 :

Object Name	Fixed Support	Force
State	Fully Defined	
Scope		
Scoping Method	Geometry Selection	
Geometry	1 Face	
Definition		
Type	Fixed Support	Force
Suppressed	No	
Define By		Components
Applied By		Surface Effect
Coordinate System		Global Coordinate System
X Component		0. N (ramped)
Y Component		0. N (ramped)
Z Component		15000 N (ramped)

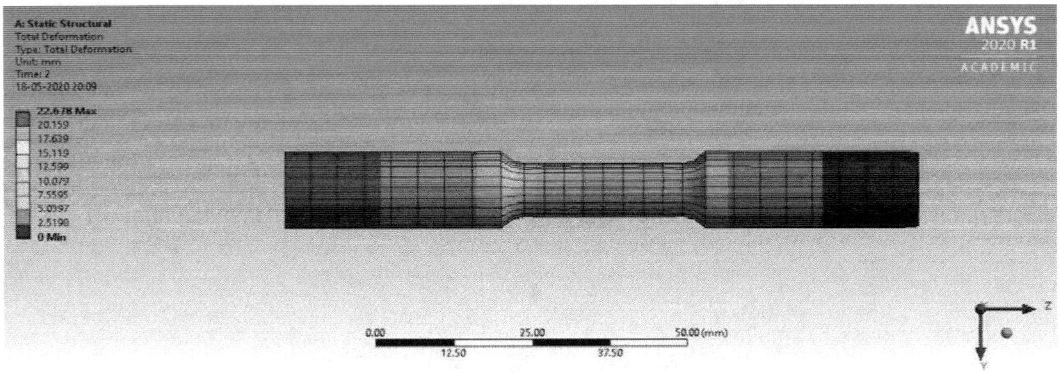

Figure 2.10 : Stress in Specimen with 5% Reinforcement

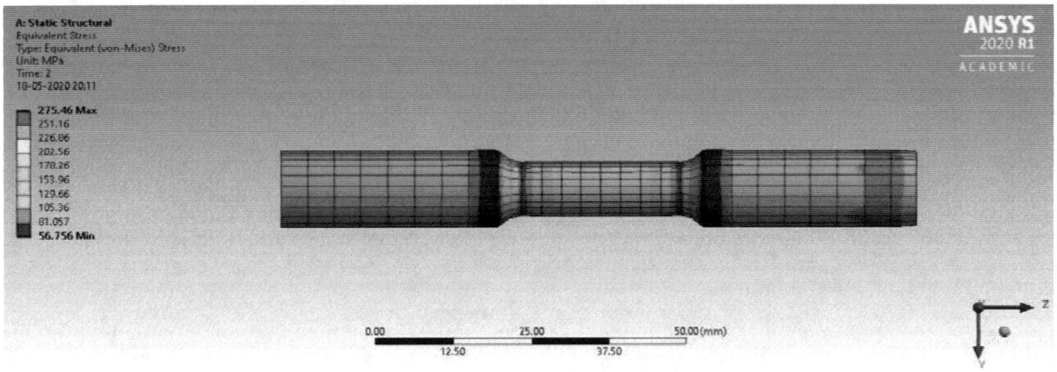

Figure 2.11: Total Deformation in Specimen with 5% Reinforcement

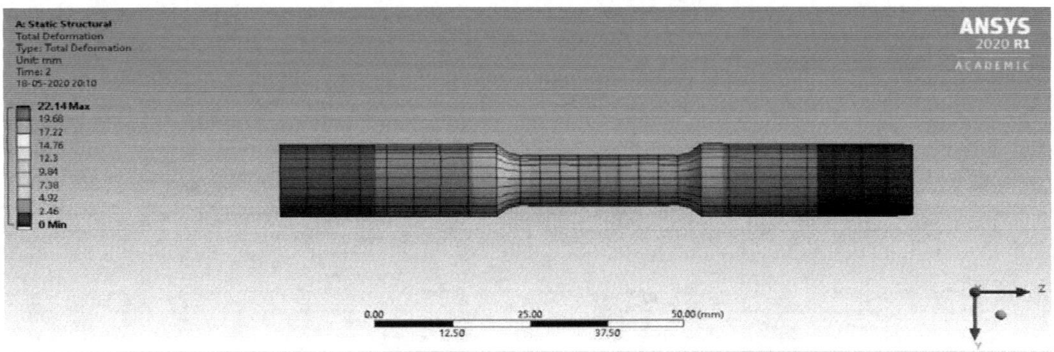

Figure 2.12: Stress Deformation in Specimen with 10% Reinforcement

IOP Conf. Series: Materials Science and Engineering 1013 (2021) 012013 doi:10.1088/1757-899X/1013/1/012013

Figure 2.13 Deformation in Specimen with 10% Reinforcement

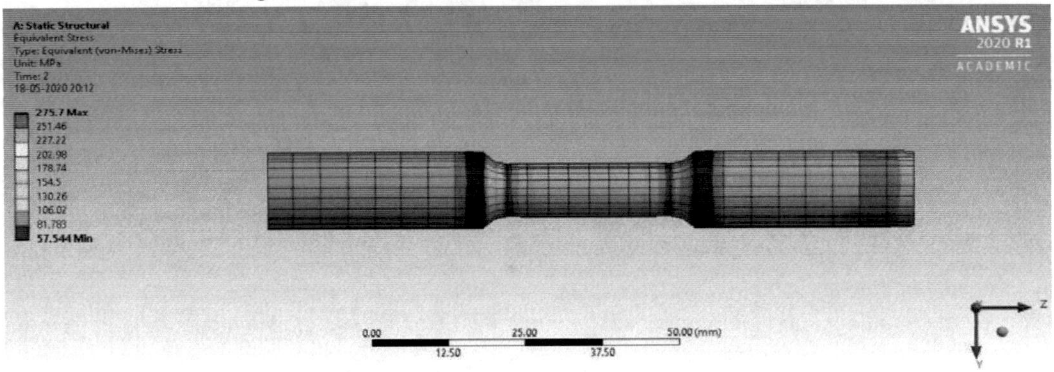

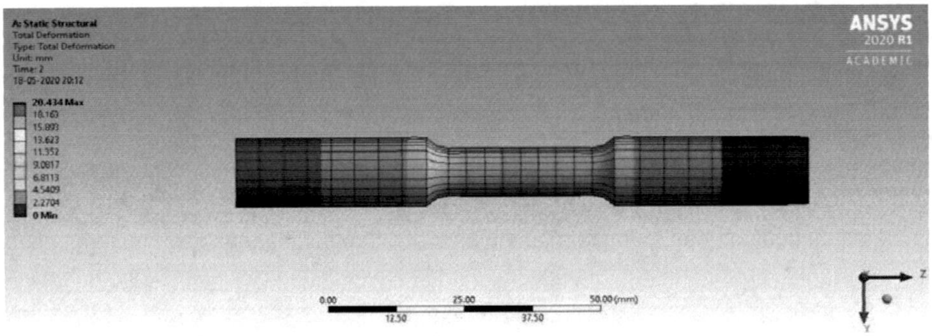

Figure 2.14: Stress Deformation in Specimen with 15% Reinforcement

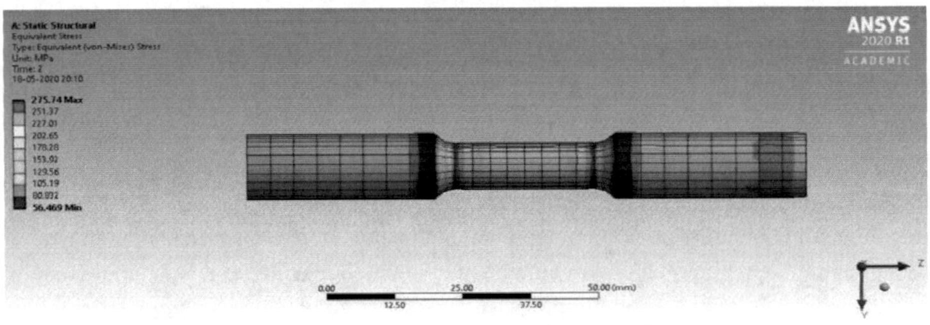

Figure 2.15: Total deformation 15% Reinforcement

3 Results and Discussions

Compared with that of the non-reinforced Al6061, strength (UTS) and Young's modulus can be linked to various factors influencing the tensile capacity of composites. It is anticipated that the strain-hardening of composites would be affected by the dislocation rate, dislocation-to-dislocation contact and plastic flow restriction due to particle resistance. The other important aspect is the high multidirectional thermal stress that is caused at the matrix– ceramic particle interface due to the disparity in thermal expansion coefficient between matrix and reinforcement. The other significant thing is the high multidirectional thermal stress induced by the matrix-the ceramic particle interface due to the variation in the coefficient of thermal expansion between matrix and reinforcement. Added; further rise of the SiC content results in decreased intensity values. The aforesaid tensile activity is observed in the variability of micro-composite tensile power. It may be the product of larger particle agglomeration and higher porosity. The sum of device region and micro porosity decreases as SiC material rises and may result in lower composite flow tension. In addition, that dislocation, i.e. defects caused around the reinforcement particles due to discrepancies in the Al(6061) and SiC thermal expansion coefficients, may contribute to the debonding of the interface and a decrease in composite UTS with more SiC volume fraction.

3.1 Tensile Test

Tensile test is also known as the tension test. During the tensile a strong force is applied in opposite directions with the specimen tightly clamped in between. It can be assumed to be a type of pulling test. After the forces are removed the material is studied to determine how it reacts to the elongation.

Table 3.1: Analytical v Experimental Results of Tensile Specimen

Percentage Composition of SiC in Aluminium	Experimental Value (mPa)	FEM Value (mPa)	Deformation(mm)
5%	218.44	221.64	21.101
10%	226.82	222.46	23.688
15%	227.65	223.89	24.532

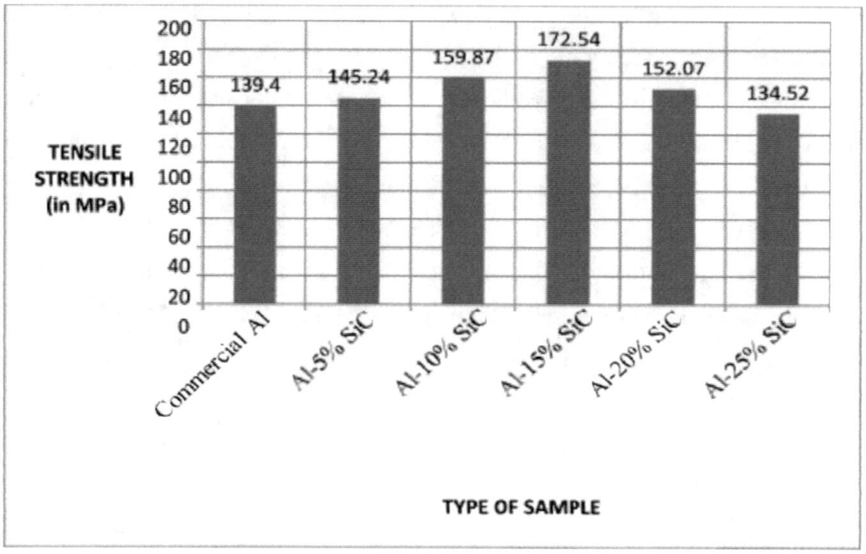

Figure 3.1: Tensile Strength

Tensile Tests of the whole arranged example were completed on UTM. At first for widely used commercial aluminum, tensile strength was 139.40 MPa which is lower than Al-5% SiC composite. It is obvious from the above figure that the composite example with 15% SiC exhibits maximu m tensile strength of 172.54 MPa. Increasing percentages of SiC in the increments of 5% showed decrease in tensile strength.

3.2: Compressive Test

Table 3.2: Compression- Analytical v Experimental Results

Weight Percentage Volume	Equivalent Stress (mPa)		Total Deformation (mm)	
	Maximum	Minimum	Maximum	Minimum
5%	275.7	57.544	20.434	0
10%	275.46	56.756	22.14	0
15%	275.74	56.469	22.678	0

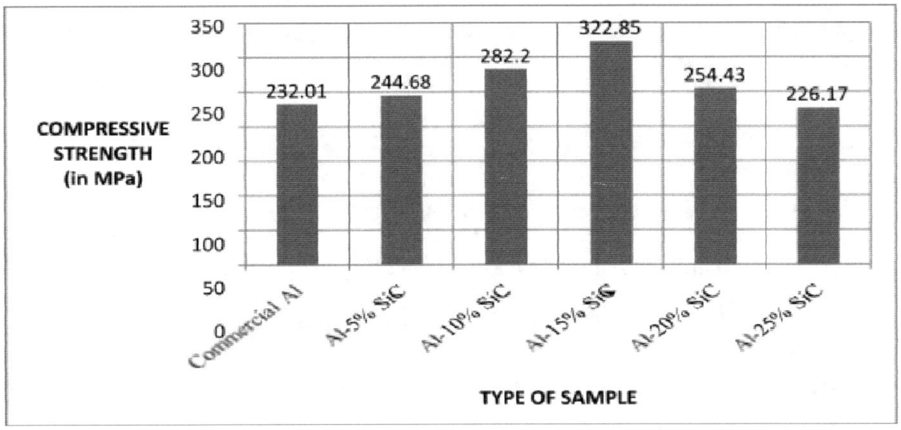

Figure 3.2: Compressive Strength

Compression Tests of the whole arranged examples were likewise done on UTM. Al-15% SiC composite material exhibited the highest value of 322.85 Mpa. The results are shown in the figure 3.2.

3.3 Hardness Test

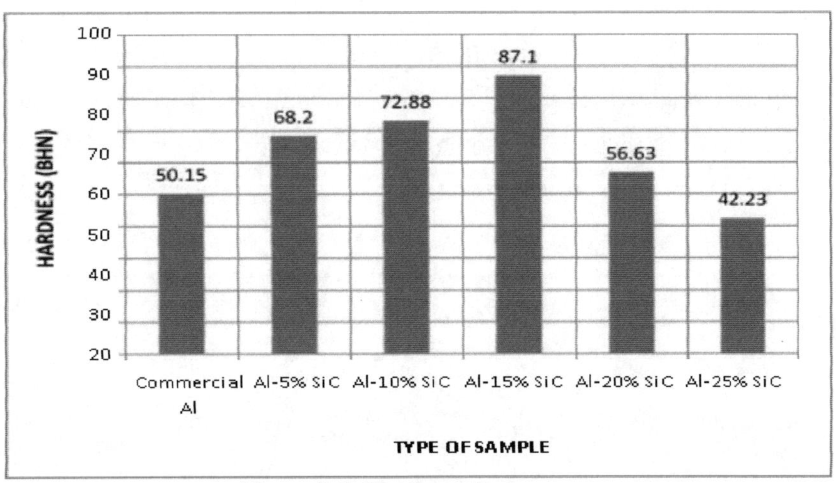

Figure 3.3: Hardness Test

IOP Conf. Series: Materials Science and Engineering 1013 (2021) 012013 doi:10.1088/1757-899X/1013/1/012013

Sample hardness increases with a rise in the weight percentage of hardened nano particulate matter and a decrease in particle size. This is due to the impediment of motion of dislocations due to increased concentration of particles and decrease in grain size.With rising reinforcing material, the stiffness improved. Hardness test was completed on Brinell Hardness Tester. In each example, four impressions were made by steel indenter balls. Average was taken to figure the Hardness value in BHN. The greatest value of hardness was seen as 87.1 BHN for Al-15% SiC composite. Above figure 3.3 shows that the option of SiC up to 15% in Al which expands the hardness estimation of the readied composites and if SiC is included over 15%, hardness value diminishes.

4. Conclusion

The experimental results of the aluminium metal matrix composite reinforced with Silicon Carbide is compared with the FEM analysis results. As observed there is a deviation in reading when compared to the actual values to the analysis values but there is a similar pattern if the results in both the analysis and experimental results the composite material tends to increase its mechanical properties like tensile strength, compressive strength and hardness at certain volume fraction of silicon carbide. The most effective volume fraction of silicon carbide to improve the mechanical properties of the aluminium metal matrix composite is 15% of Silicon Carbide. After the 15% of silicon carbide there is reduction in the mechanical properties. Therefore it can be concluded that 15% of reinforcement added in the form of silicon carbide is a major advantage to the mechanical properties of the composite material.

5 References

[1] Sidney H.Avner.,(1986); Introduction to physical metallurgy, McGraw-Hill.

[2] Introduction to Composite Materials, F.C. CampbellStructural Composite Materials.

[3] Traina , A. Hernandez , T. Scheck , C. Etuk , J. Peters ,E.W. Lee , J. Ogren , O.S. Es-Said (2005) On the correlation of mechanical and physical properties of 7075-T6 Al alloy, Engineering Failure Analysis

[4] H. Warren, Hunt,Darrell R. Herling(2004), Textbook,Aluminium Metal Matrix Composites, Advanced Materials and Processes.

[5] A.seshappa,G.K.kiranKumar,C.Bhavani Shankar "Analysis of mechanical characteristics of silicon carbide reinforced with aluminum metal matrix composites",International journal of scientific development and research (IJSDR), ISSN :2455-2631, Volume3 Issue 11, pp. 125-131,November 2018.

[6] Pradeep K. Rohatgi (1993) Metal-matrix Composites,Defence Science Journal,

[7] Rupa Dasgupta(2010) The Stretch, Limit and Path Forward for Particle Rein-forced Metal Matrix Composites of 7075 Al-Alloys Scientific research Engineering.

[8] N. Mahathaninwonga, T. Plookphola, J. Wannasina, S.Wisutmethangoonb(2012) T6 heat treatment of rheocasting 7075 Al alloy ,Materials Science and Engineering.

[9] Jin-feng l, zhuo-weipeng, chao-xing li, , zhi-qiangjia, wen-jingchenzi-qiaozheng (2008) Mechanical properties, corrosion behaviors and microstructures of 7075 aluminium alloy with various aging treatments ,Transactions of Nonferrous Metals Society of China,Volume 18(4)

EFFECT OF DISPERSION AND IMPROPER BONDING OF NANOFILLER POLYMER COMPOSITES

Manjunatha G[1], Nagesh S N[2] and Raji George[2]
1 Department of Mechanical Engineering, R.R Institute of Technology, Bengaluru, India.
2 Department of Mechanical Engineering, Ramaiah Institute of Technology, Bengaluru, India.

Email: gmanjubelaku@gmail.com

Abstract: - Nanofiller is used to enhance the properties in composite materials, with at least one phase of nanomaterials that can be referred as nanocomposites. The nanomaterial due to uniform dispersion recently used in advanced polymer nanocomposites, to mix effectively in the polymer matrix. The production of polymer composites with nanofiller is limited due to the unresolved issues such as agglomeration during processing, high production cost and limited availability of good quality. Graphene filled nanocomposites is a novel material and considered in different applications. The properties of the nanocomposites mainly depend on the arrangement of graphene layers with the polymer matrix and interfacial bonding between them. Here an attempt is made to develop polymer composites using graphene as reinforcements and epoxy as matrix materials. Sonication is a process of applying ultrasound energy to disperse nanoparticles in epoxy. The sound waves that are created in sonication are generally large ultrasound waves and during this process, considerable heat is produced. If the sonication treatment is too long, nanoparticles were seriously damaged, particularly when a probe sonicator is used. The localized damage causes nanoparticles to deteriorate the mechanical properties of the polymer composites. The study from XRD and SEM showed that the use of graphene at higher weight % resulted in a decrease in strength of the composites. Hence, it is not a better choice to produce nanocomposites at increased addition of nanofiller.

Keywords: Nanofiller, Graphene; Sonication; Composites;

1. Introduction

The discovery of graphene used as a nanofiller for the production of lightweight, low cost, and high-performance composite materials for various applications. Advanced composites are composite materials that are traditionally used in the aerospace components, parts of racing cars, transport vehicles applications[1]. Choice of fabrication method depends on matrix properties and the effect of matrix on properties of reinforcements. The incorporation of nanoparticles into polymers exhibit behavior different from conventional composite materials with microscale structure, due to the small size of the structural unit and the high surface to volume ratio[2].

In future a large number of new graphene-based polymer nanocomposites using thermoplastic, thermosetting is used due its unique properties. Synthesis of graphene-polymer composites, may have technical challenges which need to be addressed for mass production. The key to the successful development of polymer nanocomposites is to achieve homogeneous dispersion of nanofillers in the polymer matrix[3]. The different processing methods used for combining graphene with polymer matrix are in-situ polymerization, solution processing, and melt processing[4].

The ultra sound energy agitates particles using an ultra-sonic bath results in the separation of individualized nano particle from the bundles. Nanocomposites consisting of a polymer and nanofiller possess improved properties due to strong interfacial interaction between the matrix and nanofiller[5].

2. Literature review

Literature survey is required to understand the specific objective of the research work. From the literature studies gaps has been identified to manufacture nanopolymer composites. The first, prevention of agglomeration problem need to be addressed properly in preparation of nanocomposites and second, lack of information to understand several key factors influences the interaction effect, crystallinity and mechanical properties of graphene as reinforcing filler in epoxy resin[6]. The aim of the present investigation was to study the nanofiller reinforced and unreinforced composites fabricated using solution casting method. The dispersion of nanofiller in epoxy matrix will be analyzed by using X-rays diffraction and Scanning Electron Microscope (SEM).The novelty in this study is related to preparation of nanocomposites using different mixing methods to improve dispersion effect of epoxy with the graphene nanofiller.

3. Experimental details

3.1 Materials

Graphene (GR): The Graphene (GR) used as reinforcement in this study was supplied by United Nanotech Pvt Ltd Hoskote, Bangalore. Normally, Chemical vapour deposition was used to produce the GR with high aspect ratio, high purity of 96- 99% and surface area of 323-600 m^2/g.

Matrix: Epoxy liquid resin based on bisphenol was used as matrix in this study was supplied by Atul ltd polymer division Gujarat. The clear epoxy with density 1.16 g/cm^3 and yellow liquid amine based K-6 is used as curing agent.

3.2 Preparation of graphene based polymer composites

In fabrication of nanocomposites the graphene of 0.5wt%, 1.0wt%, 1.5wt% and 2.0wt% was added to epoxy material using solution casting method. The Figure 1 shows the steps involved in preparation of polymer nanocomposites.

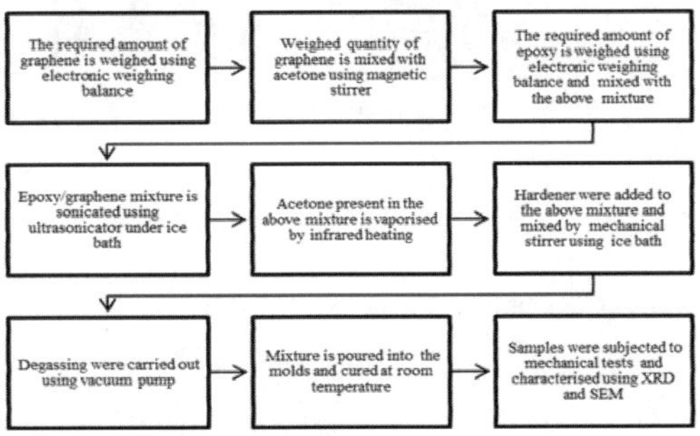

Figure 1. Preparation of polymer nanocomposites

3.3 Tensile test

Tensile test was performed for five solution casted samples according to ASTM D638 standard. An abrasive water jet cutting machine was used to cut dog bone shaped tensile specimen. This test was performed at a constant loading speed / crosshead speed of 5.2 mm/min and a strain rate of 0.1 mm/min using a multi testing machine from Mecmesin. The ultimate tensile strength, strain and Young's modulus were measured using software.

$$\text{Ultimate tensile strength } (\sigma) = \frac{P}{b \times d} \; N/mm^2 \qquad (1)$$

Where, P - Maximum load (N), b - Width (mm), d - Thickness (mm)

3.4 Characterization Technique

The diffractions are produced at a voltage of 30 kV and current of 20mA at ambient temperature, diffracted rays corresponds to coherent reflection satisfies Bragg's law in turn used to describe interference pattern of X-ray scattered by crystals. The test samples prepared was used for analysis were recorded by diffractometer and using Scanning Electron Microscope (SEM) fractured surface were examined.

4. Results and Discussions

4.1 Tensile properties

The results of tensile test showed in Figure 2, the addition of graphene with epoxy indicates improved tensile strength and decreased strength with higher addition of graphene filler. The presence of GR in polymer composites causes to increase the tensile strength and when GR content was 1.0wt% the maximum strength was 70MPa and Young's modulus was increased and found to be maximum 8 GPa. This improved in strength is due to good interfacial interaction between matrix and reinforcement material. Further increase in wt% of GR caused reduction in ultimate tensile strength to 53Mpa this is due agglomeration effect.

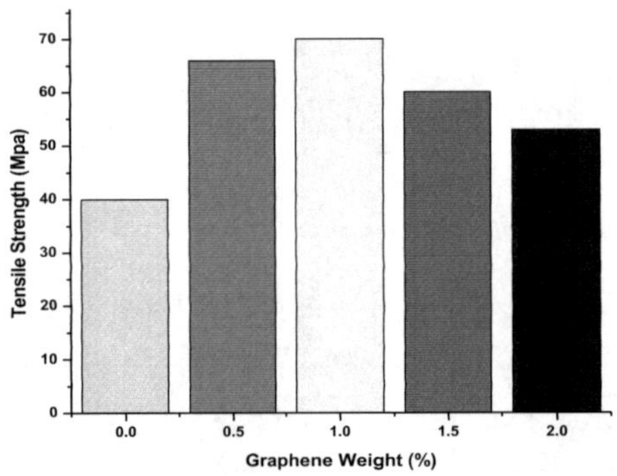

Figure 2 Tensile strength of polymer nanocomposites at different graphene contents

4.2 XRD Analysis

The X ray diffraction pattern obtained for 0.5wt% and 2wt% addition of graphene are shown in Figure 3 and Figure 4.The polymer nanocomposites material showed separate peak corresponds to graphene at $2\theta = 26.56^{o}$ and for epoxy peak obtained at $2\theta = 18.01^{o}$ for 2wt% addition of graphene filler. The different sharp peak and broad peak indicates improper binding of higher wt% of graphene nanofiller but at lower wt% both epoxy and graphene peaks are merged together with each other which indicates good dispersion.

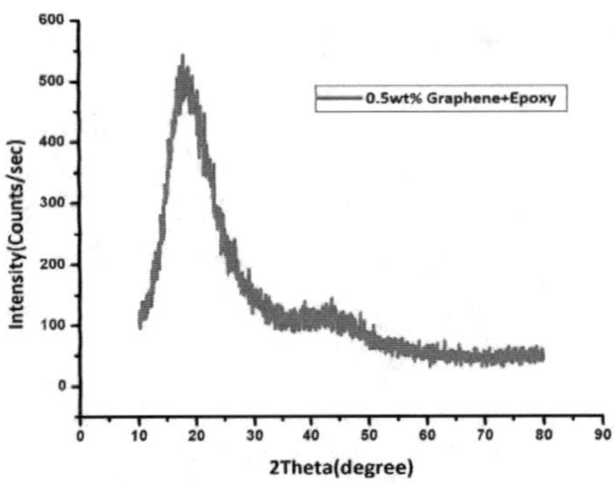

Figure 3 XRD patterns of polymer nanocomposites for 0.5 wt% graphene

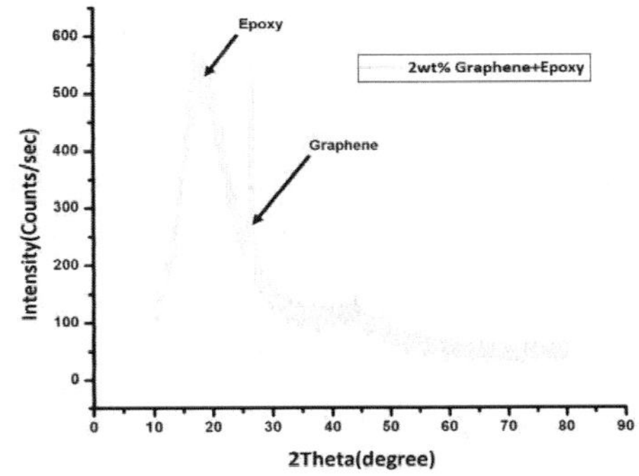

Figure 4 XRD patterns of polymer nanocomposites for 2 wt% graphene

4.3 SEM Analysis

The SEM images of fracture surfaces of polymer nanocomposites are shown in Figure 5. The fracture sample at 5000X magnification and 50000X magnification at 2wt% addition of graphene shows improper adhesion with epoxy matrix. Adding more GR fracture surface becomes cloud like structure with rough brittle was indicated. The formation of cluster in GR network causes to reduce strength of composites. It is clear from SEM image weak interface which finally effect for failure of the material.

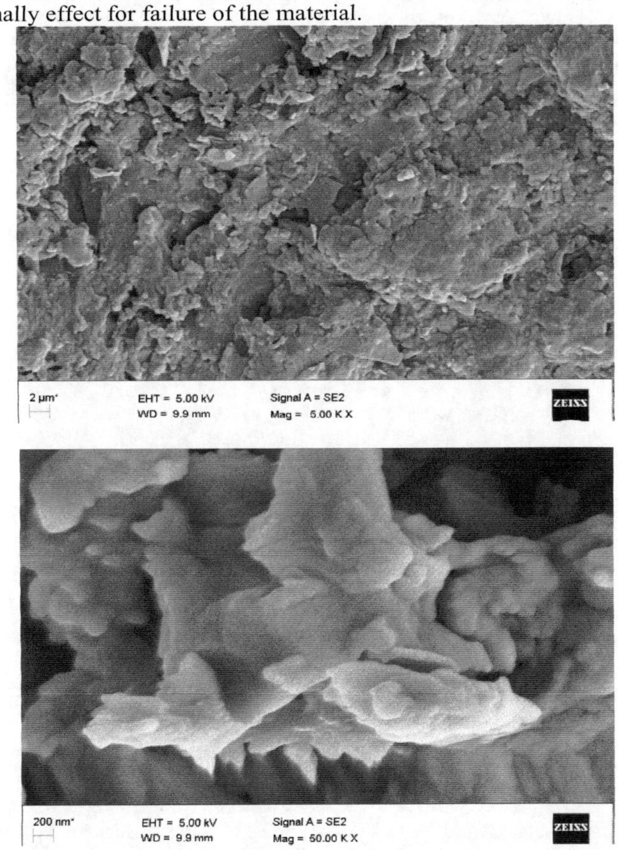

Figure 5 SEM of polymer nanocomposites for 2 wt% graphene at 5 KX and 50 KX

IOP Conf. Series: Materials Science and Engineering 1013 (2021) 012014 doi:10.1088/1757-899X/1013/1/012014

5. Conclusions

Graphene has excellent mechanical properties at 1wt% addition of graphene nanofiller due to high surface area. The tensile strength is increased progressively upto 1wt% presence of graphene compared to neat epoxy. XRD used to study dispersion effect of graphene with epoxy at different loading and found merging of peak indicated increased strength of composite material. SEM images weak interface between matrix and filler material. Finally, considering their improved properties graphene filled polymer composites can be used in different applications.

References

[1] J. Wei, T. Vo, and F. Inam, "Epoxy/graphene nanocomposites - processing and properties: a review," RSC Adv., vol. 5, no. 90, pp. 73510–73524, 2015.
[2] M. G. Prolongo, C. Salom, C. Arribas, M. Sánchez-Cabezudo, R. M. Masegosa, and S. G. Prolongo, "Influence of graphene nanoplatelets on curing and mechanical properties of graphene/epoxy nanocomposites," J. Therm. Anal. Calorim., vol. 125, no. 2, pp. 629–636, 2016.
[3] M. M. Gudarzi and F. Sharif, "Enhancement of dispersion and bonding of graphene-polymer through wet transfer of functionalized graphene oxide," Express Polym. Lett., vol. 6, no. 12, pp. 1017–1031, 2012.
[4] O. Vryonis, T. M. Harrell, T. Andritsch, A. S. Vaughan, and P. L. Lewin, "Solvent mixing and its effect on epoxy resin filled with graphene oxide," 2018 IEEE 2nd Int. Conf. Dielectr. ICD 2018, no. 642771, pp. 1–4, 2018.
[5] U. Kilic, M. M. Sherif, and O. E. Ozbulut, "Tensile properties of graphene nanoplatelets/epoxy composites fabricated by various dispersion techniques," Polym. Test., vol. 76, pp. 181–191, 2019.
[6] S. I. Abdullah and M. N. M. Ansari, "Mechanical properties of graphene oxide (GO)/epoxy composites," HBRC J., vol. 11, no. 2, pp. 151–156, 2015.

The study of Elastic Modulus and Thermal Conductivity of Different Fiber Cross-section of Fiber-Reinforced Composite in ANSYS

Rohit Vankadhar, Devendra Dandotiya, Bhaskar Pal

Department of Mechanical Engineering, School of Engineering, Presidency University Bangalore, Karnataka, India * *corresponding author e-mail*: vvsrohith9966284338@gmail.com

ABSTRACT

This work aims to predict the elastic modulus and effective thermal conductivity of the fiber-reinforced polymer composites. A representative volume element (RVE) of size 420x420x420 micron has been created in Solid-works with 10%, 17%, 27%, 40%, and 54% of fiber volume. The fiber of circular, square, and hexagonal cross-sections are considered in this analysis. The symmetric boundary conditions are applied to the RVE for determining the elastic modulus of the fiber-reinforced polymer. It has been observed from the ANSYS that the results are not exactly similar to the rule of mixture. The results are varying with a change in the cross-section of the fiber. It has been observed that elastic modulus is increasing with the increase in fiber density. It is also noticed that the elastic modulus is increased when the cross-section changes from circular to square to a hexagon with all said fiber volume. It is attributed that due to an increase in the fiber-matrix contact surface area, the adhesion between fiber and matrix increased. Surprisingly it observed that in all the cases the elastic modulus has decreased after 40% of volume fraction. This may be endorsed that due to higher fiber volume fraction the adhesion between fiber-matrices is not appropriate. Further study has revealed that the modulus of the composite is in accordance with the rule of the mixture when RVE has been prepared with a combination of circle, square, and hexagon. The effective thermal conductivity of the same RVE model has been carried out in ANSYS. It has found out that effective thermal conductivity is in reasonable agreement with the rule of mixture.

Keywords: Polymer Composites, Volume fraction, elastic modulus, different cross sections.

Introduction

Fibre reinforced composites are known as building blocks of structural material. Generally analysis is done at micro-scale level. The micro-scale defines the arrangement of fibres into the matrix. In the framework of micro-scale simulation of composite materials, the simulation, including both analytical methods and numerical methods, predicts the effective stiffness and thermal conductivity of the composite for any engineering structural materials.

The micromechanical analysis is usually performed in a representative volume element (RVE) of the fibre reinforced composite. The researchers have developed numerous methods for predicting the mechanical and thermal properties of the fibre reinforced composite. An integrated computer coded analyser developed to calculate the stiffness of the composite using a geometry [1]. Developed a unit cell that continuously reinforced with unidirectional fibre and the analysis has been done by imposing the displacement and traction continuity conditions at the interfaces [2]. The normal elastic module is predicted by considering a cylindrical fibre surrounded by tube shaped matrix [3]. Micromechanics equations are developed for the composite unit cell and then predict the mechanical properties of the composite [4]. Presented the importance of the boundary conditions on the RVE and predicted the mechanical property of the composite by using finite element analysis [5]. A computational homogenisation framework presented for predicting the mechanical properties of fibre reinforced composite [6]. Most of the micromechanical analysis is considered the RVE as a periodic structure of fibres with deterministic and ordered distribution. However, the realistic distribution of the fibres in the polymer matrix is random. Therefore, the methods being used by the researcher cannot predict the accurate result of elastic modulus of fibre reinforced composite. The effect of non-uniform fibre density on the volume of the composite has been investigated extensively [8-10]. A RVE model of an E-glass particulate reinforced composite built with randomly oriented fibres and applied iso-stressed loading for finding the upper/lower bound of Young's modulus [11]. It has been demonstrated that random models are best suited for the simulation of local phenomena [12]. The irregularities of local fibre array on the analysis of thermal stress has been reported [13]. It is found that rectangular and hexagonal periodic models show inconsistency in the result then the random model [14]. The current work is concentrated on the distribution of different volume fractions of cross-sectional fibre in the square RVE and evaluates their elastic and thermal properties. The work has been extended to model the RVE with a different percentage of combination of circular, square and hexagonal fibres.

Methodology:

The current work is to evaluate the modulus of elasticities and the thermal conductivity of the fibre reinforced composite by using finite element analysis software. The representative volume element (RVE) is a unit cell of the current analysis modelled directly in ANSYS. A three-dimensional cuboidal geometry of 0.42mmX0.42mmX0.42mm has been created. This unit cell has been modelled by aligning the fibre in the longitudinal direction of the square prism as shown in figure 1. The cross-section of fibre in the composites is circular, square and hexagonal, shown in figure 2. The fibres are distributed in line inside the matrix in an uniform manner. The volume fraction of fiber is calculated by considering the volume of fiber in the matrix related to the total volume of the composite in the RVE. The volume fraction of the fibre in the fibre reinforced composite is a very important parameter which characterises the stiffness and rigidity of the composite. The volume fraction of fibre in RVE is 10%, 17%, 28%, 40%, and 54% for this current study. The distribution of circular, square and hexagonal fibres in a RVE as shown in figure 3 and the distribution of combination of all fibres in a RVE has shown in figure 4. The fibre material is polypropylene and the matrix is polypropylene co-ethylene, the properties of the material has been shown in table 1. The stiffness of the fibre is more than the stiffness of the matrix. It has been observed that with increase in volume fraction of fibre into the matrix resulting in increase of modulus of elasticity.

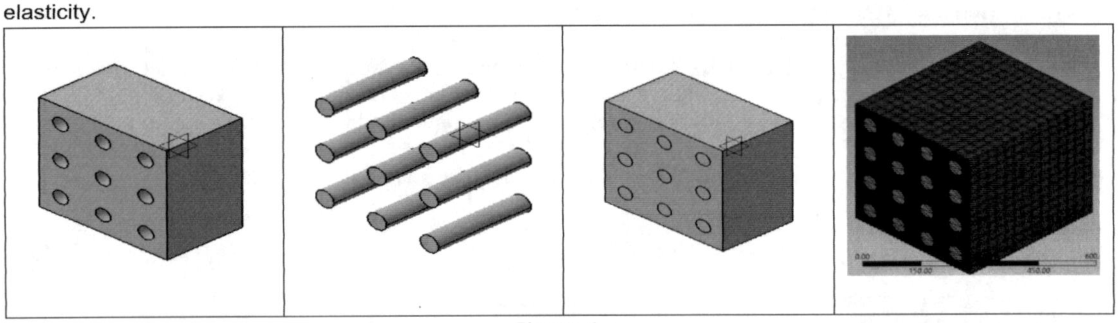

Figure 1

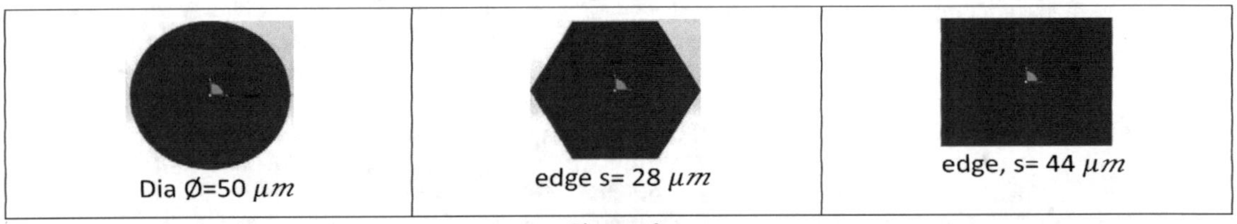

Figure 2

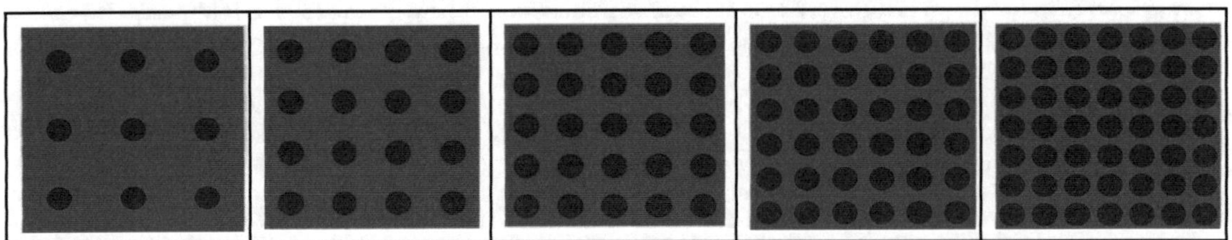

10% Circular Fibre	17% Circular Fibre	28% Circular Fibre	40% Circular Fibre	54% Circular Fibre

IOP Conf. Series: Materials Science and Engineering 1013 (2021) 012015 doi:10.1088/1757-899X/1013/1/012015

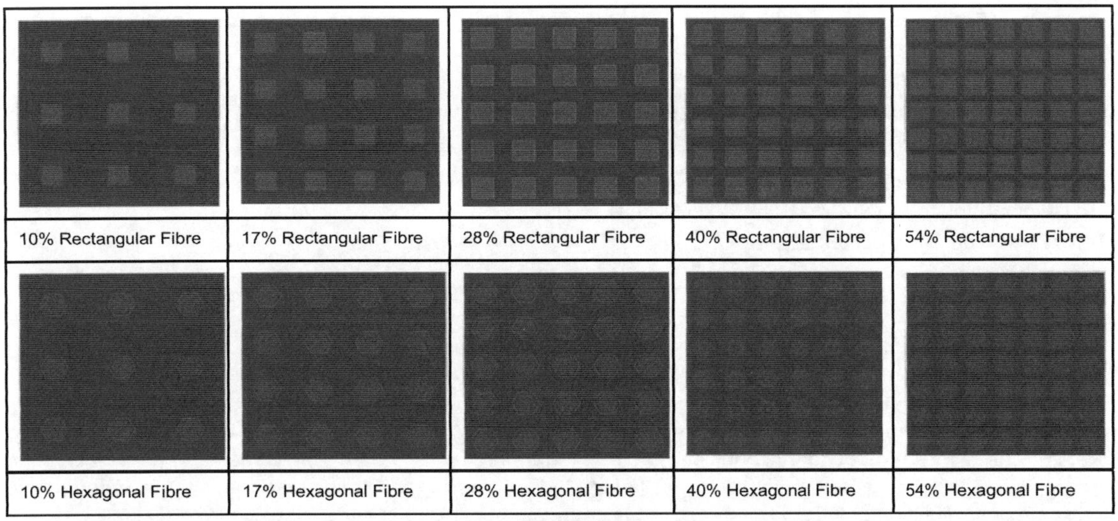

Figure 3

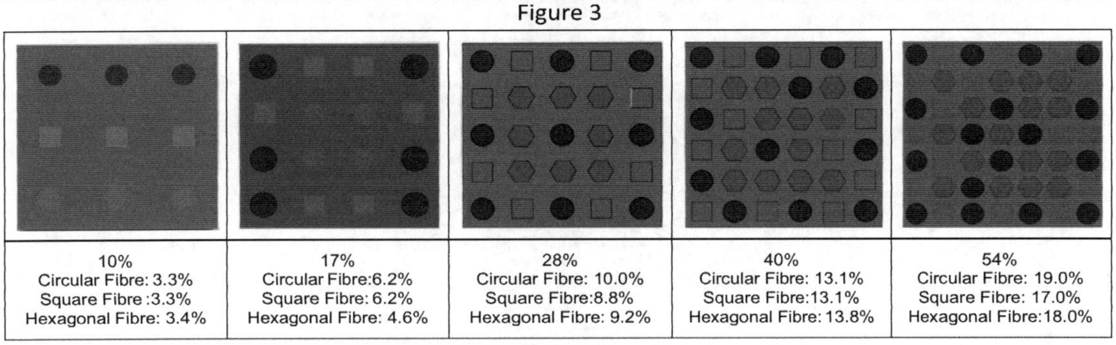

Figure 4

Table 1

Material Properties	Polypropylene co ethylene (PPE) Matrix	Polypropylene (PP) Fibre
Longitudinal Modulus E_1	1.05 Gpa	4.5 Gpa
Transverse Modulus E_2	1.05 Gpa	4.5 Gpa
Poisson's Ratio	0.33	0.2
Thermal conductivity	0.2w/µm-k	0.27w/µm-k

There are a lot of numerical models available to predict the mechanical properties of fibre reinforced composite. Rule of mixture, derived from the basic mechanics of the equation of the composites, is one of them. This analytical equation will predict the modulus of elasticity of the composite with different volume fraction. It is also predicted that the micromechanical analysis can be done by finite element software like ANSYS.

A orthogonal frame of reference, as shown in figure 5, is placed in the unit cell where the 'Z' axis is aligned with the fibre direction. The longitudinal Young's modulus is measured in the 'Z' axis direction. In linear elasticity, the longitudinal Young's modulus is one of the elastic moduli available to describe isotropic homogeneous material. A unidirectional composite with a square array of fibers can be transversely isotropic because the properties are the same along any plane which is normal to the fiber direction. The transverse Young's modulus is measured as a right angle to the longitudinal direction.

- The Z-axis is aligned with the fiber direction.
- The X-axis is in the plane of the layer and perpendicular to the fibers.
- The Y-axis is perpendicular to the plane of the layer and thus also perpendicular to the fibers.

The numerical questions for rule of mixture

Longitudinal Young's modulus, $E_z = E_f V_f + E_m V_m$

Transverse Young's modulus, $\frac{1}{E_x} = \frac{V_f}{E_f} + \frac{V_m}{E_m}$

E_m – Young'smodulus of matrix

V_m – Volume fraction of matrix

E_f – Young'smodulus of fibre

V_f – Volume of fraction of fibre

The finite element method has become a powerful tool for the numerical solution of a wide range of engineering problems. Here finite element analysis is carried out using ANSYS version 16.0 running on an Intel i3 processor system. The symmetric boundary conditions are applied to the model on X Y and Z faces. The displacements are applied with respect to the condition that is longitudinal loading and transverse loading. To determine E_z along the fiber direction of the following symmetric boundary conditions applied as indicated in Table 2

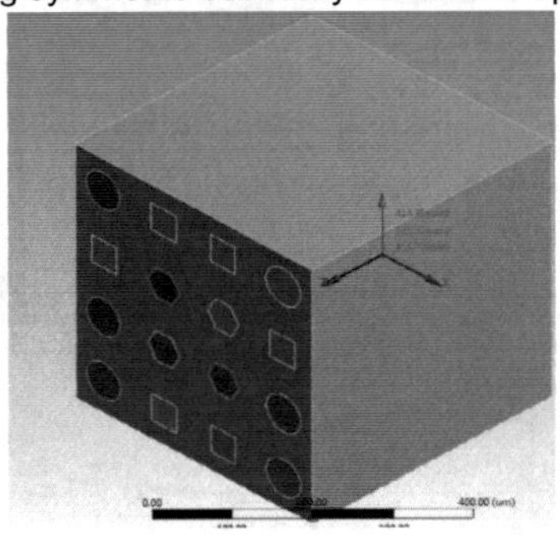

Figure 5

Table 2

X -ve Face			X +ve Face			Y -ve Face			Y +ve Face			Z -ve Face			Z +ve Face		
X	Y	Z	X	Y	Z	X	Y	Z	X	Y	Z	X	Y	Z	X	Y	Z
0	NC		No Constraints (NC)			NC	0	NC	No Constraints (NC)			NC	NC	0	NC	NC	0.05

Result and Discussions:

Determination of the effective elastic properties is one of the basic requirements in the mechanics of composite materials. It is difficult to predict the elastic moduli of fibre reinforced composite by experimentation. Thus, the numerical techniques are extensively used to calculate these elastic properties. These numerical techniques are using the analysis of a representative volume element which is packed with periodic fibre sequence. There are a few issues that have been considered carefully when carrying out such analyses. Firstly, the correct RVE corresponding to the assumed fiber distribution must be isolated. Secondly, correct boundary conditions need to be applied to the chosen RVE to model different loading situations. In the present study, the composite is macroscopically homogeneous, linear elastic, stress-free (no thermal stress); the fibres are homogeneous, linearly elastic, isotropic, regularly spaced, perfectly aligned; and the metrics is homogeneous, linearly elastic, isotropic. The finite element analysis is carried out at 10%, 17%, 28%, 40%, and 54% volume fraction and compared with the rule of mixture.

The longitudinal and transverse Young's modulus have been evaluated with the help of ANSYS. The convergence of these elastic modulus does not depend on the fixed mesh size; rather it is required to change the size of an element on different cross sections. All the values have been tabulated in table 3 and plotted in figure 6 and 7. It has been observed from figure 6 that the longitudinal Young's modulus is in good agreement with the rule of the mixture for circular cross-sectional fibre. It is predicted that up to 40% of hexagonal and square fibre, the error between the role of mixture and longitudinal young modulus is negligible. As the volume fraction increases beyond 40%, the error goes beyond 10%. This can be attribute it that due to high volume fraction of fibre, the adhesion between fibre-matrix interfaces is not appropriate, the stress concentration on matrix is accumulated along the edge of square and hexagonal fibre, and the composite become more brittle that resulted in decrease in Young's modulus. To overcome this problem, the RVE has been modelled with different cross-section of fibres into the same unit cell which has been indicated in table 4. The percentage of volume fraction of each cross-sectional fibre in every unit cell has been denoted in table 4. The longitudinal and transverse Young's modulus of modified RVE has been analysed in ANSYS by applying the same symmetric boundary conditions. It is noticed that the elastic properties of modified RVE for all volume fraction of fibre is in good agreement with the rule of mixture, indicated in figure 6. Due to proper and periodic distribution of circular, square and hexagonal fibre into the unit cell, it is attributed a good agreement with the rule of mixture in all the volume fractions. The same procedure is followed to predict the transverse modulus of fibre reinforced composite. It is observed that the transverse modulus has shown good agreement with the role of mixture when analysed the RVE filled with different percentages of circular square and hexagonal fibre as shown in figure 7.

Result & discussion

Table 3

Type of fibre	m_m	ROM, m_m	$ANSYS,$ m_m	% of error	ROM, m_m	$ANSYS,$ m_m	% of error
Circular	10	1395	1417	1.6	1301	1215	-6.6
	18	1636	1670	2.1	1454	1341	-7.8
	28	1981	2011	1.5	1627	1542	-5.2
	40	2430	2391	-1.6	1922	1747	-9.1
	54	2913	2862	-1.8	2035	2057	1.1
Square	10	1395	1413	1.3	1301	1228	-5.6
	18	1636	1637	0.1	1454	1364	-6.2
	28	1981	2026	2.3	1627	1616	-0.7
	40	2430	2446	0.7	1922	1941	1.0
	54	2913	2558	-12.2	2035	2079	2.2
Hexagonal	10	1395	1442	3.4	1301	1412	8.5
	18	1636	1634	-0.1	1454	1331	-8.5
	28	1981	2030	2.5	1627	1561	-4.1
	40	2430	2459	1.2	1922	1864	-3.0
	54	2913	2322	-20.3	2035	1842	-9.5

IOP Conf. Series: Materials Science and Engineering 1013 (2021) 012015 doi:10.1088/1757-899X/1013/1/012015

Table 4

Orientation of Fiber	ROM, E_z	$ANSYS, E_z$	% of Error	ROM of thermal conductive	ANSYS of thermal conductive	% of erro
	1395	1410	1.07	0.207	0.238	-14.97
	1618	1636	1.11	0.211	0.231	-9.47
	1969	1918	-2.59	0.218	0.229	-5.04
	2245	2430	7.61%	0.228	0.227	0.4
	2883	2913	1.04	0.237	0.232	2.109

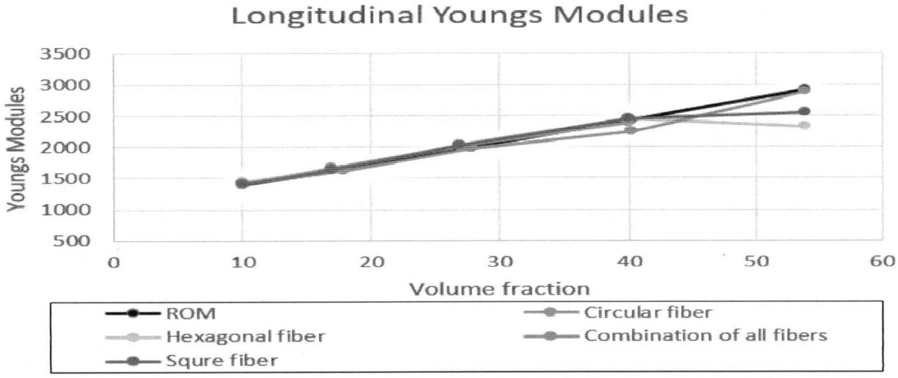

figure 6

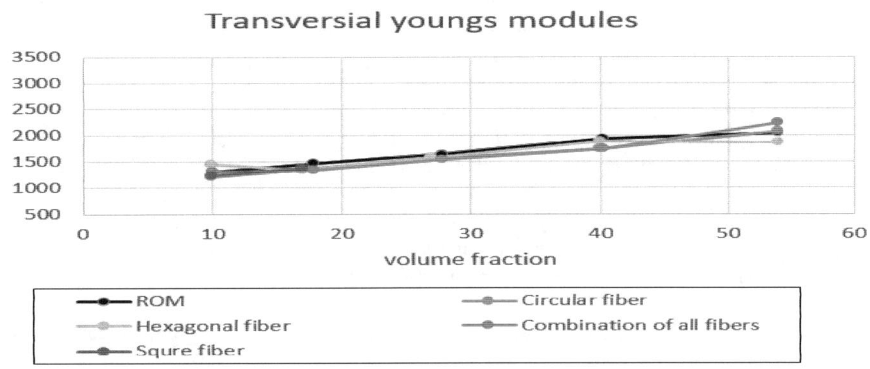

Figure 7

Figure 8 the thermal conductivity for different combinations of geometries. It is noticed that the results obtained using ANSYS FEM model are closer to the Rule of Mixture (ROM) value as the variation . In comparison with ROM it was found that errors associated with the FEM model vary from 6-14% for different cross section fibers. The variation in thermal conductivities with respect ROM decreases as the volume fraction decreases as the fiber contact area increases. This analysis can be used to predict the thermal conductivity of various Fiber Cross-section of Fiber-Reinforced Composite in ANSYS.

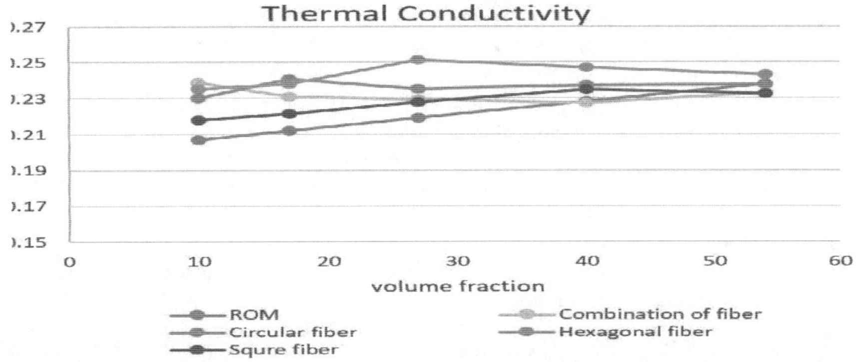

Figure 8

Conclusions:

1. In this article, the main contribution is to predict the elastic properties and thermal conductivity of fibre reinforced composite (figure 1) with different volume fraction of fibre. The cross-section of fibres are circular, square and hexagonal as shown in figure 2.
2. Two types of representative volume element (RVE) are created in this work. Type I: the unit cell consists of a matrix and a fraction circular or square or hexagonal fibres as indicated in figure 3. Type II: the unit cell consists of a matrix and fraction of combination of circular, square and hexagonal fibres as indicated in figure 4.
3. The materials for fibre reinforced composite are Polypropylene co ethylene (PPE) for matrix and Polypropylene (PP) for fibre and their properties have been indicated in table 1.
4. The longitudinal, transverse Young's modulus and thermal conductivity have been evaluated with the help of ANSYS. The symmetric boundary conditions have applied on the RVE as indicated in table 2.
5. The numerical method for type I RVE model presents a good match with the rule of mixture for circular fibre but for square and hexagonal fibre up to volume fraction of 40% showed good agreement with rule of mixture. The longitudinal modules started decreasing when the volume fraction went beyond 40%. This can be attribute it that due to high volume fraction of fibre, the adhesion between fibre-matrix interfaces is not appropriate, the stress concentration on matrix is accumulated along the edge of square and hexagonal fibre, and the composite become more brittle that resulted in decrease in Young's modulus as shown in figure 6. It is found that the transverse module showed no error with the rule of mixture as shown in figure 6.
6. To overcome this issue, Type II RVE modelled developed. The results indicated that the elastic moduli are in good agreement with the role of mixture as indicated in figure 6.
7. Further study on thermal conductivity of the RVE showed 6 to 14% error with rule of mixture as indicated in figure 8.

Reference:

1. Chamis, C.C.; Murthy, P.L. ICAN: Integrated composites analyzer. J. Compos. Technol. Res. 1986, 8, 8–17.
2. Aboudi, J. Micromechanical analysis of composites by the method of cells-update. Appl. Mech. Rev. 1996, 49, S83–S91.
3. Goldberg, R.K.; Roberts, G.D.; Gilat, A. Analytical studies of the high strain rate tensile response of a polymer matrix composite. J. Adv. Mater. 2004, 36, 14–24.
4. Herakovich, C.T. Mechanics of Fibrous Composites; John Wiley & Sons, Inc.: New York, NY, USA, 1998.
5. Sun, C.; Vaidya, R. Prediction of composite properties from a representative volume element. Compos. Sci. Technol. 1996, 56, 171–179.
6. Sonon, B.; Massart, T.J. A Level-Set Based Representative Volume Element Generator and XFEM Simulations for Textile and 3D-Reinforced Composites. Materials 2013, 6, 5568–5592.
7. Brockenbrough, J.; Suresh, S.; Wienecke, H. Deformation of metal-matrix composites with continuous fibers: Geometrical effects of fiber distribution and shape. Acta Metall. Mater. 1991, 39, 735–752.
8. Pyrz, R. Quantitative description of the microstructure of composites. Part I: Morphology of unidirectional composite systems. Compos. Sci. Technol. 1994, 50, 197–208.
9. Romanov, V.; Lomov, S.V.; Swolfs, Y.; Orlova, S.; Gorbatikh, L.; Verpoest, I. Statistical analysis of real and simulated fibre arrangements in unidirectional composites. Compos. Sci. Technol. 2013, 87, 126–134.
10. Gusev, A.A.; Hine, P.J.; Ward, I.M. Fiber packing and elastic properties of a transversely random unidirectional glass/epoxy composite. Compos. Sci. Technol. 2000, 60, 535–541.
11. Sun, C.; Saffari, P.; Ranade, R.; Sadeghipour, K.; Baran, G. Finite element analysis of elastic property bounds of a composite with randomly distributed particles. Compos. Part A Appl. Sci. Manuf. 2007, 38, 80–86.
12. Trias, D.; Costa, J.; Mayugo, J.; Hurtado, J. Random models versus periodic models for fibre reinforced composites. Comput. Mater. Sci. 2006, 38, 316–324.
13. Hojo, M.; Mizuno, M.; Hobbiebrunken, T.; Adachi, T.; Tanaka, M.; Ha, S.K. Effect of fiber array irregularities on microscopic interfacial normal stress states of transversely loaded UD-CFRP from viewpoint of failure initiation. Compos. Sci. Technol. 2009, 69, 1726–1734.
14. Zhang, T.; Yan, Y. A comparison between random model and periodic model for fiber-reinforced composites based on a new method for generating fiber distributions. Polym. Compos. 2015.

WEAR CHARACTERISTICS OF DOUBLE CERAMIC PARTICULATE HYBRID ALUMINIUM MATRIX COMPOSITE

Nagesh S.N[1], Manjunath G[2], Putta Bore Gowda[3] , Ramachandra CG[4], Nataraju S.N[5]
1,3Department of Mechanical Engineering, Ramaiah Institute of Technology, Bengaluru, India.
2 Department of Mechanical Engineering, R.R Institute of Technology, Bengaluru, India.
4 Department of Mechanical Engineering, Presidency University, Bengaluru, India.
5 Department of Mechanical Engineering, SJCIT, Chikkaballapur, India.

E-mail: snnagesh80@gmail.com

Abstract: - Metal matrix composites got popularity due to their light weight and high strength and replacing metal parts from various applications like aerospace, automobile and biomedical. Aluminum matrix composites (AMC) are suitable for the aerospace industries and still the research is going on to use it for other applications. In this research work, a double ceramic particulate AMC was synthesized. The ceramic powders used were alumina (Al_2O_3) in wt. % of 0.5, 1.0 and 2 and silicon dioxide (SiO_2) in wt. % of 0.5 and 1.0. The synthesis of composites was done by stir casting method. SEM analysis showed that porosity was reduced and the distribution of reinforcement particles were uniform throughout the matrix. Wear test was carried out using pin on disc wear testing machine, results showed reduction in wear. SEM analysis of worn out specimen showed transition from adhesive wear to abrasive wear for hybrid composites.
Keywords: AMC, Al_2O_3, SiO_2, Stir casting

1. Introduction
In the present scenario, a composite is a multiphase material can be prepared artificial or natural way by adding the constituent phases, which is chemically dissimilar and have separate distinct interface. Basically, the composite materials have been identified by two phases namely, matrix and reinforcement. Matrix phase is characterized by continuous phase whereas other phase is reinforcement which is dispersed phase [1]. The development of metal matrix composites made significant change in the aerospace and automotive industries. MMCs have been made by using most of the metals and their alloys as the matrix material, but only few metals and alloys are available for high and low temperature applications. In air craft industries, the material selected should be light weight and high strength. The popular matrix materials such as titanium, magnesium and aluminum and its alloys are popular materials suitable for aircraft components. The reinforcement material selected for making MMC should have high modulus and the resulting composite material possess better properties than the most of the alloys. The suitability of composites at different service temperatures depends on its melting point, physical and mechanical properties. Some metals have low melting point and the reinforcement materials should improve the melting point. Aluminum and its alloys can be a one good choice as matrix material which is light in weight and high strength to weight ratio. Discontinuous reinforced aluminium matrix composites (DRAMCs) are the composites in which discontinuous fibers are used like whiskers, particulates etc., which strengthen the aluminum matrix. Most of the researchers concentrated on the overall performance of the composites by selecting the proper reinforcing material. The performance of DRAMCs can be improved in three different ways. First way is to find the cheaper reinforcement material to reduce the overall production cost and to tackle the limited availability of the ceramic reinforcement material. This can be achieved by using the materials from industrial wastes and agro waste derivatives. Some of the investigations showed the usage of alternative reinforcements in the preparation of composites exhibits improved properties than

IOP Conf. Series: Materials Science and Engineering 1013 (2021) 012016 doi:10.1088/1757-899X/1013/1/012016

the unreinforced alloy. However, the developed composites show inferior properties than the conventional synthetic reinforced DRAMCs [2]. Second way the small size ceramic materials selected for the preparation ranges from micron scale to nanoscale. The nano size ceramic particulates shows better fracture toughness and ductility of DRAMCs reported in some studies, but the major limitation is high cost and availability of the reinforcement materials. Also, there is still inconclusive evidence to substantiate the mechanisms of ductility and fracture toughness improvement in nano particulate reinforced composites [2]. Third way is to use two or more reinforcing materials to produce DRAMCs hybrid composites. The objective of this approach is to produce DRAMC at lower cost coupled with optimized property. Some authors show the way to produce better performance hybrid AMCs at lower cost when compared with single reinforced AMCs [2].
Kenneth Kanayo Alanemeetal et al. prepared Al-Mg-Si alloy matrix hybrid composites by using rice husk- alumina as reinforcement and its corrosion and wear behavior have been studied. The method used was double stir casting. The percentage of alumina used was 2, 3 and 4 wt% while the Rice husk ash (RHA) was added in such a manner that the total reinforcement addition remains 10%. Corrosion and wear behavior were analyzed by open circuit corrosion potential and potentiodynamic polarization measurements methods respectively. Scanning electron microscope was taken as for characterizing corrosion and wear mechanisms. Superior corrosion resistance was seen by single reinforced Al-Mg-Si/10 wt.% Al_2O_3 composites from the other hybrid composites in 3.5% NaCl solution. Also increase in corrosion rates was due to increase in RHA reinforcement. The reason for increasing corrosion rates was increase in population of reinforcements. In wear test also, wear rate and coefficient of friction was seen to be increased because of the increased percentage of RHA reinforcement [3]. K. Umanath et al. investigated metallic worn surface of Al 6061 hybrid composites produced by stir casting method. Aluminium oxide and silicon carbide reinforcements were varied together from 5 to 25% vol. Fractions. Wear behavior showed much rougher surfaces of composites than the as cast alloy indicates abrasive wear. SEM results also revealed uniform mixing of reinforcements inside the aluminium matrix [4]. T. Hariprasad et al. investigated wear characteristics of Al_2O_3 and B_4C reinforced with Al 5083 metal matrix based hybrid composites. Al_2O_3 with 5% and four different proportions of B_4C with (0 wt%, 3 wt%, 5 wt%, 7 wt%) were used as reinforcements. Composites were synthesized by stir casting method. Wear test results concluded improved wear resistance properties because of high percentage of boron carbide and alumina particles [5]. A. Saravanakumar et al. investigated mechanical behavior of AA6063 Al_2O_3-1%, Gr (3, 6, 9, 12 wt.%) hybrid composites which was fabricated by stir casting method. One set is taken in as caste condition while other one was treated with hardening and testing was carried out on both materials. Mechanical properties were improved as addition of alumina had increased but only up to 6%. At higher percentage of alumina, agglomeration was seen as stated by SEM micrographs. The maximum improved properties were seen in 6% of alumina addition [6]. Aniruddha V. Muley et al. highlighted mechanical and tribological studies on nanoparticles reinforced hybrid aluminum based composite. Reinforcements were SiC and Al_2O_3 nanoparticles. LM 6 alloy was considered for the study. For fabrication stir casting was used and reinforcements were added in equal proportion for wt% of 0.5, 1.0, 1.5 and 2. Improvements in tensile and hardness were characterized by optical and SEM images. Hardness was improved by 17% while tensile strength was improved by 39%. Both were seen for 2wt% of reinforcements. Wear reduces by 80% compared to LM 6 alloys [7].

2. Experimental details

2.1 Materials

In the present work, AA 2014 was taken as a matrix material and the chemical composition of AA 2014 is presented in the Table 1. Alumina (Al_2O_3) particles of size 70-240 mesh and Silicon Dioxide (SiO_2) particles of size 40-150 mesh were selected as reinforcement materials for composite preparation. X-ray diffraction (XRD) test was performed to check the distribution of Al_2O_3 and SiO_2 particles. XRD phase evaluation was performed by Analytica006C X' Pert3 Powder XRD machine and confirmed that the used Al_2O_3 particles were γ- Al_2O_3.

Table 1. Chemical composition of As Cast AA2014 Al alloy

Elements	Cu	Mg	Si	Fe	Mn	Ti	Zn	Cr	Al
%Wt.	4.723	0.354	0.755	0.197	0.771	0.012	0.048	0.062	REM

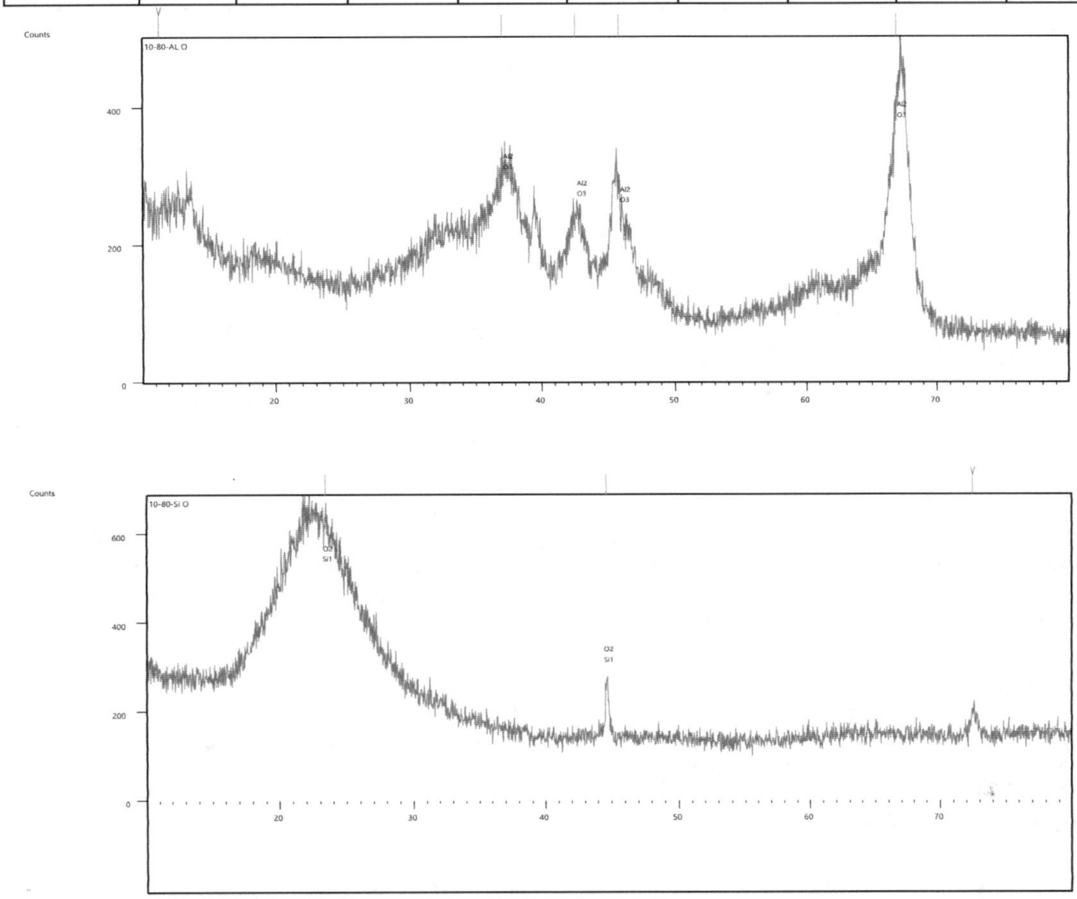

Figure 1. XRD images of Alumina and Silicon dioxide particles

2.2 Fabrication of hybrid composites

The hybrid composites of Al_2O_3/SiO_2 AA 2014 were synthesized by stir casting. For comparison AA 2014 was also synthesized. The reinforcement materials Al_2O_3 and SiO_2 were kept in small pot and preheated to 240^0C in a muffle furnace for 30 minutes. This improves wettability of the reinforcement with the matrix and remove the moisture content. After preheating, the molten metal of AA2014 was produced at 750^0C in a graphite crucible using electric resistance furnace. The unwanted gases entrapped in the melt were removed by using degassing tablets of Hexachloroethane (C_2Cl_6). The reinforcements were added to the molten matrix and stirred completely for 5 minutes at 300 rpm. During stirring, complete mixing of the reinforcement particles of Al_2O_3 and SiO_2 inside the vortex was formed. The prepared molten mixture was poured into the preheated permanent moulds.

Table 2. Composition of hybrid composites prepared

S.No.	Castings	Codes
1	AA 2014	Ascast
2	AA 2014 + 0.5% Al_2O_3 + 0.5% SiO_2	A
3	AA 2014 + 0.5% Al_2O_3 + 1.0% SiO_2	B
4	AA 2014 + 1.0% Al_2O_3 + 0.5% SiO_2	C
5	AA 2014 + 1.0% Al_2O_3 + 1.0% SiO_2	D
6	AA 2014 + 2.0% Al_2O_3 + 0.5% SiO_2	E
7	AA 2014 + 2.0% Al_2O_3 + 1.0% SiO_2	F

2.3 Wear test

Dry sliding wear was carried out on specimen to know the wear behavior. Standard wear specimen of 8 mm diameter and 30 mm height were prepared as per ASTM standard. The wear tests were conducted at a fixed 1200 rpm of disc for track radius of 50.4 mm for 5 minutes, 10 minutes and 15 minutes.

2.4 Scanning electron microscope (SEM)

Scanning electron microscopy (SEM) was used to characterize the synthesized composites. The specimens were polished using standard metallographic. The hybrid composite and as-cast specimens were examined using Tescan Vega-3 LMU High Resolution Scanning Electron Microscope (HR-SEM) to study the distribution of reinforcement in the matrix.

3. Results and Discussions

3.1 Microstructure analysis

Scanning electron microscope images show uniform mixing of reinforcements inside the matrix.

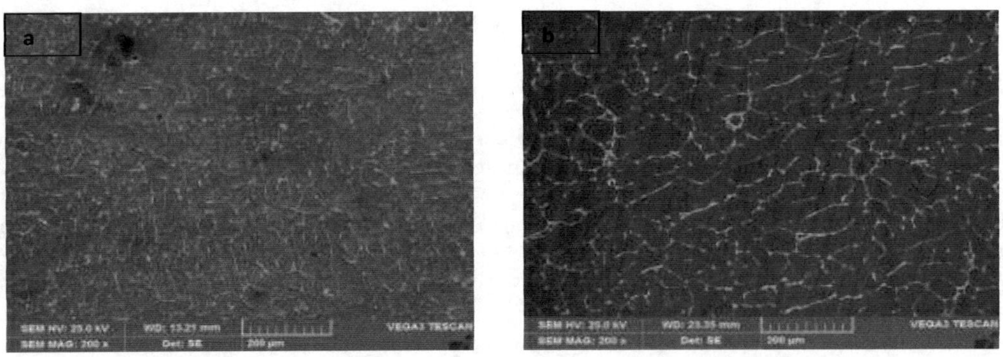

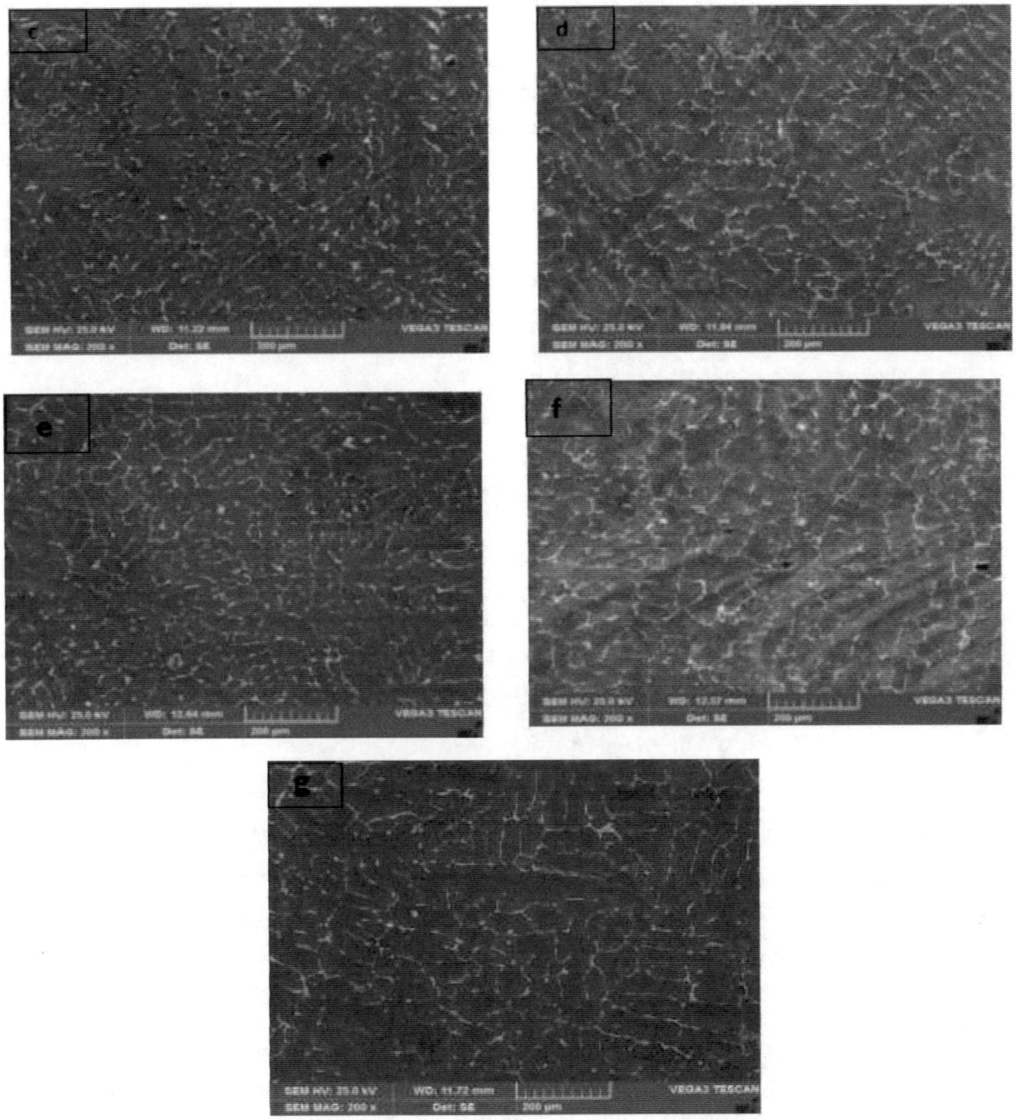

Figure2. SEM images of (a) Ascast (b) Composite A (c) Composite B (d) Composite C (e) Composite D (f) Composite E (g) Composite F, at magnification 200X.

Figure 2 shows the SEM micrograph of the specimens at 200X magnification. Agglomerated line structures are formed as shown in images of hybrid composites. Porosities can be observed in Ascast alloy. Image of hybrid composites show that reinforcements are more uniformly distributed in matrix than aluminium alloy. Hybrid composites A, B and C images show less segregation of line structure than hybrid composites D, E and F because of less reinforcement content. Line – line structure may be due to agglomeration of lesser density of SiO_2 particles.

3.2 Wear behavior

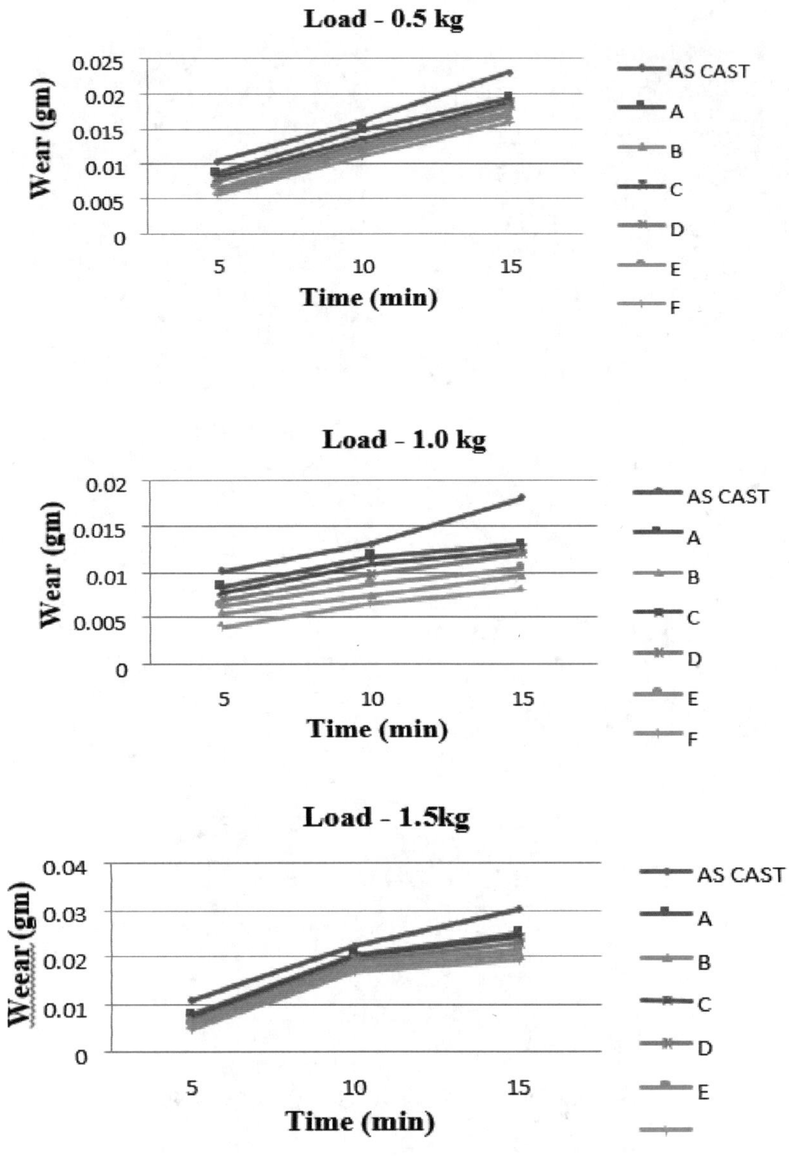

Figure 3. Wear V/s time for specimens at 1200 rpm

Figure 3 shows variation of wear with time at 1200 rpm for Ascast and hybrid composites. It can be observed from the graph that wear increases with increase in time. Ascast specimen wears out faster than hybrid composites. Low wear was observed for hybrid composites having better combination of reinforcements. Similar trend can be observed for the loads of 0.5kg, 1.0kg and 1.5kg.

3.3 Wear surface analysis

Figure 4 shows the SEM micrograph of the specimens at 500X magnification. Ascast image shows adhesive wear because of friction force given by hard EN-31 steel disc. Composite A image shows abrasive wear mechanism observed as hard particles are introduced inside matrix. Small grooves can be also observed. Composites B and C images shows track lines of wear test and reinforcement particles are forced out from matrix due to abrasive action. Composites D, E and F images shows accumulation of wear debris and partial welding of debris on surface of composite [8].

IOP Conf. Series: Materials Science and Engineering 1013 (2021) 012016 doi:10.1088/1757-899X/1013/1/012016

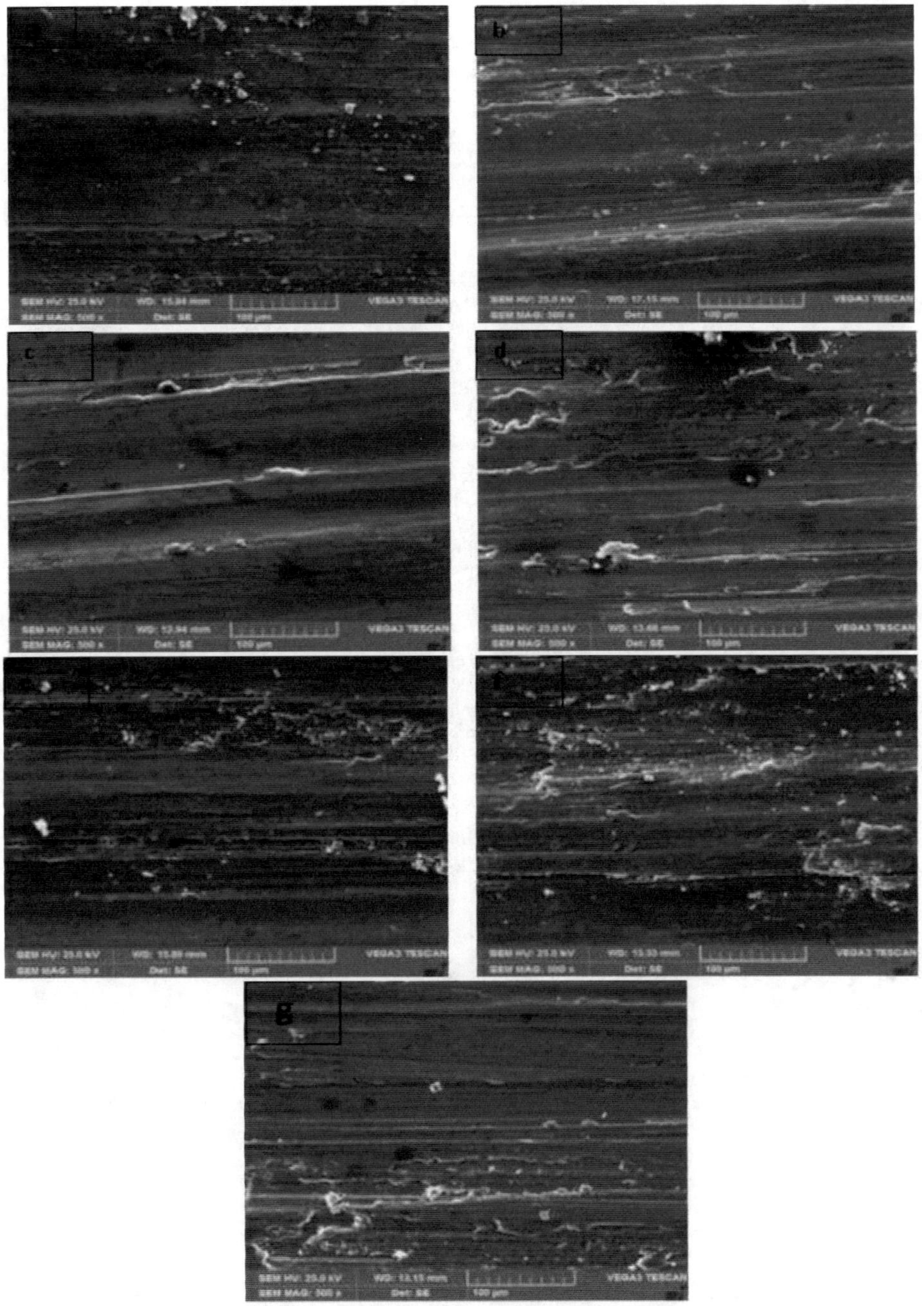

Figure 4. SEM images of worn out specimens(a) Ascast (b) Composite A (c) composite B (d) composite C (e) composite D (f) composite E (g) composite F, at magnification 500X

4. Conclusions

In the present work a hybrid aluminium matrix composite has been synthesized. The effects of reinforcements (alumina and silicon dioxide) with different percentages have been examined for wear properties of aluminium 2014.

- SEM images show successful incorporation of reinforcements inside the metal matrix.
- Decreased trend for wear was noticed for more addition of reinforcements at higher loads and higher speeds.
- The scanning electron microscope images of worn out specimens showed transition from adhesive wear to abrasive wear as reinforcement percentage increases.

References

[1] William D. Callister, Jr.David G. Rethwisch, Materials Science and Engineering an introduction, John Wiley & Sons, Inc 2018.

[2] Michael Oluwatosin Bodunrina, Kenneth Kanayo Alanemea, Lesley Heath Chown, Aluminum matrix hybrid composites: a review of reinforcement philosophies; mechanical, corrosion and tribological characteristics, j mater res technol. 4(4):434–445, 2015.

[3] Kenneth KanayoAlaneme, Peter Apata Olubambi, Corrosion and wear behaviour of rice husk ash—Alumina reinforced Al–Mg–Si alloy matrix hybrid composites, j mater res technol. 2(2):188–194, 2013

[4] K. Umanath, S.T. Selvamani, K. PalanikumarK and D. Niranjanavarma, Metal to Metal Worn Surface of AA6061 Hybrid Composites Casted by Stir Casting Method, Procedia Engineering 97, 703 – 712, 2014.

[5] T. Hariprasad, K. Varatharajan, S. Ravi, Wear Characteristics of B4C and Al2O3 Reinforced with Al 5083 Metal Matrix based Hybrid Composite, Procedia Engineering 97, 925 – 929, 2014

[6] A. Saravanakumar, P.Sasikumar, S.Sivasankaran, Synthesis and mechanical behavior of AA 6063-x wt. % Al2O3 -1% Gr (x = 3, 6, 9 and 12 wt. %) hybrid composites, Procedia Engineering 97 (2014) 951 – 960

[7] Aniruddha V. Muley, S. Aravindan, I.P. Singh, Mechanical and tribological studies on nano particles reinforced hybrid aluminum based composite, Manufacturing Review, 2, 26, 2015.

[8] Veeravalli Ramakoteswara Rao, Nallu Ramanaiah, Mohammed Moulana Mohiuddin Sarcar, Mechanical and tribological properties of AA7075–TiC metal matrix composites under heat treated (T6) and cast conditions, j mater res technol. 5(4):377–383, 2016.

Wear Characteristics of Hybrid Aluminium MMCs Subjected to Extrusion Process

Ashok Kumar M S[1*], Ajit Prasad S L[2]
[1]Assistant Professor, Department of Mechanical Engineering, DSATM, Bangalore-82
[2]Professor, Department of Mechanical Engineering, PESCE, Mandya-03
E-mail: ashok-me@dsatm.edu.in

Abstract: Aluminium alloys are extensively used due to their excellent mechanical and thermal properties, However the performance of these materials are enhanced with the addition of ceramic reinforcement phase into the aluminium matrix phase, forming a newer material having improved mechanical and tribological properties. The processing technique of these materials plays a vital role in obtaining required properties of the materials. Among the available processing techniques, liquid state processing (Stir Casting) technique is widely used because of its processing feasibility. However, the problem associated with the process is agglomeration of reinforcement particles, uneven distribution, blowholes, porosity which challenges the researchers in obtaining the required properties. To overcome these limitations, Secondary processing technique like Extrusion is used which enhances the properties of processed materials. In the present work is carried out by selecting Aluminium A356 as base material to which10% of 23μm size Al_2O_3 is reinforced. Hybrid composite material is processed by adding 1%, 3% & 5% MoS_2 particulates of 5μm size along with Al_2O_3. The sliding wear characteristics of aluminium based hybrid metal matrix composites have been studied in the present work and the sliding wear resistance of the material is found to improve with addition of MoS_2 particulates.

Keywords: A356, Al_2O_3, MoS_2, Extrusion, Wear Study

1. Introduction

Metal Matrix Composite materials are new class of materials which are intended for high performance applications in the field of automotive and aerospace applications. Among the available MMCs, Aluminium metal matrix material reinforced with ceramic materials like SiC, Al_2O_3 exhibits excellent mechanical properties compared to base alloy materials. MMCs are reinforced with self lubricating materials like Graphite and MoS_2 which leads to anti-frictions effect by forming lubricating film between mating surfaces. The presence of ceramic and lubricating particles (Hybrid Reinforcement) in the matrix material helps in obtaining the required mechanical and wear resistance properties. The main challenge is to reinforce these materials into the matrix with better distribution of particles. The cast materials normally exhibit uneven distribution of reinforcement particles and porous in nature which can be reduced by subjecting it to extrusion process.

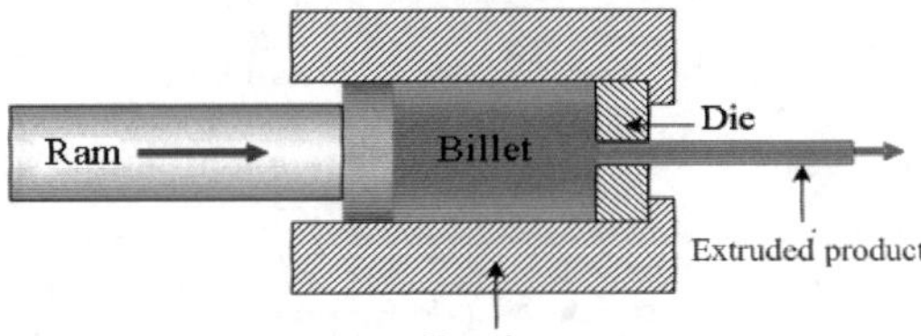

Figure 1.1: Schematic of Extrusion Process

2. Literature review

Extrusion process [Figure 1.1] is used to create objects of fixed cross-sectional profile where in the material is forced or drawn through a die of the desired cross-section [1]. Research work carried on secondary processing indicates that extrusion process can lead to break up of particle agglomerates, porosity elimination and improved bonding, all of which contribute to improve properties of MMCs. Extrusion process parameters like extrusion ratio, extrusion temperature and structure of die plays a vital role in improvement of microstructure and properties of the materials [2]. Research works carried out on extruded materials indicates that the grain growths will be easier at higher extrusion temperature and increase of extrusion ratio will cause the deformation to be increased and more uniform [3, 4].

Research work related to effects of weight fractions of reinforcement and extrusion ratios on mechanical properties of MMCs indicate that, the porosity increases with increase in the weight fractions of particulate reinforcement. Increased extrusion ratio results in significant decrease in porosity from over 2 % to 0.5 %. The higher extrusion ratios showed relatively more homogenous particle distribution within the matrix. A preferred particle orientation to extrusion axis was also observed [5].

Wear properties of Al6061–SiCp composites processed by stir casting followed by hot extrusion at a temperature of 550^{0}C has been studied under different extrusion ratios. Presence of reinforcement particles and extrusion improve the mechanical properties of the extruded composites which can be attributed to increase in dislocation densities at the interfaces of reinforcement and the matrix alloy. Increased dislocation densities leads to obstruction to the movement of the generated dislocations themselves thereby reducing the extent of plastic deformation. Wear rates of the test materials were found to decrease with increase in extrusion ratios [6].

Extrusion ratio also influence on the Surface roughness of the extruded materials. During extrusion when the material is allowed to flow in a die with high extrusion ratio whole material has to pass through the small outlet dia of the die, Since the ram speed is kept constant and as the extrusion ratio increases the time available for the material to flow through the die decreases there by increasing the resistance to flow leading to the surface roughness [7].

The present study involves processing of Al MMCs using A356matrix and addition of 23μm Al_2O_3 (10%) particles with varying weight percentage (1, 3, 5%) of 5μm MoS_2 particles by stir casting technique. The cast materials are extruded with varying extrusion ratios of 12.25, 16.0 and 20.25. The cast and extruded materials are subjected to sliding wear test under varying load conditions of 10N to 60N, for a constant sliding distance of 5000m and sliding velocity of 2m/s.

3. Experimental details: Materials and Method

Table 3.1 Salient properties of A356

Properties	Values
Elastic Modulus (Gpa)	70-80
Density (g/cc)	2.7
Poisson's Ratio	0.33
Tensile Strength in Mpa	220
Hardness (Vickers)	99
Melting Temperature	660^{0}C
Thermal Conductivity	151 W/m-^{0}K

Table 3.2 Salient properties of Al_2O_3

Properties	Values
Elastic Modulus (Gpa)	300
Density (g/cc)	3.69
Poisson's Ratio	0.21
Tensile Strength (Mpa)	300
Hardness (Vickers)	1365
Melting Temperature	2072^{0}C
Thermal Conductivity	30 W/m-^{0}K

Table 3.3 Salient properties of MoS_2

Properties	Values
Elastic Modulus (Gpa)	270
Density (g/cc)	5.06
Poisson's Ratio	0.27
Tensile Strength (Mpa)	230
Hardness (Vickers)	29.1
Melting Temperature	2375^{0}C
Thermal Conductivity	34 W/m-^{0}K

Table 3.4 Specimen Designation

Specimen	Designation
Base Alloy	A356
A356 + 10% Al_2O_3 (23μm)	23P
A356 + 10%Al_2O_3 + 1%MoS_2	1 M
A356 + 10%Al_2O_3 + 3%MoS_2	3 M
A356 + 10%Al_2O_3 + 5%MoS_2	5 M
Extrusion Ratio - 12.25	E1
Extrusion Ratio - 16.0	E2
Extrusion Ratio - 20.25	E3

Table 3.1, 3.2, 3.3 shows the salient properties of base material A356, Al_2O_3 & MoS_2 respectively, Different test samples are designated based on their composition and extrusion ratio as shown in the Table 3.4.

Hybrid composites (A356 + Al_2O_3+MoS_2) are fabricated by the stir casting process (Figure 3.1) with varying percentages of MoS_2 reinforcements. Initially base material is charged into the furnace containing graphite crucible and heated up to liquid state (700^{0}C). Once the required temperature is attained degassing tablets are added to remove the entrapped air inside the liquid material followed by slag remover which removes the impurities present inside the cast element. Reinforcing particles (Al_2O_3) are preheated and added to the molten material followed by stirring with the aid of stirrer rotating at 300rpm and suitable percentage of MoS_2 particles are then added followed by continuous stirring. The molten material is then passed on to the mould of the required shapes. The cast material matrix composites are subjected to secondary extrusion process with extrusion ratios of 12.25, 16.00 and 20.25 at constant temperature of 300^{0}c, Cast specimen of 35mm, 40mm and 45mm diameters were reduced to 10mm diameter by extrusion process with the above mentioned extrusion ratios respectively.

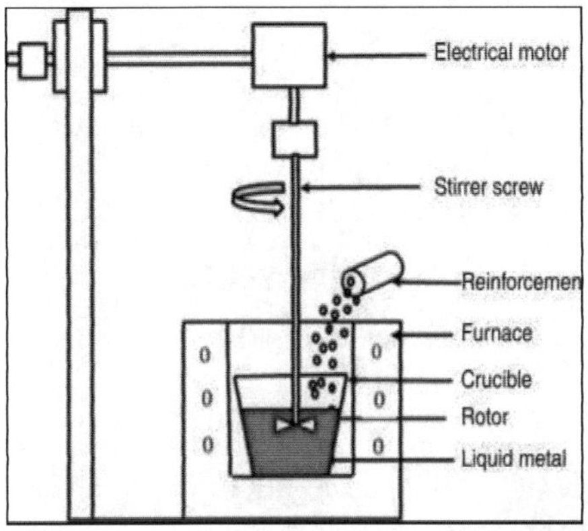

Figure 3.1: Schematic of Stir Casting Setup

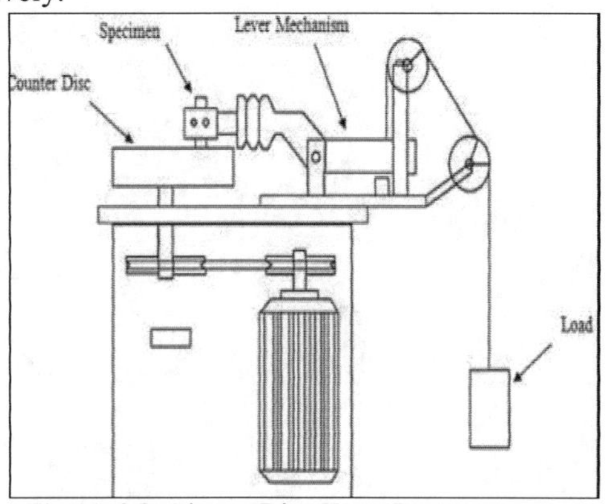

Figure 3.2: Pin-on-Disc Wear Test Apparatus

Figure 3.3: ASTM G99 Standard Wear Testing Sample

Figure 3.2 shows the schematic of wear testing apparatus used for the present investigation. The test samples are prepared as per the ASTM G99 standard for sliding wear test. Figure 3.3 shows the sample specimen for wear testing.

4. Results & Discussion

4.1 Microstructural studies

SEM images of reinforcement materials Al_2O_3 and MoS_2 are shown in the Figure 4.1 and Figure 4.3 respectively. Figure 4.2 and 4.4 shows the corresponding EDAX analysis. The SEM image as cast Al-Al_2O_3 specimen is shown in Figure 4.5, it can be observed that the reinforcement particles are uniformly distributed over the matrix materials and the pores present in the cast materials can also be visualized. Figure 4.6 indicates the SEM image of extruded material with uniform distribution and reduced porosity. Figure 4.7 shows the hybrid composites with agglomerates of reinforcement particles, Figure 4.8 shows the extruded hybrid material extruded where in reinforced particles are getting aligned in the direction of extrusion process with uniform distribution.

Figure 4.1 SEM Image of Al_2O_3 Particle

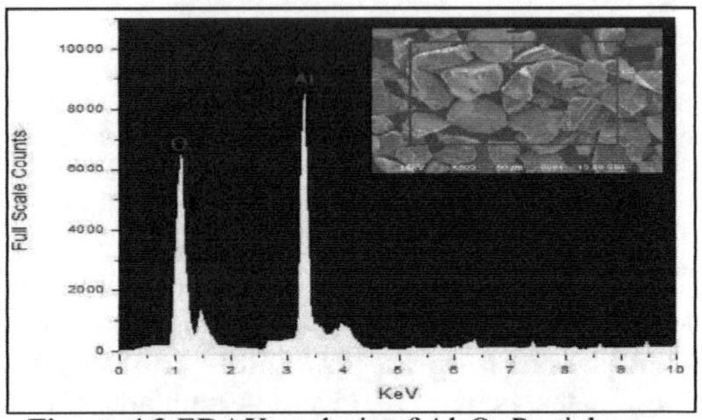

Figure 4.2 EDAX analysis of Al_2O_3 Particles

Figure 4.3 SEM Image of MoS_2 Particles

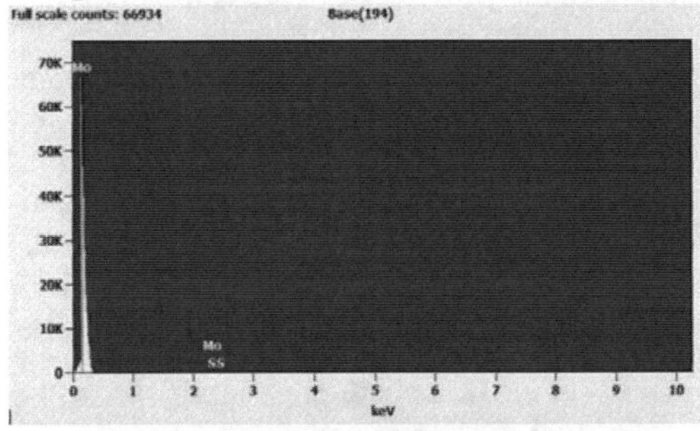

Figure 4.4 EDAX analysis of MoS_2 Particles

Figure 4.5 SEM of as cast Al–Al_2O_3 specimen

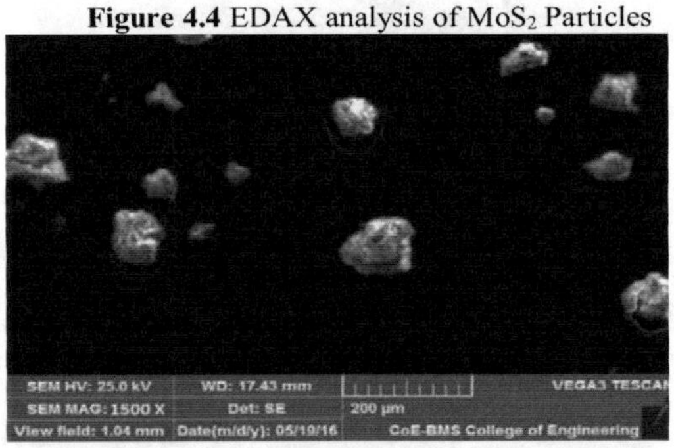

Figure 4.6 SEM of Extruded Al–Al_2O_3 specimen

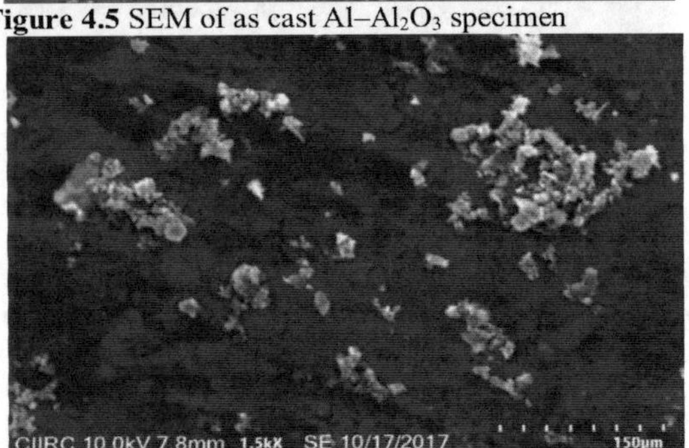

Figure 4.7 SEM of as cast specimen 5% MoS_2

Figure 4.8 SEM Image of extruded specimen (A356 + 10% Al_2O_3, 5% MoS_2, Ext. Ratio 20.25)

SEM images of worn surface of cast Al–Al_2O_3 and extruded specimen are shown in Figure 4.9 and 4.10 respectively. In cast specimens, grooves that are generated along the direction of sliding due to plastic

IOP Conf. Series: Materials Science and Engineering 1013 (2021) 012017 doi:10.1088/1757-899X/1013/1/012017

deformation can be visualized. In the extruded conditions, it can be noted that there is lesser plastic deformation due to the presence of reinforcement particle.

Similar results are also obtained with addition of MoS_2 as additional reinforcement member along with Al_2O_3 particles. Presence of MoS_2 material acts as a solid lubricant that reduces the friction generated at the pin and the disc interface during the wear test. Results obtained indicates that the friction and wear generated during rolling and sliding motions can be minimized with the aid of self lubricating materials which enchants the tribological properties of the materials by forming a tribo layer which reduces the friction between the mating components.

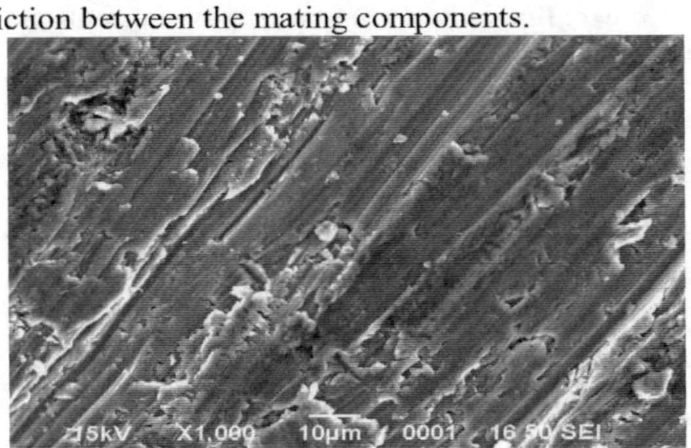

Figure 4.9 SEM of Worn Surface (Al–Al_2O_3, As Cast)

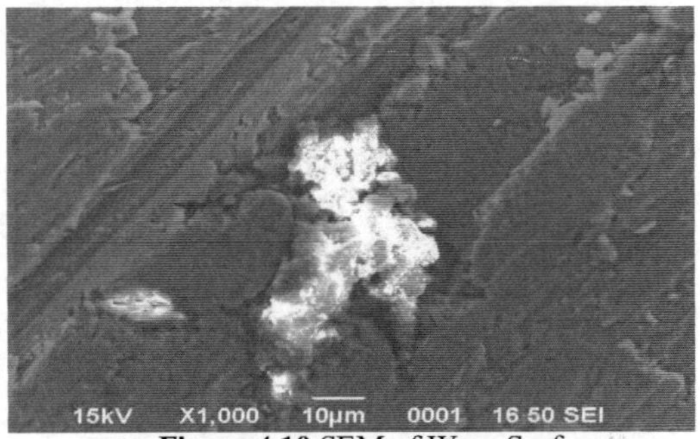

Figure 4.10 SEM of Worn Surface (Al–Al_2O_3, Extruded, Ext. Ratio 12.25)

SEM images of worn surface of cast Al–Al_2O_3 and extruded specimen are shown in Figure 4.9 and 4.10 respectively. In cast specimens, grooves that are generated along the direction of sliding due to plastic deformation can be visualized. In the extruded conditions, it can be noted that there is lesser plastic deformation due to the presence of reinforcement particle.

Similar results are also obtained with addition of MoS_2 as additional reinforcement member along with Al_2O_3 particles. Presence of MoS_2 material acts as a solid lubricant that reduces the friction generated at the pin and the disc interface during the wear test. Results obtained indicates that the friction and wear generated during rolling and sliding motions can be minimized with the aid of self lubricating materials which enchants the tribological properties of the materials by forming a tribo layer which reduces the friction between the mating components.

4.2 Wear **v/s** Normal Load

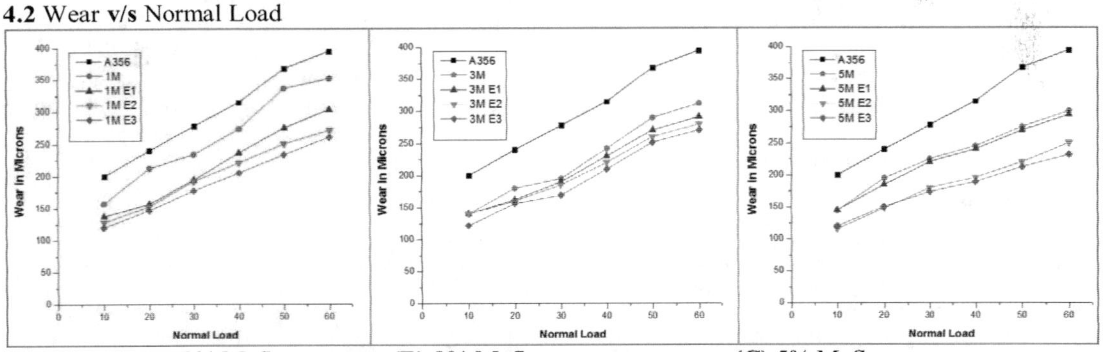

(A) 1% MoS_2 **(B)** 3% MoS_2 **(C)** 5% MoS_2

Figure 4.11 Wear v/s Normal Load

Figure 4.11 shows the Wear rate obtained with respect to normal loads of 10N to 60N for a sliding distance of 5000m at a constant sliding velocity of 2m/s. Wear pattern of 1M, 3M & 5M materials in comparison with base alloy and with different extrusion ratios are shown in Figure 4.2 A, B & C respectively. Presence of MoS_2 aids in extruding the materials easily by reducing the friction between the surface of die and the material by forming a solid lubricant layer and minimizes the degree of shear stress transferred to the sliding material underneath the contact area resulting in decreased plastic deformation in the subsurface region thereby improving the wear resistance property. Increase in MoS_2 percentage in the matrix material helps in reducing the forces required to extruded the material from the die. Increase in extrusion ratio leads to recrystallization and better distribution of the particles in the matrix material, uniformly distributed Al_2O_3 particles helps in improving the mechanical properties where as the MoS_2 helps in improving the frictional properties. Results obtained indicate that hybrid composite material reinforced with 5% of MoS_2 has experienced lower wear rate in both as-cast and extruded condition.

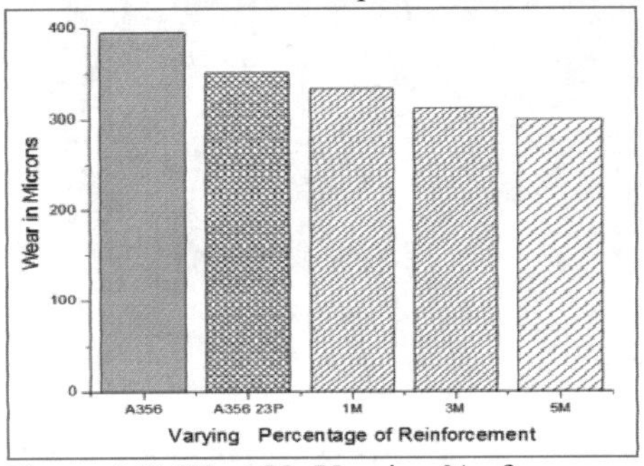

Figure 4.12 Wear Vs Varying % of Reinforcement (As Cast- Load 60N, 5000m SD)

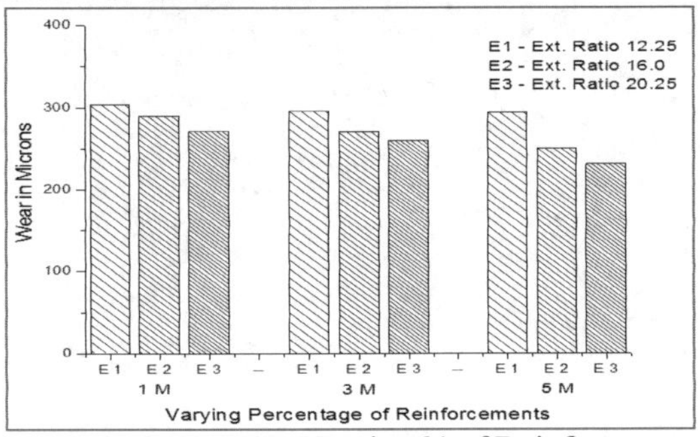

Figure 4.13 Wear Vs Varying % of Reinforcement (Varying Extrusion Ratio - Load 60N, 5000m SD)

Figure 4.12 shows the wear loss experienced by as–cast samples of different compositions at end of each trial. Results indicate that the wear loss is higher in base alloy sample compared with reinforcements. Further, hybrid composites having different percentage of MoS_2 experience lesser amount of wear when compared to those samples without MoS_2. Addition of Al_2O_3 particulate reinforcements enhances the strength of the material where as MoS_2 particulates enhances the self lubricating properties of the material resulting in reduced wear loss.

Figure 4.13 shows the variation in wear loss experienced by different hybrid composite materials when subjected to secondary processing with different extrusion ratios. From the results obtained it can be observed that there is marginal decrease in wear loss of each composite material with increase in extrusion ratio. When the material is subjected to extrusion, grain refinement of the matrix material takes place strengthening the material structure. Also because of the large amount of plastic deformation, dislocation density of the material increases, enhancing the hardness. Higher extrusion pressures also cause breaking up of agglomerates and closure of pores which are present in as cast material resulting in improved structure of the material. These beneficial influences of extrusion result in improved wear strength of the material as indicated by the experimental results.

5. Conclusions

Materials with self lubricating properties are making remarkable changes in the mechanical and tribological properties of the composite materials. Improvement in friction and wear properties leads to increase in demand for the materials especially for tribological applications where lubrication is preferred. In this regard present work is an attempt made to evaluate the wear characteristics of hybrid composites reinforced with Al_2O_3 and varying percentage of MoS_2 for both as cast and extruded conditions. Result of SEM analysis indicates the presence of reinforcement particulates that are uniformly distributed over the matrix material. Similar results are also obtained for the extruded materials with even better distribution of reinforcement particles in the matrix materials. Presence of reinforcement particles with better bonding yields improved wear resistance properties. Increase in addition of reinforcement percentage of MoS_2 showed improved wear characteristics due to the formation of lubrication film. Extrusion process has led to changes in the microstructure of the processed materials which results in improvedwear resistance properties.

References

[1] C.S Ramesh, Adarsha Hirianiah, Harishanad K.S, Naveen Prakash Noronha "A review on hot extrusion of Metal Matrix Composites (MMC's)" International Journal of Engineering and Science ISBN: 2319-6483, ISSN: 2278-4721, Vol. 1, Issue 10 (December 2012), PP 30-35.

[2] Jiang Bin, Gao Liang, Huang Guang-jie, Ding Peidao, ang Jian "Effect of extrusion processing parameters on microstructure and mechanical properties of as-extruded AZ31 sheets" Transactions of Nonferrous Society of China, Science Press, 18(2008) S160 –S164.

[3] Wang L Y, Fan Y G, Huang G J, Huang G S "Flow stress and softening behavior of wrought magnesium alloy AZ31B at elevated temperature" Transactions of Nonferrous Metals Society of China, 2003, 13(2): 335−338.

[4] Wang L Y, Fan Y G, Huang G J, Huang G S. Grain refinement of wrought AZ31 magnesium alloy" The Chinese Journal of Nonferrous Metals, 2003, 13(3): 594−598.

[5] Ghanaraja.S, Vinuth Kumar.K.L, Ravikumar.K.S, Madhusudan.B.M "Mechanical Properties of Al_2O_3 Reinforced Cast and Hot Extruded Al Based Metal Matrix Composites" 5th International Conference of Materials Processing and Characterization (ICMPC 2016) Materials Today: Proceedings 4 (2017) Pp 2771–2776.

[6] C.S. Ramesh, R. Keshavamurthy, G.J. Naveen "Effect of Extrusion Ratio on Wear Behavior of Hot Extruded Al6061–SiCp (Ni–P Coated) Composites" Elsevier Publication Wear 271 (2011) Pp 1868– 1877.

[7] Ambati Vijay Kumar, Chalamalasetti Srinivasa Rao, Damera Nageswara Rao "The Influence of Extrusion Ratio on The Properties of Nano Sic Reinforced 6061 Aluminum Alloy In Cold Extrusion" International Journal of Engineering Research & Technology (IJERT) Vol. 3 Issue 1, January – 2014.

IOP Conf. Series: Materials Science and Engineering 1013 (2021) 012018 doi:10.1088/1757-899X/1013/1/012018

Impact of Alumina Particulates Addition on Hardness and Wear Behaviour of 2014Al Metal Matrix Composites by Vortex Method

Bharath V[1*], V Auradi[1], Madeva Nagaral[2], Samuel Dayanand[3] and Satish Babu Boppana[4]

R & D Centre, Department of Mechanical Engineering, Siddaganga Institute of Technology, Tumakuru, Visvesvaraya Technological University, Karnataka, India
[2]Aircraft Research & Design Centre, HAL, Bangalore, Karnataka, India
[3]Department of Mechanical Engineering, Government Engineering College, Raichur, Karnataka, India
[4]School of Engineering, Presidency University, Bengaluru, Karnataka, India

*Corresponding author: Bharath V; Email – bharathv88@gmail.com, Contact No: +91-9742222523

Abstract:- In this study, 2014Al matrix alloy reinforced with a 9 wt% of alumina (20 μm) particulates composites by novel two stage melt stirring (stir casting) practice used to synthesize, describe and analyze wear behaviour. Adding up of reinforcement was kept at 0 and 9wt%. For each wt. % preheated (i.e., 250°C) alumina particles were introduced in to the molten 2014Al alloy in steps of two by creating a vortex to overcome the clustering effects. Produced composite was examined by SEM analysis in support of exploring the microstructures and chemical components. Additionally characterization of wear studies of cast 2014Al matrix alloy and 2014Al-Al_2O_3 particulate composites were analyzed. After the assessment, it had been seen that there is an improvement in the hardness of the composites prepared and this is due to the incorporation of Al_2O_3 particulates in 2014Al alloy matrix meanwhile. Additionally, the wear rate of all composites prepared diminishes with increase in sliding distance while the wear rate of the composites prepared enhances with increase in load. By using Scanning Electron Microscope (SEM) the diverse wear mechanism for different test states of various compositions were examined.

Key words- 2014Al Alloy; Alumina; Hardness; Wear; Metal Matrix Composites;

1. Introduction

Metal Matrix Composites (MMCs) are generally encouraging in accomplishing upgraded mechanical properties, for example, Young's modulus, hardness, 0.2 percent yield quality strength and extreme elasticity because of the nearness of small scale measured fortification particles in the matrix. Generally, regards to the mechanical properties, the fortifications result in increased quality and hardness, even at the cost of some malleability [1]. On account of their improved mechanical and physical properties, aluminum matrix composites (AMCs) strengthened with particles and whiskers are broadly utilized for superior applications, particulate strengthened aluminum matrix composites and whiskers are broadly utilized for predominant applications, for example, in automobiles, military, aviations and power industries [2]. By adding mainly strong yet brittle particles such as Al_2O_3 and SiC, comparatively soft alloy such as aluminum can be rendered extremely resistant in the composites. Due to their high particular consistency and firmness, aluminum compounds are popular materials among the various accessible matrix materials. Nevertheless, their applications are constrained because of their poor wear opposition.

Within Al-alloys, 2014Al-alloy broadly utilized in various designing applications together with transportation and development, where exceptional mechanical properties, for example tensile, hardness, strength and so forth, are basically required [1]. Particulate reinforced aluminum matrix composites are as of now being considered for their good mechanical and tribological properties over the customary compounds, and along these lines, these composites have improved broad applications in vehicle and aviation industries. Hard particles, for example, Boron Carbide (B_4C), Alumina (Al_2O_3), and Silicon Carbide (SiC) are normally used in the composites as reinforcement phases. In the automotive and aircraft industries, the application of Al_2O_3 or SiC particle strengthened aluminum matrix alloy composites is increasingly for cylinder heads, connecting rods, pistons and so forth. Meanwhile, the materials properties of tribology are significant [1–4]. MMC's mechanical properties are also delicate to the manufacturing method utilized to produce the products. By making use of science-based modeling practices to automate the manufacturing cycle will make significant improvements. A few systems have been utilized for the composite preparation, involving powder metallurgy; squeeze casting and melting practice [2, 3]. Be that as it may, powder metallurgy gives off an impression of being the favoured procedure taking into account its capacity to give progressively uniform scatterings. In general, hot extrusion will be used as post treatment to take benefits from simultaneous compressive force and high temperature application [5].

MMC's liquid state production involves two approaches relying upon the temperature, where the particles are brought into the melt. The particles are incorporated in the melt stirring phase over the temperature of the liquid molten metal, although the particles are introduced in compo-casting system at the metal's semi-solid slurry temperature. The vortex is used for injection of reinforcement particles in both systems. Moreover, the melting procedure has two significant issues (i) the liquid metal matrix usually does not wet the ceramic particles and (ii) the particles appear to drift or descend as per their relative density to the metal liquid. Therefore ceramic particle scattering is not uniform. The pressure casting such as die and squeeze casting practices are required to reduce the composite material porosity [6]. While powder metallurgy delivers enhanced properties of mechanical in MMCs, there are several major improvements to melt processing i.e., better matrix-particle bonding, simpler matrix structure control, flexibility, low processing cost, closer net shape and large material selection [1–4].

Be that as it may, small data is accessible as respects to the microstructure, friction and wear behaviour of 2014Al strengthened with Al_2O_3 particulates treated by two-stage stir casting technique. The aluminum–Al_2O_3 composites play an important role with the growing competition for light weight materials applications in evolving industries. Keeping the above perceptions in seen, it is recommended to develop 2014Al-Al_2O_3 composites with 9 wt. % of Al_2O_3 particulates. The core aim of this study is to explore the impact of alumina on various factors such as processing method on the microstructure, hardness and wear behavior of 2014Al combination based composites with Al_2O_3 particulates by utilizing two stage stir casting technique.

2. Experimental details

2.1 Selection of Materials

For the present experiment 2014Al alloy is used as matrix material whereas, alumina is used as a reinforcing material. 2014Al matrix alloy chemical composition is shown in Table 1 and matrix properties and reinforcement properties are shown in Table 2. Alumina (Al_2O_3) particle with a size 20 μm is used for synthesis of composites.

Table 1. Chemical composition of 2014Al alloy in weight percentage

Elements	Si	Fe	Cu	Mg	Cr	Zn	Ti	Mn	Al
Contents	0.7	0.2	0.45	0.63	0.01	0.19	0.06	0.83	Balance

Table 2. Chemical composition of 2014Al alloy in weight percentage [7]

Material/ Properties	Density (gm/cc)	Hardness (HB500)	Strength (Tensile /Compressive) (MPa)	Elastic modulus (GPa)
Matrix- 2014Al	2.8	135	483	70-80

2.2 Preparation of composites

To manufacture the casted composites as mentioned below, the 2-stir casting practice has been used. After successful degassing with solid hexa-chloro ethane (C_2Cl_6), preheated 20 μm size of alumina particle are inserted into the vortex of the molten alloy. Before transferring into the melt, alumina particles were preheated (250^{0}C). In steps of 2 the degree to which Al_2O_3 particles were integrated in the matrix alloy was accomplished. i.e., the overall quantity of reinforcement needed has been determined and is transferred into the melt two stages instead of presenting it at once. Mechanical stirring is carried out for a period of 10 minutes for the molten alloy and it was accomplished by the use of steel impeller coated with zirconia at either point of before and after the adding of strengthening particles. Until plunging into the melt, the stirrer was preheated, positioned roughly at a depth of 2/3 height from the bottom of the molten metal and worked at a 200 rpm. Once the pouring temperature (750^{0}C) is accomplished, the liquid molten composite is poured into the cast iron molds and allowed for complete solidification for some time. Thus, composites containing 0, and 9 wt. percent alumina particles were accomplished as chambers of distance across 12.5mm and length 125mm.

2.3 Evaluation of produced composites

Obtained samples were subjected to machining operation i.e., samples were cut to suitable size of 5 mm thickness and 15 mm diameter then processed for polishing in order to examine the microstructure of the produced samples. At first, the cut examples were cleaned with emery paper up to 1000 grit size followed by cleaning with Al_2O_3 suspension on a cleaning plate utilizing velvet material. This was trailed by cleaning with a diamond paste (0.3 microns). The cleaned surface of the samples was engraved with Keller's reagent lastly exposed to microstructure concentrate under the examining electron magnifying instrument. Micro- hardness tests were conducted on the cleaned surface of the specimens utilizing micro-vickers hardness measuring system with a load of 2N for a dwelling time of 30 seconds, ten sets of measurements were taken at various locations of the cleaned surface of the specimen and the average was taken according to ASTM E10. As per ASTM G99 standard [8] wear test is carryout by the utilization of pin on disc machine (DUCOM, TR-20LE). Dry sliding wear tests were executed on samples of 8 mm diameter and 30 mm height. The counter circle material was of EN31 steel. Previous to testing, the pin and surface disc was dirt free with acetone. The examinations were driven at a consistent sliding speed of 1.25m/s and total time is 40 minute under room temperature over a varying load of 10N, 20N and 30N and a sliding distance of 500 m to 3000 m. In the midst of testing the pin samples was kept fixed and opposite to the disc while the indirect plate was rotated. Electronic estimating machine with the precision of 0.0001 g., was used to quantify the fundamental load of the models. During every single test, the circle counter face was dirt free with acetone. Test is followed by weight loss method.

3. Results and Discussions

3.1 Microstructural studies

Fig. 1a and b demonstrates the SEM micrographs of as- cast 2014Al matrix alloy and 9 wt. % of alumina strengthened with 2014Al alloy composite. These two analyzed samples were obtained from the middle segment from the cylindrical specimens. The microstructure of as cast 2014Al combination involves fine grains of aluminum strong arrangement with an adequate scattering of between metallic precipitates. Besides the standardized uniform scattering of estimated alumina particles with absence of agglomeration and bunching in the composites, it also shows the great holding between the structure and the fortification. This is due in part to the viable mixing operation that was achieved through two phases of expanding the fortification. All over the grain limit of the lattice the alumina particles hinder the grain improvement and resist the grain separation production in the midst of stacking.

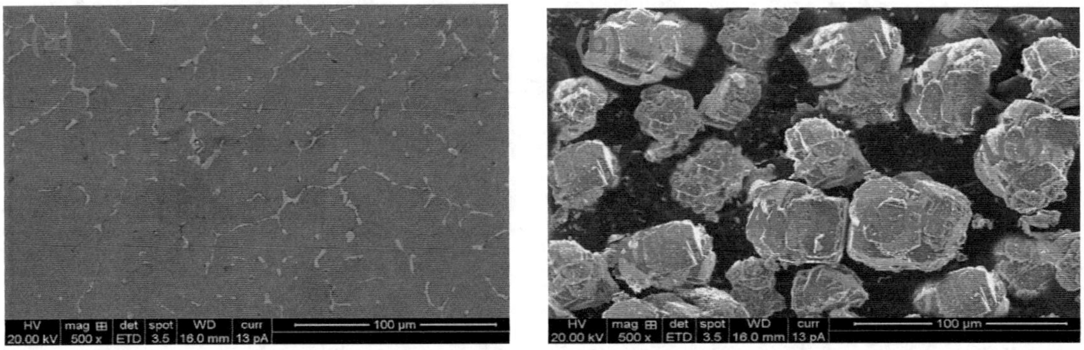

Fig. 1 SEM photographs of (a) as-cast 2014Al alloy (b) 2014Al-9 wt. % Al_2O_3 composites.

3.2 Hardness measurement

Fig. 2 demonstrates the difference in hardness when 9 weight percent of alumina is added to the 2014Al alloy. A material's hardness is a mechanical parameter showing the capacity to withstand nearby plastic deformation. With the incorporation of 9 wt. % of alumina particles, the hardness of 2014Al-Al_2O_3 composite is found to increase. This rise is observed for Al composites, from 93.67 VHN to 104.7 VHN. This can be traced essentially to the closeness of tougher alumina particles in the lattice and also to the maximum constraints of localized deformation of matrix throughout the indentation, due to the presence of tougher phase [9, 10 and 11]. Moreover, alumina, like different fortifications, reinforces the matrix by making of high dislocation density between cooling and room temperature, because of the differentiation of coefficient of thermal expansion among the alumina and the 2014Al compound. Disparity strains formed involving the reinforcement and the matrix inhibits the development of dislocations, thus improving the composite hardness.

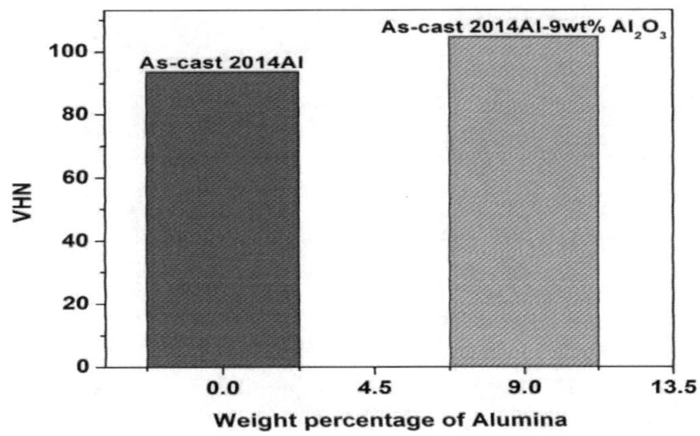

Fig. 2 Hardness of 2014Al alloy and Al_2O_3 composites

3.3 Wear behaviour of the produced composites:

Fig.3 displays the wear rate of as-cast 2014Al compound and 2014Al-9 wt. % Al_2O_3 composites. Wear rate diminishes after incorporation of alumina particles contrasted with 2014Al matrix with increment in distance of sliding varied from 500 m to 3000 m. The wear rate reduction with increment in the distance of sliding is because of (i) the improved hardness of the scattered alumina particulates as a load-bearing part (ii) the proximity of alumina particulates as a solid lubricant [12], (iii) Alumina particulates released on the erode surface keeps from the prompt metal to metal contact and serves as a lubricant consequently decreasing the coefficient of friction among the steel disk and the arranged composite pin [12], (iv) Rigid alumina particulates in the 2014Al matrix resist these rigid steel plugging operation and increases wear resistance [13]. The wear opposition of the produced composites altogether upgraded due to the presence of the alumina particulates. The outcomes as showed from Fig. 3 shows the diminishing pattern of wear rate after the introduction of 9 wt. % of alumina particles.

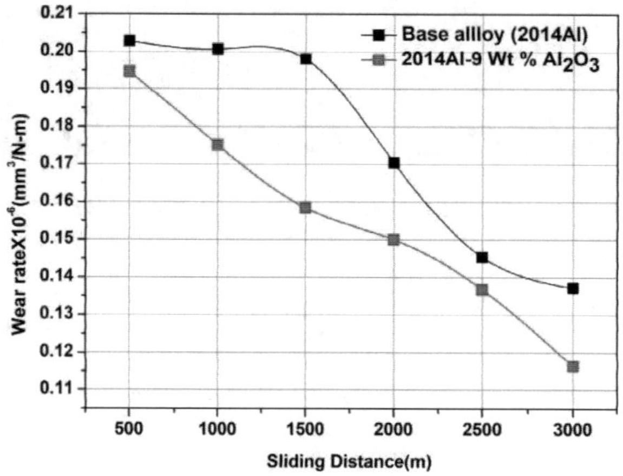

Fig.3 Variation of wear rate with respect to distance of sliding (As- cast 2014Al matrix alloy and 2014Al-9wt% Al_2O_3 composites)

Fig. 4 shows the impact of load applied on 2014Al alloy wear behavior and its composites. As the usual load rises from 10N to 30N the wear rate is increased and is lower for composites that are strengthened with alumina. Maximum wear rates are identified at higher loads for 2014Al matrix alloy and the composites. The temperature of the sliding surface and pin is greater than the critical value at optimum loads and as the load increases on the plate, the wear rate of both the matrix alloy and the alumina composites eventually increases. Wear rate of the produced composites is observed to decrease with 9 wt. % of alumina additions in alloy matrix. Enhancement in the wear opposition of the produced composites with 9 wt. % of fortification can be credited to the high hardness of alumina particulates which serves as the hindrance for the material misfortune [14].

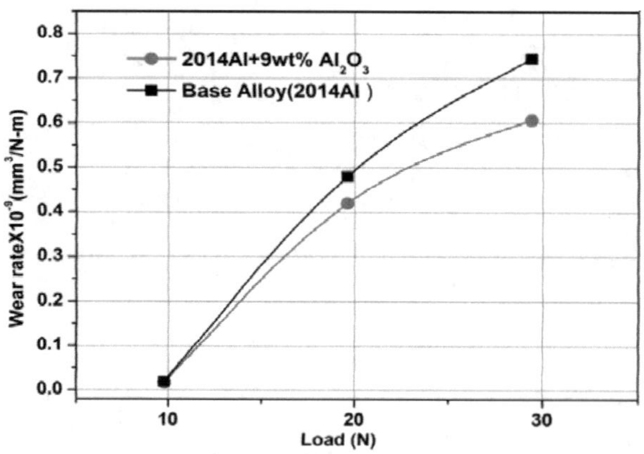

Fig.4 Wear rate of 2014Al alloy and 2014Al-9 wt. % Al_2O_3 composites at different loads

3.4 Worn Morphology

Using scanning electron micro-photographs wear surface analysis of 2014Al alloy and Al_2O_3 reinforced composites are studied. Figure 5 displays the wear worn surfaces of matrix material 2014Al alloy and 9 wt. % of Al2O3 particles strengthened composite at a load of 30N and a sliding speed of 1.25 m / sec.

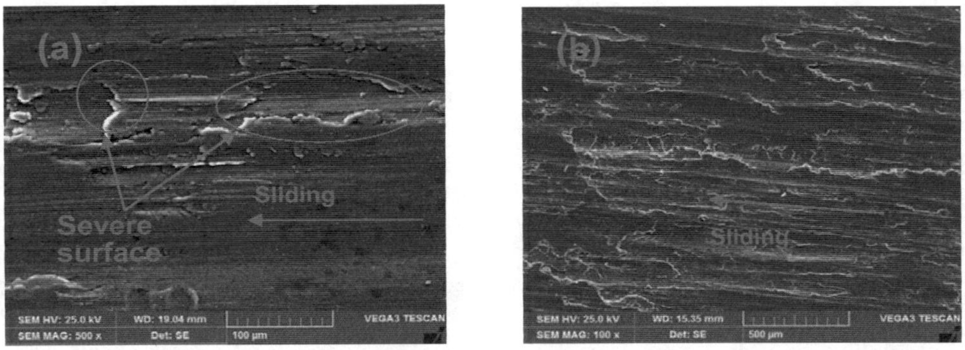

Fig.5 SEM microphotographs of worn surfaces of (a) as-cast 2014Al matrix alloy (b) 2014Al-9% Al_2O_3 at a load of 30N and 1.25m/sec

Fig. 5a displays specific depressions and edges moving in sliding direction parallel to one other. From the micrograph it can be seen that the furrows in the grid combination 2014Al are more significant and deeper as compared with the composites tried with practically identical conditions. Furrowing on the worn out surface of the 2014Al-9 wt. % of alumina can also be seen in fig. 5b, which may be due to slipping of the composite oxide molecule. A thick layer could be seen due to the composites, which shields the basic matrix from being in contact with the sliding partner and diminishes the volumetric wear misfortune along those lines. This trade layer surrounded on the composites gives the protective cover to hidden material and thus suppresses the metal- metal contact.

4. Conclusions

In this study, composites 2014Al/Al_2O_3 were manufactured using a process of 2-stage stir casting by taking 9 wt. % of alumina reinforcement. The microstructure, hardness, and wear conduct of prepared samples are examined. The matrix is nearly pore-free and uniform dispersion of alumina particle which is clear from SEM microphotographs. The micro- hardness estimation of 2014Al-9 wt. % alumina composite is better than those of unreinforced material. The wear obstruction of 2014Al-9 wt. % alumina composite is extensively higher than that of the 2014Al alloy. In addition, the wear rate of 2014Al matrix alloy and 2014Al-9 wt. % alumina composites diminished with increment in distance of sliding and wear rate of the composite produced is less as contrasted to as-cast 2014Al matrix alloy with increment in load and further, these composites can be utilized for aviation basic applications and furthermore can be utilized as a high load transfer components.

References

[1] Vencl A, Bobic I, Arostegui S, Bobic B. Structural, mechanical and tribological properties of A356 aluminium alloy reinforced with Al2O3, SiC and SiC + graphite particles. Journal Alloys Compound, 506:631–9, 2010

[2] Sajjadi SA, Ezatpour HR, Beygi H. Microstructure and mechanical properties of Al–Al2O3 micro and nano composites fabricated by stir casting. In: Proceedings of 14th national conference on Materials Science and Engineering,

Tehran, Iran, p.325, 2010.

[3] Sajjadi SA, Torabi Parizi M, Ezatpour HR, Sedghi A. Fabrication of A356 composites reinforced with micro and nano Al2O3 particles by a developed compo-casting method and study of their properties. Journal of Alloys Compound, accepted for publication. 2011

[4] Mazahery A, Abdizadeh H, Baharvandi HR. Development of high-performance A356/nano-Al_2O_3 composites. Mater Sci Eng A, 518:61–4, 2009

[5] Kok M. Production and mechanical properties of Al_2O_3 particle-reinforced 2024 aluminum alloy composites. J Mater Process Technol, 161:381–7, 2005

[6] S.A. Sajjadi, H.R. Ezatpour, H. Beygi, "Microstructure and mechanical properties of Al– Al_2O_3 micro and nano composites fabricated by stir casting, Materials Science and Engineering A 528 8765– 8771, 2011

[7] ASM Handbook; Volume 2; p 62-122, 1990

[8] Sekar K, Allesu K, Joseph MA, Effect of heat treatment in tribological properties of A356 aluminium alloy reinforced with Al2O3 nano particles by combination effect of stir and squeeze casting method. Applied Mechanics and Materials 592:968-971, 2014

[9] Bhaskar Chandra Kandpal, Jatinder Kumar, Hari Singh. Fabrication and characterization of Al2O3-aluminium alloy 6061 composites fabricated by stir casting. Materials Today Proceedings 4:2783-2792, 2017

[10] Marcin, Pietrzak Katarzyna. Processing, microstructure and mechanical properties of Al2O3-Cr nanocomposites. Journal of European Ceramic Society 27:1273- 1279, 2007

[11] Bharath, V., Madeva Nagaral, Auradi, V., & Kori, S.A. (2014).Preparation of 6061Al-Al2O3 MMC's by stir casting and evaluation of mechanical and wear properties.' Procedia of Material Science. 6, pp.1658-1667. doi.org/10.1016/j.mspro.2014.07.151

[12] DevarajuAruri, Kumar Adepu, KumaraswamyAdepu,and KotiveeracharіBazavada, "Wear and mechanical properties of 6061-T6 aluminium alloy surface hybrid composites [(SiC+Gr) and (SiC+Al2O3)] fabricated by friction stir processing", Journal of Material Research and Technology, Vol. 4, pp.no 362-369, 2013.

[13] Bharath V, MadevNagaral, V. Auradi and S. A. Kori, "Studies on dry sliding wear characteristics of ceramic Al2O3 particulate reinforced 6061Al matrix composites", Advanced Materials Research, Vol. 985, pp.no 319-325, 2014.

[14] Kaveh Edalati, Maki Ashida, Zenji Horita, Toshaiki Mastui, Hirotaka Kato. Wear resistance and tribological features of pure aluminium and Al-Al2O3 composites consolidated by high pressure torsion. Wear, 310:83-89, 2014

IOP Conf. Series: Materials Science and Engineering 1013 (2021) 012019 doi:10.1088/1757-899X/1013/1/012019

Microstructure Evaluation of Si_3N_4 reinforced Al6082 Composites subjected to Severe Plastic Deformation

Nishanth B N[1], Guruchannabasavaiah N G[1], Virupaxi Auradi[1], Bharath V[1], Satish Babu Boppana[2], Madeva Nagaral[3]
1R & D Centre, Department of Mechanical Engineering, Siddaganga Institute of Technology, Tumakuru, Visvesvaraya Technological University, Karnataka, India.
2School of Engineering, Presidency University, Bengaluru, Karnataka, India.
3Aircraft, Research and design centre, HAL, Bengaluru, Karnataka, India.

E-mail: nishanth.bn1996@gmail.com

Abstract: - In this research, work is carried out on Al6082 alloy because of its thermal stability and stiffness which can be used for strong structural application, here the hard ceramic particle of Si_3N_4 are reinforced with Al6082 for further enhancing its strength, stiffness. Stir casting technique is used to fabricate aluminum metal matrix composition with varying wt. % of Si_3N_4, this technique involves cost effective process and also results in homogenous mixture between the matrix and the reinforcement. The percentage of reinforcement varies from 0 and 9wt%. When subjected to severe plastic deformation it tends to change the grain size from micron to submicron level size in the grain structure by the application of load. The scanning electron microscope reveals the distribution of Si_3N_4 on the surface of the sample. The hardness value increased as compared to as- cast Al6082 alloy.

Keywords: Al6082, Si_3N_4, Severe plastic deformation, Hardness, Stir casting;

1. Introduction

Metal grid composites (MMCs) are developing consistently. Metal is used as a matrix in this, whereas ceramic or other organic compounds are used as reinforcements. Aluminum, magnesium and titanium are light weight and are also commonly used as a matrix material [1- 3]. Ceramic reinforced composites are discovered to be quite popular materials with the end goal of basic applications because of its phenomenal properties such as high quality, good malleability, durability and high modulus; which is the product of the interaction of its constituent process, i.e., ceramic reinforcement and matrix metal [4–5]. Form of the composite strengthening and manufacturing process plays a significant role in affecting the composites mechanical and physical properties [6]. Reinforcements may be in the form of fibre, particulate or whiskers. An appropriate collection of reinforcement and matrix and the strengthening form is extremely valuable for improving the property of the composites [7-9]. Due to its high strength-to-weight ratio, ease to produce, and low cost of production, aluminum and its alloys have become the most commonly used non-ferrous metals in recent years and have become more common [10,11,12]. Composites of aluminum metal matrix (AMMCs) strengthened with ceramic particles are quick and effortless to process and have the ability to provide customized combinations of properties [13]. These unique combinations of properties make composites more encouraging for aviations and vehicle applications [14-16]. High hardness and low thermal expansion coefficient developments are some of the basic properties of ceramic reinforcements that offer ascent to immediate and indirect reinforcement of the ultimate composite of aluminum [17, 18, and 19]. Components of automobiles that operate under friction and temperature conditions are produced by hard ceramic composites and are commonly used to produce AMMCs for this SiC, B_4C, Al_2O_3, Si_3N_4 [20-27].

From the literature, the nitride ceramic reinforcement (AIN, Si_3N_4) are favorable materials for the production of AMMCs because they have good heat conductivity, specific modulus, low density and high temperature stability. Researchers also led to many methods of fabricating AMMCs, such as stirring, mechanical alloying and pinch casting. However, in an effort to yield products that are less cost effective and mass produced, stirring is the furthermost popular procedure used by numerous researchers. Many researchers recognized the introduction of ceramic particles to develop mechanical characteristics. Ceramic reinforced with matrix material is the most common method of generating MMC.

Through the last few years, studies have been based on the efforts on enhancing the thickness of the grain material using different SPD methods. Such as (i) Equal Channel Angular Pressing (ECAP), (ii) Accumulative Roll Bonding (ARB), (iii) Large Strain Hot Rolling (LSHR), (iv) High-Pressure Torsion (HPT), (v) Extrusion and (vi) Constrain groove pressing (CGP) and others [31, 32]. Among these Constrain groove pressing is one of the effective severe plastic deformation process which is mainly used for producing an ultra fine grin structure throughout shape and size of the specimen. The whole method of Constrain groove pressing (CGP) the material is mounted in a symmetrically formed groove and then smoothed by the load operation. It is necessary to replicate the cycle in order to achieve the broad effective strain essential to refine the rough grain structure. By contrast to the commonly employed ECAP method for structural grinding, the benefit of the CGP procedure is that immense plastic deformation can be applied to metal by shape of sheet or plate. The Hall Petch effect is measured to critically enhance the properties of metallic materials by abating grain size [28, 29]. Besides, the reduction in grain size it produces a super plastic resistance at elevated temperatures [30]. Usually, SPD activates at higher compressive load to obtain ultra grain pre arrangement with a high-angle grain size. The variation of grain size can be applied to produce products with the vital properties. Physical and mechanical properties will expressively help from a reduction in grain size. The evolution and character of the new interfaces appears as a very significant property when studying microstructure in SPD materials as regards evaluating their effect on the mechanical properties. Given the deformation processing state, the heterogeneity in the formation of microstructures was often observed across the bulk specimen depending on the strain introduced [33, 34]. This dissertation examines the processing problems in anticipating commercialization attempts. Next, a set of major SPD processes will be presented (i.e. constrain groove pressing). Later on in the article, the contribution of highly deformed aluminum to microstructural evolution and hardness improvement is addressed in short.

2. Experimental details

2.1 Material

The material used was Aluminum AA6082 alloy of purity 96.55% with composition as shown in Table II. Silicon nitride (Si_3N_4) reinforcement involves 50nm size of powder particles; the purity of silicon nitride particles is 99.7%. The silicon nitride powders were pre heated to obtain homogenous mixing before adding it into the molten mixture. Chemical composition and physical properties of Al6082 is shown in table I and table II respectively. In the present study, the addition level of Si_3N_4 is selected as 0 and 9 wt% to know the effect of higher weight fraction of the Si_3N_4 on microstructural characterization and mechanical properties of produced composites (Al6082-Si_3N_4).

Table 1. Chemical composition of Al6082

Element	Si	Fe	Cu	Mn	Mg	Zn	Ti	Cr	Al
% present	0.7-1.3	0.0-0.5	0.0-0.1	0.4-1.0	0.6-1.2	0.0-0.2	0.0-0.1	0.0-0.25	Balance

Table 2. Physical properties of Al6082

Properties	Density	Melting point	Thermal Expansion	Modulus of Elasticity
Value	2.71 g/cm^3	555°C	24 x10^-6 /K	70 GPa

2.2 Method

Aluminum 6082 metal block is first sliced into small parts and carefully washed to eliminate salt, dirt, etc. so that the slag can be minimized. Such aluminum strips are placed inside a crucible and placed in a furnace. A 350g 6082 aluminum alloy sample was melted using a 6kW electric furnace. The ceramic materials were then preheated to 250^{0}C to increase the wettability and remove the gases deposits on the surface of the particles. Pre-heating of the reinforcement improves the wettability of the heated aluminum alloy and the reinforced substance while retaining the temperature of the mix, reducing the surface impurities and gasses involved with the agglomeration of the metal. The casting obtain by stirring process is one of the cost effective process. To prevent the mixing of the molten metal with iron, the stirrer used must be a stainless steel stirrer covered with zirconium. The AA6082 was charged to a graphite crucible, furnace is heated at 750^{0}C. The preheated Si_3N_4 was supplied to the molten metal and stirred with a mechanical stirrer rotating at a speed of 300 rpm to generate a complete vortex. Si_3N_4 is now slowly added in to the fine vortex created by mechanical stirrer. The duration of the stirring was 5min and the surface of the furnace is again tightly closed and waited till the temperature reaches to 750^{0}C and then the fluid is permitted to pass around the edge of the die. Finally, the melt is poured into a rectangular cavity (70mm x 35mm x 50 mm) container. Table III demonstrates the process parameters used in the present studies to produce the composites.

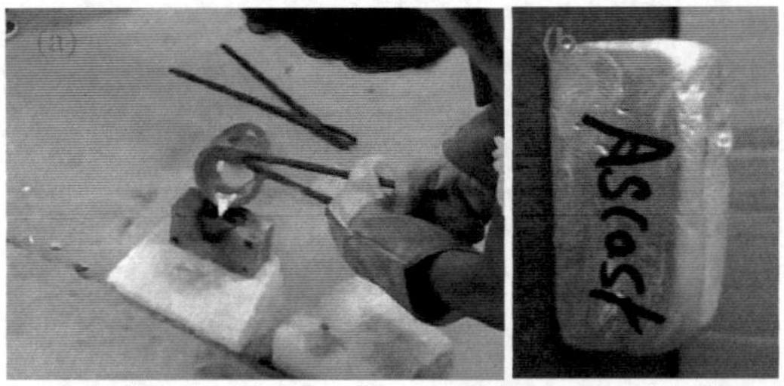

Figure 1. (a) Molten metal poured into the mould cavity. (b) Specimen obtains from die.

Table 3. Process parameter of Stir casting

Process parameter	Stirring temperature	Stirring speed	Stirring time	Preheat temperature of reinforced particles	Preheat temperature of Permanent method
Value	750°C	200rpm	5min	250°C	250°C

2.3 Steps involved in SPD (Constrain groove pressing)

The composite samples has prepared by using stir casting technique and the same samples is subjected to severe plastic deformation using flat and corrugated as shown in fig. 2(a) with the application of the load (50KN and 60KN) for which the material plastically deform and again the plastically deform specimen is placed between the flat die and the same amount of load is applied (50KN and 60KN) to obtain fine grain structure throughout the volume of the material. Fig 2(b) demonstrates deformed specimen with the help of corrugated die and fig 2(c) demonstrates deformed specimen under flat die.

Figure 2. (a) Severe plastic deformation die (i.e., flat and corrugated die). (b) Sample deformed under corrugated die. (c) Sample deformed under flat die.

3. Results and Discussions

3.1 Microstructural studies

Olympus microscope is used to examine the microstructure. To get the microstructure of the material, the material should be polished by using different size of grit papers (example 200, 400, 600, 800, 1000, 1200 and 2000). It is then polished with velvet cloth to get a mirror finish. Before placing the samples in the Olympus microscope the sample should be thermally etched. It is then placed in between the lenses to get the micrograph of the required samples by using elevating screw while focusing on the sample.

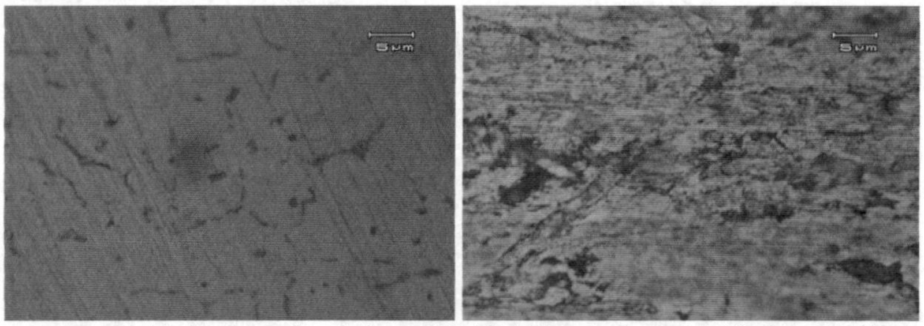

Figure 3. (a) As-cast aluminium (b) Al6082 reinforced with 9wt % Si_3N_4.

3.2 SEM analysis

Scanning electron microscope (SEM) reveals the distribution of silicon nitride powder in a metal matrix composition. The microscope reveals the microstructure of each polished specimen at different location on the surface of the specimen and the image shows proper distribution of Si_3N_4 on the surface of the metal. The particles are clearly visible on every increasing magnification in the SEM (example 100X, 500X 1000X, 1500X).

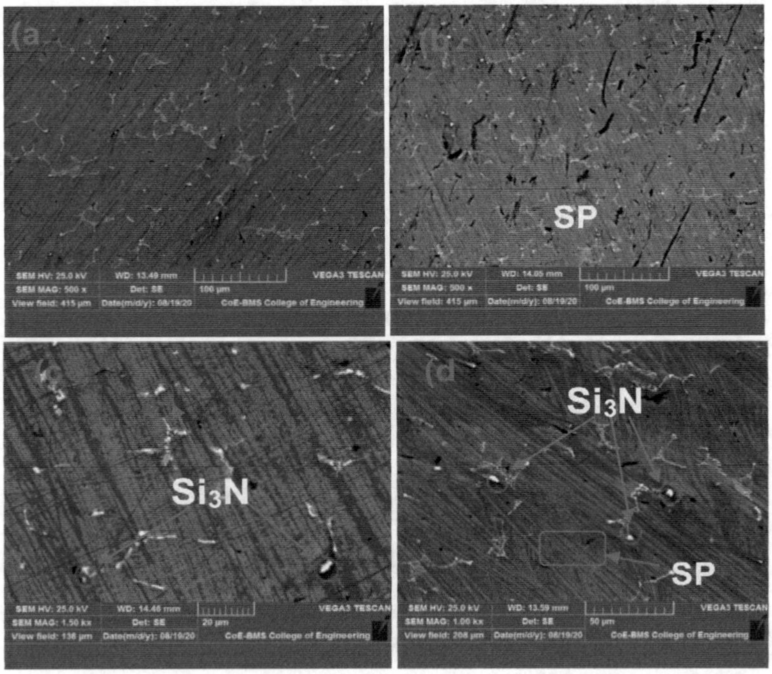

Figure 4 (a) As-cast 6082Al alloy without SPD (b) As-cast 6082Al alloy with SPD (c) 6082Al- 9 wt% Si_3N_4 without SPD (d) 6082Al- 9 wt% $Si_3N_4$with SPD

SEM image shows the microstructure on the surface of the specimen. Figure 4 (a) shows the as-cast aluminum alloy without severe plastic deformation, the grain size remains itself because the material doesn't undergo under any deformation so the length of the grains size will be standard. In figure (b) the same as-cast aluminum alloy under goes a sever plastic deformation, here the grain size increased as compared to as-cast aluminum alloy, the high compressive load (50KN) rubs the grains on the surface of the material and tends to increase grain length. Figure (c) reveals micrograph of the 6082Al alloy reinforced with 9wt% of Si_3N_4. The material will not undergo any deformation, while the white particles represent the Si_3N_4 particles on the surface of the specimen. In figure (d) the reinforced material undergoes a severe plastic deformation, which increases the length of the grain size under the load of 50KN.

3.3 Evaluation of Microvickers hardness

The Microvickers hardness was measured on a flat surface of the specimen. Variation of hardness values were observed by increasing Si_3N_4 in Al6082 alloy, it clearly shows that increasing wt. % of reinforcement in matrix material tends to increase the hardness value. The test was carried out with a specific load of 0.2HV for every specimen. The homogeneous distribution of Si_3N_4 led to increasing hardness values in the composites materials. Many factors are affected by the hardness value like particle size, percentage reinforcement, and the method of preparation of composites.

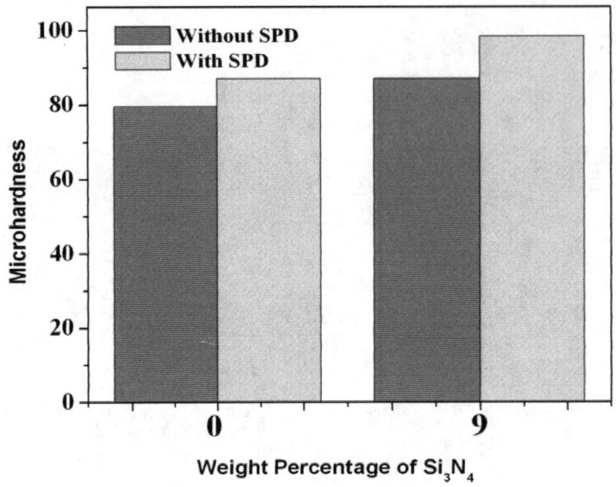

Figure 5. Microvickers hardness value of Al6082 with 9w.t% Si_3N_4

Table 4. Microvickers hardness values

Microvickers Hardness		
	Without SPD	With SPD
As-cast	76.65	87
9wt%	92.6	98.5

4. Conclusions

Microstructure evaluation of Al6082 alloy reinforced with Si_3N_4 particles subjected to severe plastic deformation led to the following conclusion:

1. Aluminum MMC with 0 and 9wt% of Si_3N_4 is successfully developed by automated stir casting technique.
2. Olympus microscope revealed the microstructure of Al6082 as-cast alloy and 9wt% Si_3N_4 reinforced Al6082 composite with and without severe plastic deformation.
3. SEM micrographs revealed the distribution of Si_3N_4 in the Al6082 matrix. The micrographs revealed the grain refinement process formed due to severe plastic deformation.
4. The hardness of as-cast Al6082 alloy without carrying out SPD is 79.65VHN while the as-cast Al6082 alloy that was subjected to SPD is 87VHN. The composite with 9wt% Si_3N_4 reinforcement without being subjected to SPD has a hardness value of 92.5VHN while the same composite that under went through SPD could achieve a hardness of 98.5VHN.

References

[1] Srivatsan TS, Ibrahim IA, Mohamed FA, Lavernia EJ. Processing techniques for particulate reinforced metal aluminum matrix composites. Journal of Material Science. 1991.
[2] Shukla M, Dhakad SK, Agarwal P, Pradhan MK Characteristic behaviour of aluminium metal matrix composites: A review. Materials Today Today Proceedings. 2018
[3] Bodunrin MO, Alaneme KK, Chown LH. Aluminium matrix hybrid composites A review of reinforcement philosophies; Mechanical, corrosion and tribological characteristic. Journal of Material Research Technology. 2015
[4] Miracle, Dan. Metal Matrix composites – From Science to Technological Significance. Composites Science and Technology, 2005
[5] Sinclair I, Gregson PJ. Structural performance of discontinuous metal matrix composites. Mater Sci Technol. 1997
[6] Rahman JF, Yunus M, Yezdani TMT. Charting of a strategy for the application of aluminium metal matrix Composites for Different Engineering Service Requirements. Int J Mod Eng Res. 2012
[7] Sahu PS, Banchhor R. Effect of different reinforcement on mechanical properties of aluminium metal matrix composites. Res J Eng Sci, 2017
[8] Mortensen A, Llorca J. Metal Matrix Composites. Annu Rev Mater Res, 2010
[9] Jain P, Soni S, Baredar P. Review on machining of aluminium metal matrix composites. Mater Sci Res India. 2014

[10] Ramnath BV, Elanchezhian C, Atreya TSA, Vignesh V Aluminum metal matrix composites - a review. Rev Adv Mater Sci 2014
[11] Rino JJ, Chandramohan D, Sucitharan KS. An overview on development of aluminium metal matrix composites with hybrid reinforcement matrix. 2012
[12] Milos K, Juric I, Skorput P Aluminium based composite materials in construction of transport means. Sci Traffic Transp, 2011
[13] Ma Kaka , Enrique JL, Julie MS Particulate reinforced aluminum alloy matrix composites-a review on the effect of micro constituents. Rev Adv Mater Sci,(2017)
[14] Sharma R, S.J. P, Kakkar K, Kamboj K, Sharma P. A Review of the aluminium metal matrix composite and its properties. (2017) IRJET
[15] Shivasankaran N. Aluminium metal matrix composites-A Review. (2017) TASTONLINE 01
[16] A. Macke, Schultz BF, Rohatgi P Metal matrix composites offer the automotive industry an opportunity to reduce vehicle weight, improve performance. Adv Mater Process, (2012)
[17] Chawla KK. Composite Materials - Science and Engineering. 2nd Ed Springer-Verlag New York (1997)
[18] Chawla KK, Metzger M. Initial dislocation distributions in tungsten fibre-copper composites. J Mater Sci (1972)
[19] Arsenault RJ, Shi N, Dislocation generation due to differences between the coefficients of thermal expansion. Mater Sci Eng (1986)
[20] Panwar N , Chauhan A Fabrication methods of particulate reinforced Aluminium metal matrix composite-A review. Mater Today Proc, (2018)
[21] Vijaya Bhaskar K, Sundarrajan S, Subba Rao B, Ravindra K Effect of reinforcement and wear parameters on dry sliding wear of aluminum composites-A review. Mater Today Proc (2018)
[22] Yashpal, Sumankant, Jawalkar CS, Verma AS, Suri NM Fabrication of Aluminium Metal Matrix Composites with Particulate Reinforcement A Review. Mater Today Proc (2017)
[23] Bharath, V., Auradi V, Nagaral M, Satish Babu Boppana. Experimental Investigations on Mechanical and Wear Behaviour of 2014Al–Al_2O_3 Composites. J Bio Tribo Corros (2020).
[24] Satish Babu Boppana, Samuel Dayanand, MR Anil Kumar, Vijee Kumar, and T. Aravinda. "Synthesis and characterization of nano graphene and ZrO_2 reinforced Al 6061 metal matrix composites." Journal of Materials Research and Technology (2020)
[25] Bharath V, Santhrust S Ajawan, Madeva Nagaral, Virupaxi Auradi and S A Kori. Characterization and Mechanical Properties of 2014 Aluminum Alloy Reinforced with Al_2O_{3p} Composite Produced by Two-Stage Stir Casting Route. J. Inst. Eng. India Ser. (2019).
[26] Madeva Nagaral, V. Bharath, and V. Auradi. "Effect of Al_2O_3 particles on mechanical and wear properties of 6061Al alloy metal matrix composites." J. Mater. Sci. Eng (2013)
[27] Boppana, S.B., Dayanand, S. Impact of Heat Treatment on Mechanical, Wear and Corrosion Behaviour of In Situ AlB_2 Reinforced Metal Matrix Composites Produced by Liquid Metallurgy Route. J Bio Tribo Corros (2020).
[28] E.O. Hall, The deformation and ageing of mild steel. Proc.Phy. Soc. B 64 (1951)
[29] N.J. Petch, The Cleavage Strength of Polycrystals," Journal of the Iron and Steel Institute,1953
[30] T.G. Langdon, The mechanical properties of super plastic materials. Acta Metall. Trans. (1982)
[31] R.Z. Valiev, Y. Estrin, Z. Horita, T.G. Langdon, M.J. Zehetbauer & Y.T. Zhu, Fundamentals of Superior Properties in Bulk NanoSPD Materials, Materials Research Letters, 2016
[32] G. Faraji and H. S. Kim, Review of principles and methods of severe plastic deformation for producing ultrafine-grained tubes, Materials Science and Technology, 2017
[33] Harsha, R. N., V. Mithun Kulkarni, and B. Satish Babu. "Severe Plastic Deformation-A Review." Materials Today: Proceedings (2018)
[34] J. Zrnik, Metallic Materials, (2006)

MICROSTRUCTURE AND WEAR BEHAVIOUR OF MATRIX Al7075 REINFORCED WITH MICRO WC-Co PARTICULATE COMPOSITE PROCESSED BY STIR CASTING METHOD

Gopal Krishna U B[1], Manoj Kumar D[2], Naveen A[3], Naveen H R[4], Nikhil H S[5], Virupaxi Auradi[6], Vasudeva B[7] and Satish Babu Boppana[8]
1,6,7 Department of Mechanical Engineering, Siddaganga Institute of Technology, Tumakuru, Karnataka, India
1,2,3,4,5 Department of Mechanical Engineering, Alva's institute of engineering and technology, Moodabidire, Dakshina Kannada, Karnataka, India
8 Presidency University, Bengaluru, Karnataka, India

E-mail: gopalkrishnaub@gmail.com

Abstract: - The present work dry sliding wear behaviour of Al7075 alloy matrix composite reinforced with 9Wt. % of WC-Co particulates is examined after prepared by stir casting. The dry sliding wear behaviour was studied for the composites with parameters of varying sliding speed, load and sliding distance. Microstructure characterization is done using SEM/EDX studies for the composite samples. From microstructural characterization even distribution of WC-Co particulates is observed in Al7075 matrix. DUCOM wear testing machine with EN32 steel disc is used to evaluate the wear rate of the composite. Worn out surface of the composite is subjected to SEM analysis to observe the layer formations. From the results of wear studies it is observed that the composite wear rate is reduced by restricting dislocation between the matrix and reinforcement interface.

Keywords: Al7075, WC-Co, SEM, Dry sliding wear

1. Introduction

The advancement of composite materials was fixated on the requirement for consolidated weight and improved execution properties in applications for military, car, aviation, and space vehicles. The advances in structure and computerized fabricating techniques have brought down their creation costs and extended their utilization for high volume applications in aviation and non-aviation sectors. [1-2]. Which incorporate development materials, for example, plain sheets, water tanks, restroom machines, home fittings, furniture industry, corrosion resistant compartments, angling vessels, mechanical items, portions of hardware, electrical industry, car parts, motor housings, vehicle seats and pedal mounting etc. Also, a few particulate reinforcements like SiC, Al_2O_3, TiC, B_4C, WC and TiO_2 are used for Al matrix as strengthening particulates [3]. In the aviation area, advancement or weight decrease assumes a significant job. Since, an airplane's presentation is straightforwardly relative to the airplane weight. It is in this manner important to create lighter structures or segments so as to give the essential quality and unbending nature in the base cross-sectional region. Metal grid composite has a significant influence in this specific region [4, 5]. A few processing methods like powder metallurgy, mechanical alloying and other various techniques including spray deposition technology have produced particulate strengthened metal matrix composites. Stir casting method is customary as a traditional process which is economic, the scope of procedures set up accessible for intermittent metal matrix composites. Its advantage lies in its effortlessness, adaptability and relevance for the preparing of huge amounts. It is the generally cheap of every single open course to produce metal network [6-7]. Mechanical examinations, for example, hardness, ductile, pressure and power of effect have been assessed particularly. A few parts utilized in a mechanical application experience sliding or pivoting movement, where the two segments slide together. Composites with more wear-resistance should be developed. The aim of this paper is to understand the effect on wear properties of Al7075 alloy by micro size WC, Co particle reinforcement. Wear experiments were carried out using pin-on - disc wear devices made by DUCOM. The influence of load applied, sliding speed, and sliding speed on the behaviour of composites Al7075-WC-Co was studied.

2. Experimental Details

For the current work, Al7075 alloy is used as the base matrix alloy and WC-Co mixture is used as reinforcing materials. The chemical composition of Al7075 alloy is tabulated in Table 1.

Table 1: Chemical composition of Al7075

Si	Fe	Cu	Mn	Mg	Cr	Zn	Ti	Al
0.4	0.5	1.6	0.6	2.5	0.15	1.5	0.2	Balance

The secondary phase materials WC and Co are blended with the Zirconia (Zr) balls to prepare the cermet mixture using planetary ball mill. During the preparation of the composite, an average size of 37μm of the WC-Co mixture is extracted using sieving analysis and used as reinforcement.

9 Wt. % WC-Co cermet particulates reinforced Al7075 alloy composites were prepared by stir casting method. A batch of calculated amount of base matrix Al7075 was melted to its liquidus temperature in a resistance furnace using graphite crucible. Degasification is performed using solid hexachloroethane (C_2Cl_6) to the liquid metal and unwanted absorbed gases are removed from the melt. Mechanical stirring was performed to form a fine vortex using zirconia coated steel rod [8]. The molten metal of Al alloy was stirred at a pace of about 300rpm for around 5 to 10 minutes. The preheated calculated amount of WC-Co mixture is added with constant stirring once the molten metal reaches a superheated temperature of 780°C. Constant stirring was maintained before and after addition of reinforcing particles and homogeneity is achieved. Maintaining a steady stirring action helps the melt to evade particle agglomeration and severance. The permanent steel mould with length 140 mm and diameter Ø15 mm is preheated and kept ready in parallel. The liquid composite mixture is then poured into permanent mould and kept under normal atmospheric temperature for solidification.

The sliding wear test at normal room temperature is performed with DUCOM wear testing equipment. The samples used to check wear are prepared using standard ASTM G99. The size of the test specimens is 8 mm in diameter and 30 mm in length. The equipment used to check sliding wear is shown in Figure 1. Digital electronic weighing machine with 0.0001gram accuracy is used for weighing the samples before and after testing.

Figure 1: DUCOM wear testing equipment

3. Results and Discussions

3.1 Microstructure observation

Figure 2(a-b) represents 9Wt. % WC-Co reinforcement added Al7075 composite Scanning Electron Microscope (SEM). The uniform homogeneous distribution of the reinforcement can be observed from the SEM images without any agglomeration forming. Si and α-Al boundaries can be observed from microstructure of the composite. Elements like Co, W, C, Zn, Al, Cu, Mg and Mn are identified in EDX elemental analysis which is represented in Figure 2(c-d).

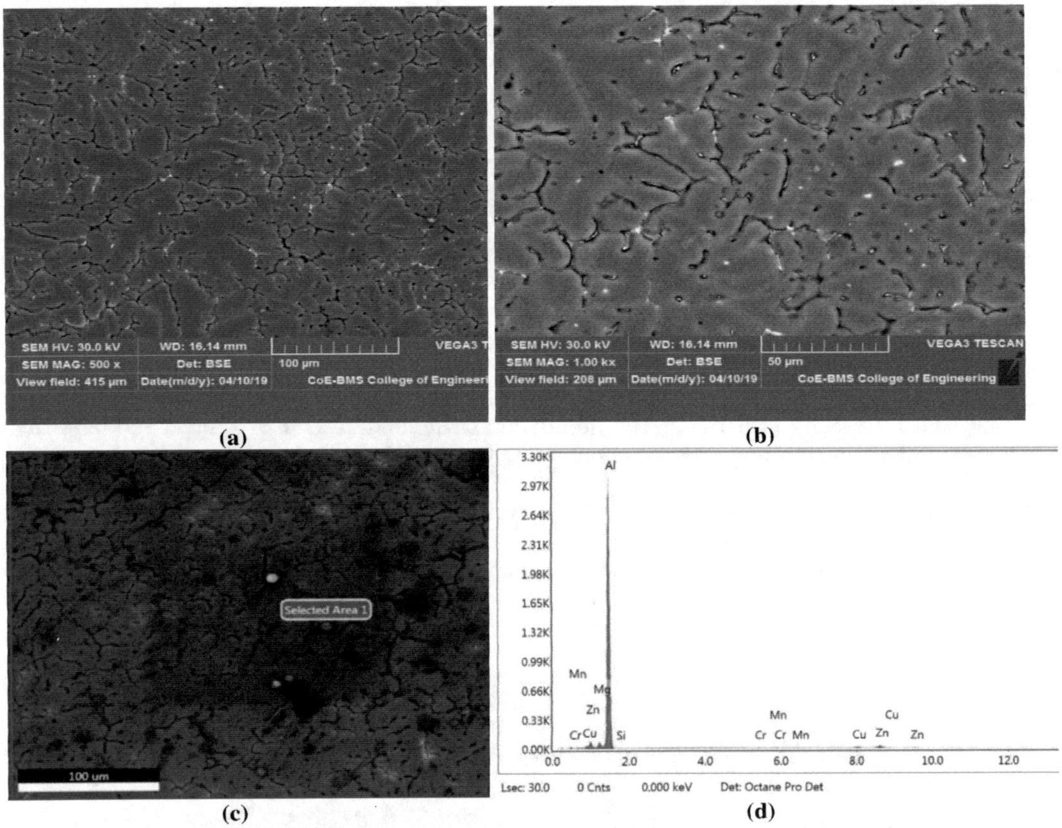

Figure 2. (a-b) SEM images of the composite, (c-d) EDX spectrum of composite

3.2 Wear rate for varying sliding velocity

Variation in wear rate of the composite prepared for 9Wt. % composite under different speed level conditions of 0.94, 1.87, 2.82 and 3.75 m/s are examined. The corresponding results obtained are represented in Figure 3. In this condition, the two parameters load and sliding distance are kept constant at 19.6N and 500m respectively. For the incremental in the speed the wear rate of the composite decreases both in matrix and composite sample. In compare with the matrix, composite wear rate is observed as decreased, this is mainly because as the speed increases the formation of wear debris which acts as a lubricating media between pin and disc which made wear rate as smoother [9]. Hence the combination of WC-Co particulates in the matrix has more wear resistance over different variables of speed.

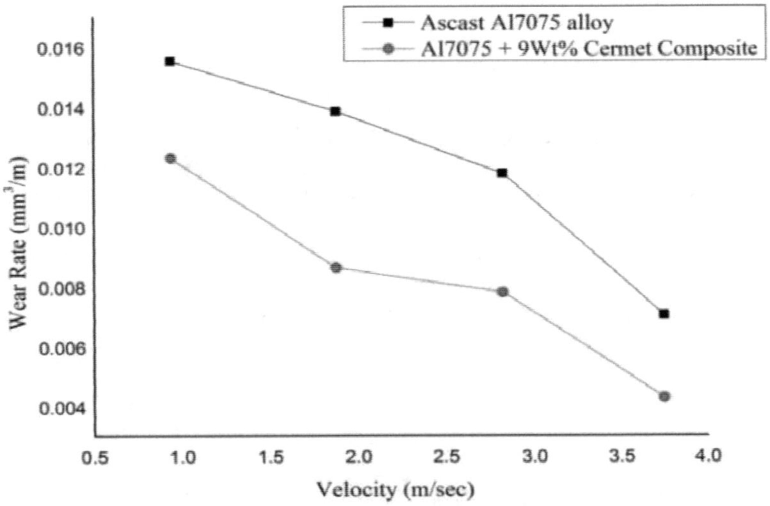

(a)

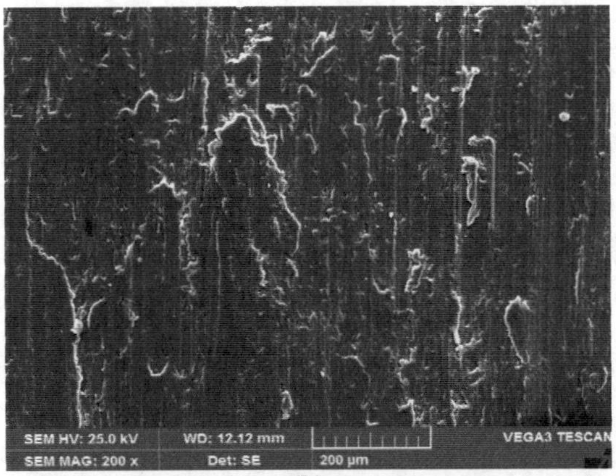

(b)

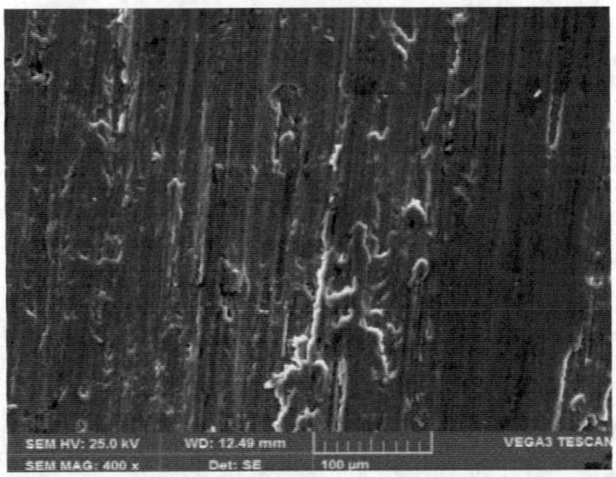

(c)

Figure 3. (a) Wear rate of 9Wt% of WC-Co reinforced with Al7075 composite for varying sliding speed, (b-c) Worn out surface SEM images of the composite

IOP Conf. Series: Materials Science and Engineering 1013 (2021) 012020 doi:10.1088/1757-899X/1013/1/012020

3.3 Wear rate for varying load

Variation in wear rate of the composite prepared for 9Wt% composite under variable loading conditions of 4.9, 9.8, 14.7 and 19.6N are examined. The corresponding results obtained are represented in Figure 4. In this condition, the two parameters speed and sliding distance are kept constant at 400rpm and 500m respectively. From Fig 4(a) it is clearly observed that as the load increases the corresponding wear rate is also increases. Wear rate of the composite is observed less in comparison with the matrix. The presence of hard ceramic particulates that accumulate at the boundary region of the matrix that makes composite to resist the pressure applied [10]. It can be observed from the SEM images as represented in Fig 4(b-c) that no formation of oxide layer in the direction of wear. The application of the pressure on the specimen which generates a clear adhesive wear on the sample surface.

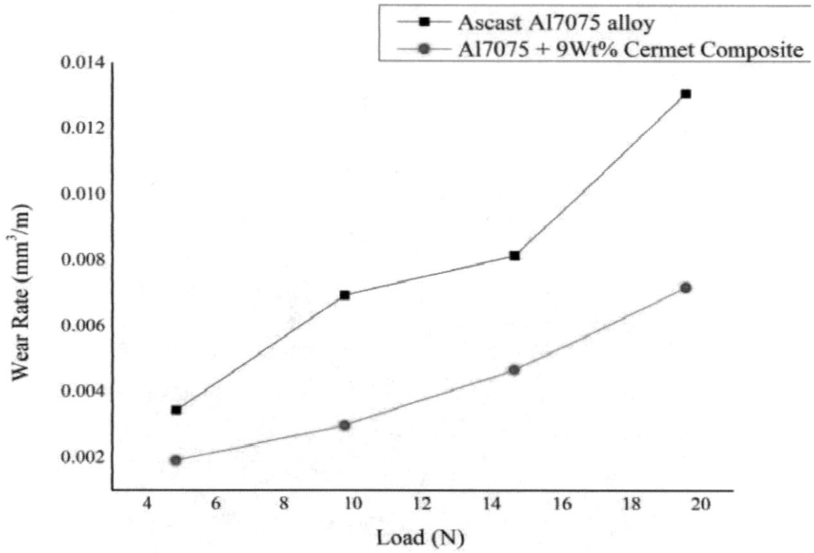

(a)

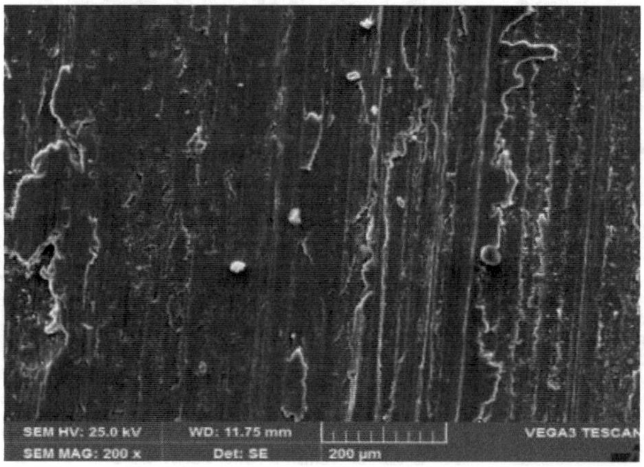

(b)

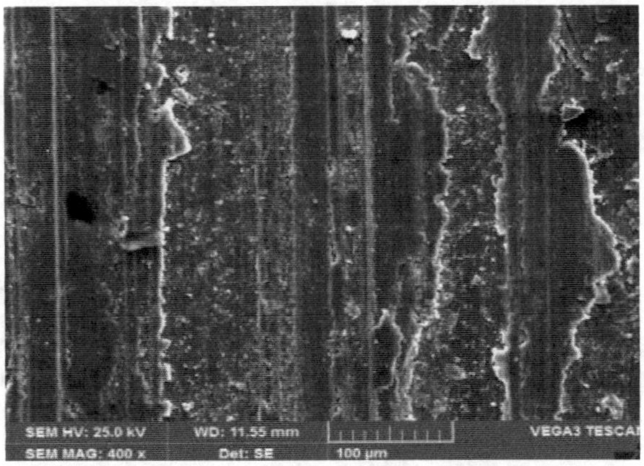

(c)

Figure 4. (a) Wear rate of 9Wt% of WC-Co reinforced with Al7075 composite for varying load, (b-c) Worn out surface SEM images of the composite

3.4 Wear rate for sliding distance

Variation in wear rate of the composite prepared for 9Wt% composite under variable sliding distance conditions of 250, 500, 750 and 1000m are examined. The corresponding results obtained are represented in Figure 5(a). The worn out regions are represented in Figure 5(b-c). The addition of WC-Co particulates in the matrix results in decrement of wear rate in the composite. The prolonged distance travel of the specimen leads to the worn surface of the disc part which is accumulated with the wear debris [11-12]. Further as the sliding distance increases the corresponding wear rate also increase.

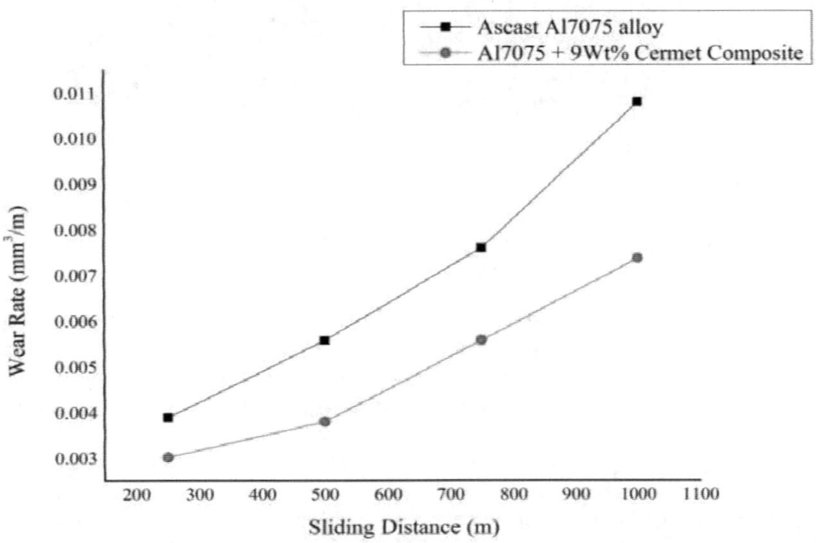

(a)

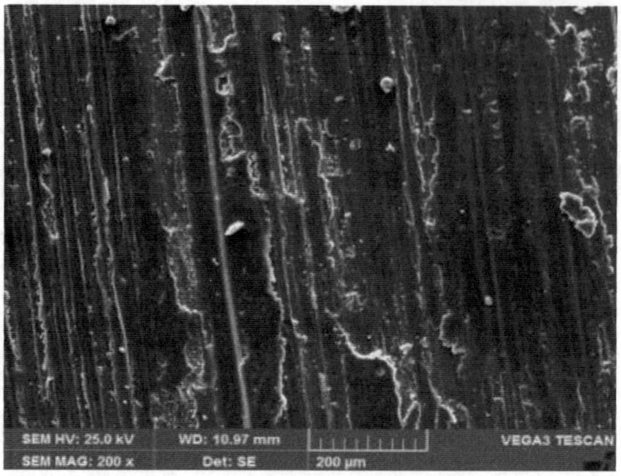

(b)

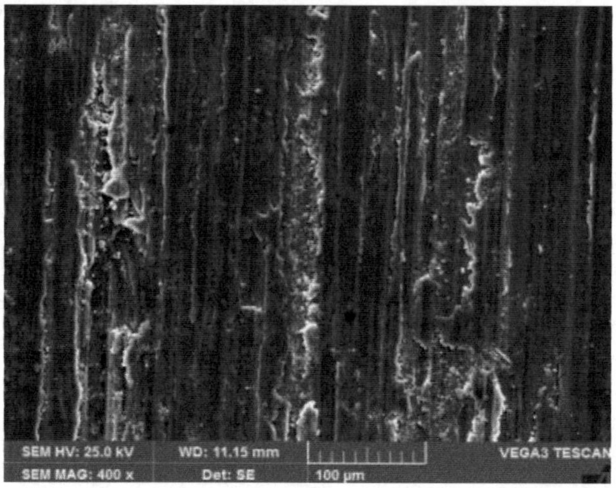

(c)

Figure 5: (a) Wear rate of 9Wt% of WC-Co reinforced with Al7075 composite for varying sliding distance, (b-c) Worn out surface SEM images of the composite

4. Conclusions

The conclusions made for the present work on the microstructure and wear characteristics of Al7075 matrix reinforced with 9Wt. % WC-Co particulate composite prepared by stir casting is as follows. The preparation of the composite using stir casting method by maintaining even distribution of the reinforcement is successful. Effective stirring action and proper maintenance of temperature during the process leads to the even distribution of ceramic particulates over the matrix. The same was observed from the Scanning Electron Microscope (SEM) images. The EDX spectrum analysis also confirms the presence of elements like Al, Co, W, C, Zn, Mg and Mn in the composite. The presence of hard ceramic particulates and other elements which lead to the decrement of wear rate with respect to matrix under different parameters of varying sliding speed, load and sliding distance. The application of the pressure on the specimen which generates a clear adhesive wear on the sample surface which leads to the decrement in the wear rate.

References

[1] H. Hocheng, S. B. Yen, B. K. Yen, "Fundamental turning characteristics of a tribology favoured graphite/aluminium alloy composite material" Composites Part A: 28A, 883-890, 1997.

[2] Gaohui Wu, Xi Wang, Longtao Jiang, Boran Ma, "A nanostructural design to produce high ductility of high volume fraction SiCp/Al composites with enhanced strength", Materials and Design, vol.61, pp. 141-145, 2014.

[3] J.H.Shin, H.J.Choi, D.H.Bae, "The structure and properties of 2024 aliminum composites reinforced with TiO_2 nanoparticulates", Materials Science and Engineering A, vol. 607, pp. 605-610, 2014.

[4] R. N. Rao, S. Das: Effect of matrix alloy and influence of SiC particle on the sliding wear characteristics of aluminium alloy composites. Materials and Design 31, 1200-1207. 2010

[5] B.Vijay Ramnath, C.Elachezhian, M.jaivignesh, S.Rajesh, C.Parswajinan, A.Siddique Ahmed Ghias, "Evaluation of mechanical properties of aluminum alloy-alumina-boron carbide metal matrix composites", Journal of Materials and Design, vol. 58, pp. 332-338, 2014.

[6] H.Pouraliakbar, A.Hosseini Monazzah, R.Bagheri, S.M.Seyed Reihani, G.Khalaji, A.Nazari, M.R.Jandaghi, "Toughness prediction in functionally graded Al6061/SiC composites produced by roll-bonding", Ceramics International, vol. 40, pp. 8809-8825, 2014.

[7] W Hartaj Singh, sarabijit, Nrip Jit, Anand K Tyagi , "An overview of metal matrix composites: Processing and SiC based mechanical properties", Journal of Engineering Research and Studies, vol. 2, pp. 72-78, 2011.

[8] R.S.Rana, Rajesh Purohit, S.Das, "Review of recent studies in Al matrix composites", International Journal of Scientific and Engineering Research, vol. 3, 2012.

[9] S. Jagatheesh Kumar, G. Santhosh, D. Nirmalkumar, A. Saravanakumar, Dr. P. Sasikumar, Dr. S. Sivasankaran, "Mechanical and dry sliding wear behavior of Al6063/Al_2O_3/Graphite hybrid composites", International Journal of Innovative Research in Science, Engineering and Technology, vol. 3, march 2014.

[10] C.S. Ramesh, A.C. Vijetha, Nirupama Mohan, Harsh R Gudi, "Slurry Erosion wear of Al6061-SiC composites developed by hybrid technique", Applied Mechanics and Materials, vol. 592-594, pp. 734-738, 2014.

[11] V.S. Aigbodion, J.O. Agunsoye, V. Kalu, F. Asuke, S. Ola, "Microstructure and Mechanical properties of Ceramic Composites", Journal of Minerals and materials Characterization and Engineering, vol.11, pp. 1085-1094, 2012.

[12] Vinod Lakshminarayana, Madeva Nagaral, Suthan Rajendran, V Auradi, "Microstructure and dry sliding wear behavior of Al2219-8 wt% of B_4C particulate Reinforced composites", International Jouranal of Applied engineering research, vol.10, 2015.

[13] Shouvik Ghoah, Prasanta Sahoo, Goutham Sutraghar, "Wear Behavior of Al-SiCp Composites and Optimization Using Taguchi Method and Grey Relational Analysis", Journal of Minerals and materials Characterization and Engineering, vol. 11, pp. 1085-1094, 2012.

Investigation on the Microstructure and Mechanical Properties of Aluminum 7075 Reinforced With Different Weight Percentage of Tungsten Carbide and Cobalt Metal Matrix Hybrid Composites

Guruchannabasavaiah N G [1], Dr. Virupaxi Auradi[2], Nandeeshaiah B M [3], Satish Babu Boppana[4] and Nishanth B N [5]
1,2,3 and 5 R & D Centre, Department of Mechanical Engineering, Siddaganga Institute of Technology, Tumkur, India
4 School of Engineering, Presidency University, Bengaluru, Karnataka, India

E-mail: gurusky.1984@gmail.com

Abstract: - In this paper an effort has made to develop and characterize Al7075 alloy reinforced with Tungsten carbide (WC) & Cobalt (Co). For uniform grain size and homogeneity of the reinforcements with matrix, Cermets (WC-Co) of grain size 50 microns were prepared with the help of ball milling operation. An experiment was conducted to evaluate the microstructure; mechanical properties like hardness test & tensile test as per the standards for different weight percentages of Cermets like 3, 6 and 9wt. %. Micrographs were studied to know the micro structure of the composite material and the result showed the, randomly dispersed reinforcement particles & fine interdendritic precipitates were seen. A comparative study was conducted for the microstructure and mechanical properties by using Cermets with varied weight percentages reinforced with Al7075 matrix. Mechanical properties results reveal that there is a considerable improvement in tensile strength and hardness as the percentage of Cermet increased from 3 to 9wt. %.

Keywords: Al7075, Tungsten carbide, Cobalt, Hardness, Cermet, Stir casting;

1. Introduction

Composites are materials made from more than one material, which has physical and chemical properties and differs widely when compared to other materials. Depending on the mechanical properties of the composite material, the elements used may be a matrix or reinforcement. The newly produced composite chosen over pure form can be processed easily, wear-resistant and resistant to corrosion in comparability to other materials. Composites are produced in order to improve mechanical and physical properties by incorporating and mixing the ingredients to the metal matrix. Aluminum matrix composites (AMCs), due to their inherent properties, are latest trending materials for different applications. We observe increased functional strength, tear, creep and fatigue compared to traditional engineered materials by the addition of reinforcing materials in the metal matrix. MMC is a highly capable material in the nonferrous family because of their outstanding properties such as low densities, high strength to weight ratio, good thermal and electric conductivity and fluidity, and it plays an important role in aerospace applications [1]. High specific strength, longevity, electrical and thermal conductivity, poor thermally expansion and wear resistance are vital advantages of aluminum matrix composites [2]. By injecting a ceramic material into a metal matrix, composite materials create a combination of attractive mechanical properties which cannot be attained by monolithic alloys [3]. Due to the rising production volume and the wider market applications one needs knowledge on the processing methods and mechanical behaviour of particulate matter MMCs. Particulate-reinforced MMCs are being increasingly focused, mainly due to the simple availability of particulate matter and economic strategies for particle-enhanced MMCs processing [4]. The main interest in strengthening alloy matrices with ceramic particulates is due to the low density, low thermal expansion and high strength of reinforcement as well as their wide availability [5]. Cermets represent a combination of processes in ceramic and metal. Cermet products such as WC-Co (tungsten carbide and cobalt) were commonly used and investigated for their high resistance against the corrosion of brittle (hard) and ductile (soften) phases and the exceptional combination of high hardness and resilience. Tungsten carbide is famous for its excellent hardness and resistance to wear. Ductile elements, such as cobalt, increase their resistance significantly in order to prevent cracks [5]. Various researchers [6-9] have successfully fabricated composites using stir casting techniques. Currently, substantial research has been conducted during the past few years in reinforced MMC (WC-Co) cermets, which have low wear tolerance, high temperature and corrosion resistance applications.

2. Experimental details

2.1 Matrix and Reinforcement:

In the present research work Al7075 alloy is used as metal matrix. The spectroscopic structure analyzed for atomic absorption of Al 7075 is tabulated in Table I. Reinforcement was produced by mixing and preparing a Cermet which is a mixture of tungsten carbide (WC) and cobalt (Co) with a mean size of 50μm processed by ball milling.

Table 1. Chemical composition of al7075 matrix used in the present study

Si	Fe	Cu	Mn	Mg	Cr	Zn	Ti	Al
0.40	0.50	1.60	0.30	2.50	0.15	5.50	0.20	Balance

2.2 Preparing Cermet (WC-CO)

30grams of reinforcements of WC and Co accurately weighed and measured was added to the ball mill and a dry milling operation was carried out. The grinding balls along with WC-Co powders were rotated with a prescribed speed of 300rpm for 3 hours. During ball milling, alumina pounding balls were used instead of steel balls to avoid the iron contamination and to maintain purity and homogeneity. Homogeneous dispersion of hard phases in flexible metallic networks occurs due to the process of cold welding. The WC-Co mixture produced has a size less than 50μm which is analyzed by sieve analysis. The WC-Co Cermet manufactured by the ball milling process has been used as reinforcement for the fabrication of matrix composites.

2.3 Preparation of Composite by stir casting

Initially aluminium which is available in the form of bar is cut into small pieces as per the melting capacity size of the crucible. The weighted amount (3, 6 & 12wt percent) of WC-Co cermets in different batches were taken through an aluminum foil. These reinforcements were preheated to 250°C in an electric oven and maintained at the same temperature by keeping it in the oven until it is added to molten aluminum matrix. The measured amount of aluminum matrix was melted in a graphite crucible at a melting temperature of 800-850°C in an electric resistance furnace with a melting temperature control unit. During the melting process to extract the oxygen content from aluminum the degasifying agent (C2Cl6-Carbon hexa-chloride) was applied to the molten Al7075 matrix. The molten aluminum was stirred for a while using mechanical stirrer till the formation of slag and then the produced slag was extracted through a metallic spoon. Then the preheated WC-Co Cermet was added slowly to compose the homogeneity with the molten aluminium and stirred continuously for 15 minutes at constant speed using a mechanical stirrer. The stirred mixture of molten metal with a preheated Cermet was then poured into the prepared cylindrical mold (125 mm in height and 15 mm in diameter) using the bottom pouring process. After the melt was poured, it was allowed to cool and solidify in the pot. For the intent of the study, few specimens were cast under identical conditions. The poured mixture is allowed for solidification and then the castings were removed from the mold and the specimens were produced using appropriate machining process as per the ASTM test specifications.

2.4 Preparation of specimens for testing

Tensile test

The prepared specimens were tested for tensile properties as per ASTM-E8 norms. These tests were carried in a tensile testing machine. Specimen is prepared as shown in Fig 1 for the tensile test.

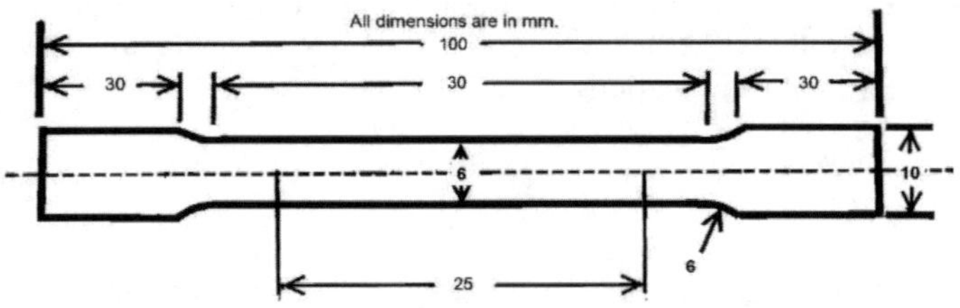

Figure 1: ASTM E8 specimen dimension

Hardness test

The hardness test is conducted as per ASTM E8 norm to one of the prepared specimens to analyze the hardness properties.These experiments were carried out in a research unit.

3. Results and Discussions

3.1 Microstructure characterization

The specimens were carefully polished for the studying the microstructures. From the optical microstructures (Fig. 2), inter-dendrite precipitates are seen in the aluminum solid solution matrix the presence of inter-metallic particles in the mixture are also noted from the micrographs. A moderate allocation of WC-Co reinforcement can be visualized in the micrographs. Figure 3 and 4 shows the microstructure image of composite 6% wt and 9% wt of reinforcement respectively.

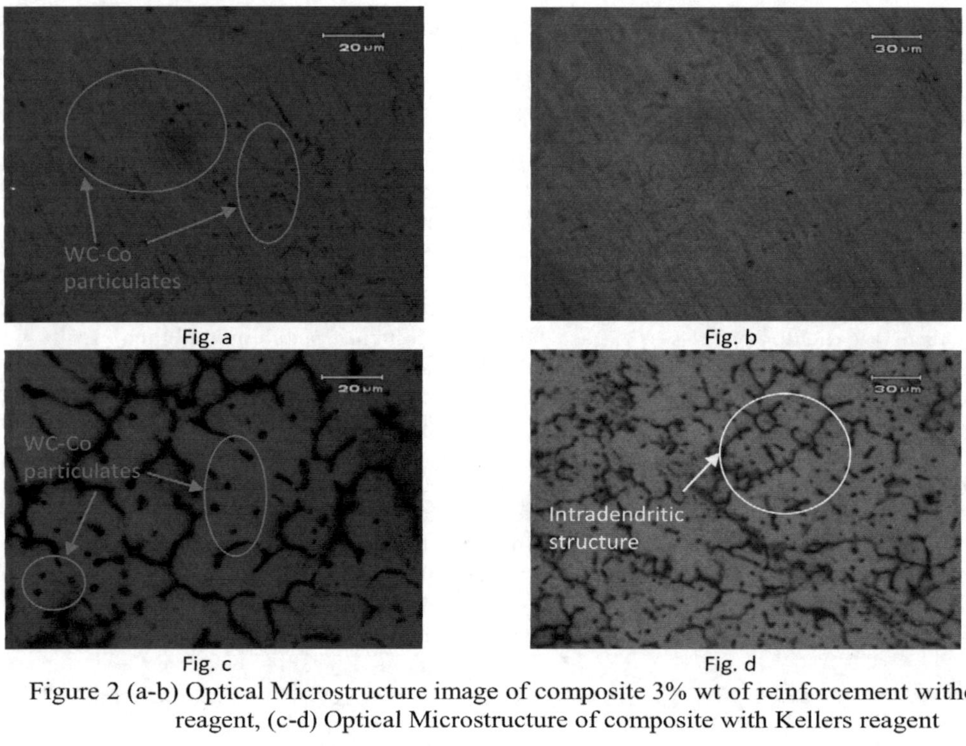

Figure 2 (a-b) Optical Microstructure image of composite 3% wt of reinforcement without keller's reagent, (c-d) Optical Microstructure of composite with Kellers reagent

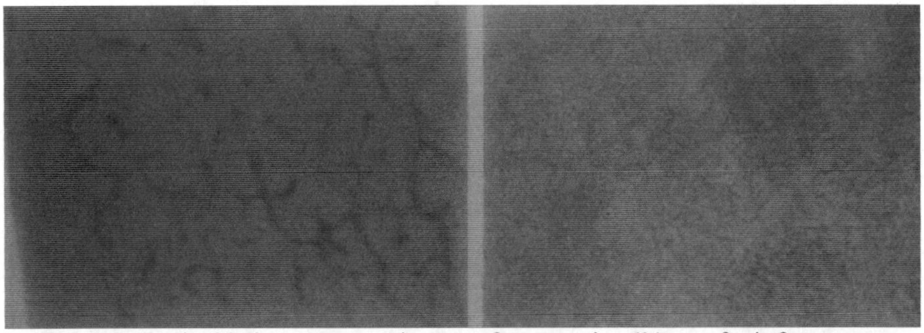

Figure 3. Optical Microstructure image of composite 6% wt of reinforcement

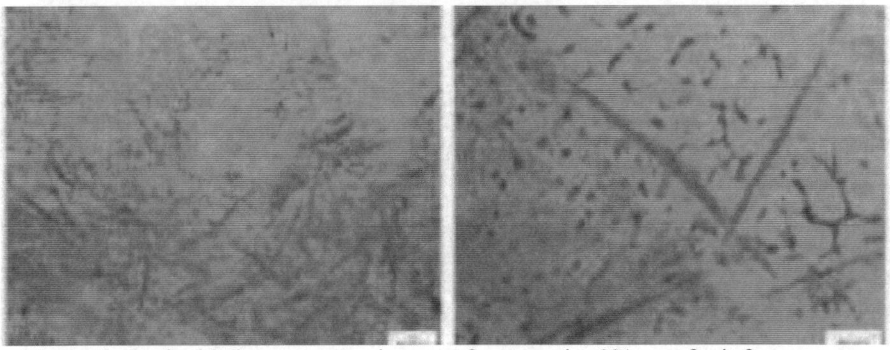

Figure 4. Optical Microstructure image of composite 9% wt of reinforcement

3.2 Tensile properties

Prepared composite materials were examined for tensile strength as per ASTM requirements and the results are shown in table II and table III. For each test, 3 samples were subjected to loading and the average result was considered as shown in the table 3. Table 4 shows an increase in the tensile strength of 47% was observed in the composite as compared to the base alloy with a decrement in its % of elongation.

Figure5 shows the variation of mechanical properties for different weight percentage of reinforcement.

Table 2. Comparison between base alloy and composites

Sample type	Tensile strength N/mm²	Yield strength N/mm²	Percentage Elongation	Compressive strength N/mm²	Hardness BHN
Only Al7075	102.26	83.54	9.45	575	95
WC-Co 50μ Al7075+WC-Co 3%	111.429	87.797	1.62	659.195	104
WC-Co 50 μ Al7075+WC-Co 6%	183.775	125.024	1.63	631.52	105
WC-Co 50 μ Al7075+WC-Co 9%	193.928	161.108	1.4	649	108

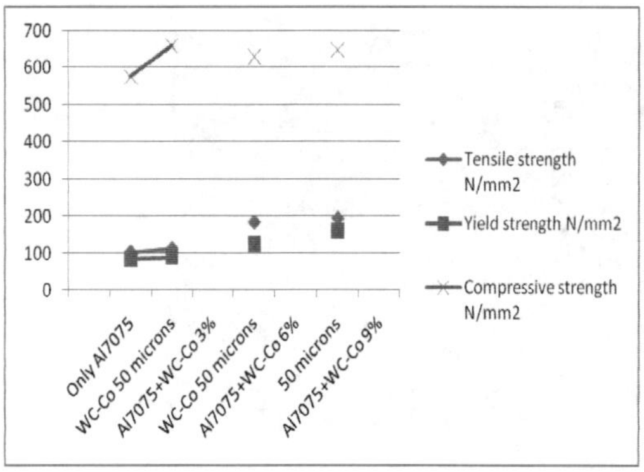

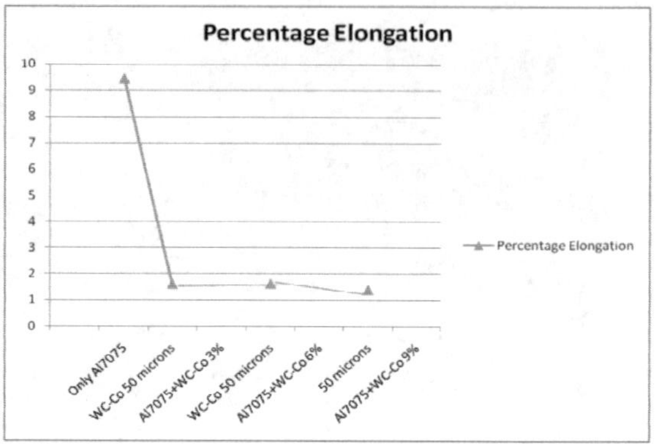

Figure 5. Variation of mechanical properties for different weight percentage of reinforcement

Table 3. Average of tensile property values of specimens

Properties	Sample 1	Sample 2	Sample 3	Average
Tensile strength (N/mm^2)	210.293	184.713	184.778	193.938
Yield stress (N/mm^2)	185.002	154.790	147.743	163.128
% Elongation	2.47	2.61	1.86	2.32

3.3 Hardness properties

The specimens were prepared according to ASTM E8 norm to analyze the hardness properties.The testing details are tabulated in the table 4. Comparative study conducted between the base metal and the composite material to observe the hardness and found slight increment in the hardness values.Figure 6 shows the variation of hardness for different weight percentage of reinforcement.

Table 4. Hardness testing details

Machine	Briness hardness testing machine
Indentor	Ball Indenter
Specification	5mm ball –Total load750kg (Pre load 250 kg) 0utside load : 500 kg
Timings	20 secs

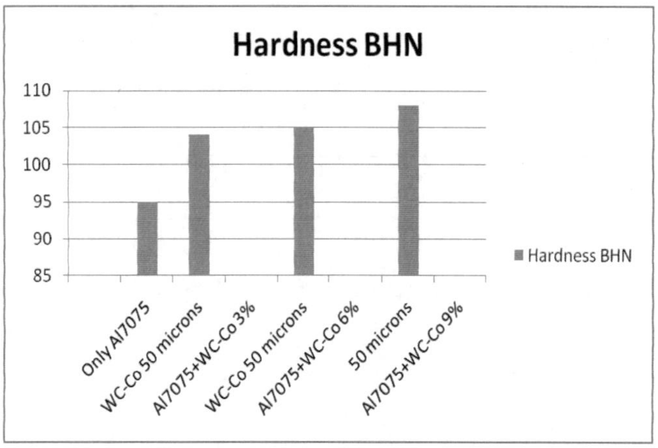

Figure 6.Variation of hardness for different weight percentage of reinforcement

4. Conclusions

i. The WC-Co mixture with a particle size 50μ is effectively processed with homogeneity through ball milling process.
ii. Composite of Cermet mixture (WC-Co) and Al7075 are easily produced using stir casting process.
iii. Microstructural characterizations carried out using optical microscope and the study showed homogeneous dispersion of the cermet particles in the Al7075 matrix.
iv. An improvement of10.52% in hardness was obtained with the addition of Cermet (WC-Co) reinforcement to Al7075 matrix.
v. Addition of Cermet (WC-Co) to Al7075 matrix increases the yield stress and tensile strength by 49% and 58% but drastically decreases the percentage elongation by 84%.

References

[1] Jacob J, Zhengui Yao, Stiglich, T S Sudarshan, "Nano-grained tungsten carbide-cobalt (WC/Co)", Material Modification Inc, VA 22031.

[2] Swami A R "Effect of Particulate Reinforcement on the Mechanical Properties of Al6061- WC and Al6061-Gr MMCs", 2011, part 10, page 1141-1152.

[3] Hari Prasada Rao, Pydi Balamurugan Adhithan, Syed Bava Bakrudeen, "Microstructure exploration of the aluminium tungsten carbide composite with different manufacturing circumstances" International Journal of soft computing and engineering, January 2013, ISSN: 2231-2307, Vol 2, Issue-6.

[4] Zhang Yuan wei Daobin Mu Chengyi Wu, "Microstructure and properties of tungsten-carbide particle-reinforced high-speed composites fabricated by spray forming Powder Technology", 1999, Vol 104, pages 100-104.

[5] Rong Zhou, Yehua Jiang, Dehong Lu "The effect of volume fraction of WC particles on erosion resistance of WC reinforced iron matrix surface composites", 2003, Vol 255, pages 134- 138.

[6] Bharath, V., Mahadev Nagaral, and V. Auradi. "Preparation, characterization and mechanical properties of Al2O3 reinforced 6061Al particulate MMC's." International Journal of Engineering Research and Technology (IJERT) (2012): 1-6.

[7] Mithun, B. R., Nagaral, M., Auradi, V., & Bharath, V. (2017). Microstructure and mechanical properties of Cu-coated Al2O3 particulate reinforced 6061 Al metal matrix composite. *Materials Today: Proceedings*, *4*(10), 11015-11022.

[8] Boppana, Satish Babu, and Samuel Dayanand. "Impact of Heat Treatment on Mechanical, Wear and Corrosion Behaviour of In Situ AlB 2 Reinforced Metal Matrix Composites Produced by Liquid Metallurgy Route." *Journal of Bio-and Tribo-Corrosion* 6, no. 2 (2020): 33.

[9] Boppana, Satish Babu, Samuel Dayanand, MR Anil Kumar, Vijee Kumar, and T. Aravinda. "Synthesis and characterization of nano graphene and ZrO2 reinforced Al 6061 metal matrix composites." *Journal of Materials Research and Technology* 9, no. 4 (2020): 7354-7362.

ASSESSMENT AND MODELLING OF CUTTING FORCES IN TURNING OF ALUMINIUM METAL MATRIX COMPOSITE USING RESPONSE SURFACE METHODOLOGY

Vivek Bhandarkar V N [1], Dr. Bhaskar Pal[2] and Dr. S.P. Shanmuganathan[3]
1,3Department of Mechanical Engineering, Dayananda Sagar College of Engineering, Bangalore, India.
2Department of Mechanical Engineering, Presidency University, Bangalore, India.
E-mail: viveckbhandarkar@gmail.com

Abstract: - The conventional materials are nowadays replaced by Metal matrix composites owing to their excellent Mechanical and machining phenomena. The growing/ up-trending growth in Automobile and aerospace sector paves / necessitates the need for accuracy / closer tolerance in machining parameters. The constraints while machining MMC's are surface roughness and cutting force parameters. The present study focuses on the cutting parameters, condition influencing the surface roughness in terms of material composition, tool wear, speed feed, during the machining of LM6/SiC composites. The mathematical model has been established using Taguchi technique (L16 orthogonal array). By using the RSM approach empirical relation for various machining attributes has been generated to assess the output parameters namely, Feed force (Ff), Cutting force (Fc), Radial force (Fd) and surface roughness (Ra). Desirability function approach is used to find the optimal machining condition. The influence of various parameters in machining of LM6/SiC composite have been analyzed in detail. The study showed that the optimized outputs for the machining of LM6/SiC composite for the input parameters 0% Reinforcement, new tool flank wear(0mm), speed in the range of 684 rpm and 0.1mm feed rate respectively are 7.827 N, 11.196N, 7.890 N and 10.786μm

Keywords: LM6/SiC Composites, Response Surface Methodology, Surface Roughness

1. Introduction

A composite material is a heterogeneous combination of matrix and reinforcing phase yielding the desired properties. Aluminium composites have become an apparent material in the industrial sector owing to superior properties such as resistance to corrosion, enhanced strength to weight ratio and improved resistance to wear.

Most of the engineering services/applications in the present scenario demands materials that exhibits varied composition and proportion of properties which cannot be substituted by conventional existing alloys/ceramic materials. The reason is obvious in the case of materials employed in aerospace sectors requiring low density and impact resistance characteristics. In order to meet the demand and surpass the shortcomings faced by industry composites have emerged as a promising candidate of interest. MMC exhibits high strength, moduli, good damping and improved resistance to wear compared to unreinforced.

MMCs machining and conventional alloys machining will be significantly different in many aspects of machining. During the machining of MMCs behaviour of the MMCs will be non-homogeneous. Machining mainly depends on the matrix, reinforcement material & its volume fraction. Proper mechanisms should be implemented for selection of proper tool based on the behaviour of the work material, geometry and tool wear. Hence proper cutting force data to be generated to check the behavior of the cutting tool for different working conditions. Since machinability cannot be changed for the given work material, however to increase the tool life, Machining parameters like speed, feed and depth of cut to be selected appropriately [1,2].

Santosh Kumar Tripathy et al [3] carried out an investigation related to the influence of process attributes like Tool wear, Tool life, Power utilization and Surface roughness in Turning of Aluminium Metal Matrix reinforced with waste derived from industrial sectors. The investigated study concluded that surface roughness and power consumption projected a declining and rising trend respectively with the spike in cutting speed and feed rate. Also decrement in performance of tool (Tool life) with increase in addition reinforcement is also reported.

Ireneusz zagorski et al [4] observed the influence of technology bound factors on assessing the roughness characteristics on assessing the roughness characteristics of aluminium alloys post machining (Turning). The observation arrived at the fact that feed rate exhibits significant influence on quality of machined surface and cutting speed factor is insignificant.

Diptikanta das and Vivek Chakraborty [5] performed machining on T6-AMMC's under Dry environment for different levels of attributes like speed, feed rate and depth of cut (DOC). The performance study showed that flank wear maximized with rising attribute levels and reduction in surface roughness was observed with increase in cutting speed and elevated with rising feed /DOC

Joardar [6] adapted Response Surface Methodology (RSM) to optimize the process parameters in turning of cast AMMC reinforced by Titanium di boride. The influence of attributes like speed, feed, DOC on surface roughness were examined with the establishment of Numerical model through ANOVA analysis. The analysis unveiled that reduction of feed rate results in improved surface finish.

Philip Hoier et al [7] studied the microstructural characteristics of superalloys 718 annd waspaloy and impact on flank wear of uncoated tungsten carbide tools on turning. The study concluded that significant progression of flank wear is noticed when machining alloy 718 in comparison to the later. The study also confines to the fact that mechanism contributing to flank wear is abrasion by the presence of inclusions.

Shubhajit das et al [8] studied the machinability characteristics of Al 2024 reinforced with B4C in terms of surface roughness and Metal Removal Rate. The study revealed that feed rate is the prominent factor affecting surface roughness and Depth of Cut being the influential attribute affecting the MRR.

Ram Singh et al [9] studied the turning characteristics of Al 5083 reinforced with B4C under dry condition. The effect of addition/dispersion of B4C in Al 5083 was analyzed and optimized to arrive at effective and efficient machining using algorithm. The study revealed that cutting and DOC has an substantial effect/influence on MRR. It also exposed that surface roughness is effected by DOC and feed rate.

The studies pertaining to the machining of Al 2024, Al 5083, super alloys 718, have been adequately carried out to assess the cutting forces using experimental and numerical methods. The present study concerning the machining of LM6 composites, subsequent analysis of impact of input process parameters like feed, speed, artificially induced tool wear on the output responses like surface roughness, cutting forces , followed by optimization of cutting parameters is scarcely reported. The scope of the present study is oriented to characterize the machinability nature of the LM6 composite aided with the analysis of the contributing factors, assessment of cutting forces and surface finish of the machined composite.

2. Materials and Methods

For the fabrication of the composite base/ matrix material used is LM6 as per BS1490:1988 standard [16]. Fabrication of the composite is carried out by using stir casting technique. In this process SiC is dispersed in the molten aluminium matrix by means of stirring. Graphite crucible is used to melt the aluminium alloy LM6 by using the coke fired crucible. Preheating is carried out for the reinforcement material SiC to a temperature of about 250°C for 1/2 hr to take away the volatile impurities on the particle surface. Preheating also oxidize the surface artificially to improves the wettability by formation of SiO2 layer. Preheated SiC partials are added to the molten metal, which is kept at a temperature of 720-7400C. Molten metal is stirred at a speed of 250rpm, and SiC particles are slowly added to the vertex to distribute inside the matrix. The mechanical properties of base metal aluminium alloy and MMC (LM6/SiC) such as Ultimate tensile strength and hardness are measured and tabulated in Table 1.

Table 1. Mechanical properties of LM6 and MMC

	LM6/SiC composite material			
	0 wt% SiC	4 wt% SiC	8 wt% SiC	12 wt% SiC
Tensile strength (N/mm2)	173	208	216	230
Hardness (BHN)	55.0	65.87	76.25	78.95

Table 2. Levels of various control factors

Code	Control Factors	Level			
		-2	-1	1	2
A	Composite Material (LM6/SiC) (%)	0	4	8	12
B	Tool wear (Flank wear) (mm)	0	0.1	0.2	0.3
C	Cutting speed(rpm)	250	500	750	1000
D	Feed rate (mm/rev)	0.1	0.2	0.3	0.4

Table 3. Measured values of Cutting force components and surface roughness.

Run	Composite Material (LM6/SiC) (%)	Tool wear (Flank wear) (mm)	Cutting Speed (rpm)	Feed (mm/ rev)	Feed force (Ff) N	Cutting force (FC) N	Radial force (Fd) N	Surface roughness Ra μm
1	0	0	250	0.1	15.14	30.18	12.48	10.5
2	0	0.1	500	0.2	11.25	20.46	10.56	11.2
3	0	0.2	750	0.3	11.89	21.48	10.89	11.6
4	0	0.3	1000	0.4	24.48	28.14	18.17	12.4
5	4	0	500	0.3	28.47	47.28	26.54	12.8
6	4	0.1	250	0.4	55.18	80.17	52.18	11.4
7	4	0.2	1000	0.1	6.14	15.48	4.15	13.8
8	4	0.3	750	0.2	20.48	40.15	18.45	12.6
9	8	0	750	0.4	45.15	44.58	44.89	12.3
10	8	0.1	1000	0.3	40.15	72.13	32.47	13.2
11	8	0.2	250	0.2	52.89	84.78	42.18	11.4
12	8	0.3	500	0.1	44.16	76.87	34.58	12.2
13	12	0	1000	0.2	69.48	98.48	57.47	14.2
14	12	0.1	750	0.1	70.58	96.18	60.48	13.8
15	12	0.2	500	0.4	150.49	178.54	135.87	11.3
16	12	0.3	250	0.3	140.58	158.47	118.47	11.8

Composite samples are machined in the conventional lathe using the tungsten carbide tool bit, which is grounded to induce the artificial tool wear. Experiment is carried out using the Taguci's L16 orthogonal array, levels of various control factors are as shown in the table 2. The force components feed force (Ff), Cutting force (Fc) and Radial force (Fd) are measured using the lathe tool dynamometer. Surface roughness values are measured for all the 16 experimental run using the Mitutoyo SJ-210 surface roughness tester. The results are tabulated in the table 3.

2.1 Response surface methodology.

By using Design-expert trail version 12 software [22], Machining variables and the RSM responses are correlated by quadratic polynomial regression model. By regression approach second order polynomial mathematical models shows the correlation between the force component (Ff, Fc, Fd) and surface roughness (Ra), with the input variables such as composite material (A), flank wear (B), cutting speed(C), and flank wear (D)

From the ANOVA table the R2 value (the coefficient of regression) obtained is 0.9532 (95.3%), 0.9324 (93.24%), 0.9399(93.99%), 0.8437(84.37%) for Ff, Fc, Fd, and Ra respectively. These R2 value indicated that model developed through experimental study is reasonably good and hence relationships are satisfactory.

$$\text{Feed force (Ff)} = +27.80+37.17A-8.71B+0.4120C+2.92D-15.85AB-7.78AC16.88AD+13.32BC-4.23BD-9.08CD+38.14A2+2.47B2-10.02C2+8.05D2 \quad (1)$$

The variation of feed force with respect to speed and flank wear on the machining characteristics of the metal matrix composites with varying reinforcement shows a positive effect. On the other hand, the interaction of composite of different reinforcement ratio with flank wear and speed over the output feed force showed a minimal effect.

$$\text{Cutting force (Fc)} = +45.11+38.87A-7.91B+0.9374C+11.55D-20.17AB-9.05AC-19.22AD+10.00BC-4.94BD-7.58CD+48.26A2-0.37362-9.63C2+3.60D2 \quad (2)$$

The variation of the cutting force with respect to flank wear, cutting speed and feed on the machining characteristics of MMC with varying reinforcement shows the significant effect. On the other hand minimal effect is exhibited in the interaction of flank wear and speed.

$$\text{Radial force (Fd)} = +25.11+33.55A-7.39B+0.2869C+2.92D-12.86AB-6.92AC12.76AD+12.16BC-4.82BD-10.66CD+30.95A2+3.64B2-10.22C2+6.13D2 \quad (3)$$

The variation of radial force with respect to flank wear and speed shows the positive effect on the machining characteristics of the MMC with varying reinforcement. On the other hand meagre effect is observed in the interaction of composite of varying reinforcement with cutting speed and feed.

$$\text{Surface roughness (Ra)} = +12.55+0.5901A+0.3592B-0.0485C+0.3982D+0.6165AB+0.4270AC-0.2530AD-0.1749BC+0.1188BD+0.0628CD-0.1882A2-0.4607B2+0.6402C2-0.2183D2 \quad (4)$$

The variation of surface roughness with respect to tool wear, cutting speed and feed on the machining characteristics of the MMC with varying reinforcement shows a positive effect. On the other hand, the interaction of composite of different reinforcement ratio with flank wear and speed over the output surface roughness showed a minimal effect

3. Results and Discussions

3.1 Evaluation of Mechanical and microstructural characteristics

From the mechanical properties Table 1, Tensile strength of the composite were increased by 33% and hardness increased to 43% with the increase in SiC reinforcement content, compared to base material LM 6. The properties of the MMC's has been changed due to addition of the SiC reinforcement. Figure 1 shows the EDS analysis report for LM6/12% SiC, proves that with the addition of the reinforcement SiC particulate will tend to increase the silicon element peak in the composite. Apart from the major element aluminium, silicon, traces of Fe, Mn, Ni, Cu, Zn, Cu and Mg were identified [10]. The SEM image in figure 2 revels that there is a uniform distribution of the SiC particles in the LM6/SiC MMC. In the micrograph some porosity regions are identified. This is principally due to the precipitation of the dissolved gas in the melt, incomplete evacuation leads to the formation of the gas entrapment, solidification shrinkage occurs because of inability to feed the voids due to insufficient pressure gradient to overcome the resistant offered by dendritic networks and incomplete infiltration. The micro porosity observed in composite was ascribed to poor nucleation at the SiC particulate sites and due to the particle clustering will hinder the flow of metal. Thereby the presence of porosity in the casting, could be attributed to the fact that, ceramic reinforcement have lower thermal conductivity it needs to be pushed to the solidifying front of a freezing melt. [10, 11].

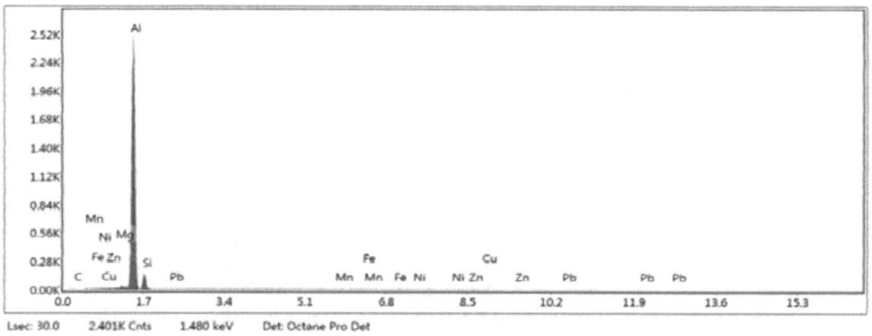

Figure 1. LM6/ 12wt% SiC-EDS analysis

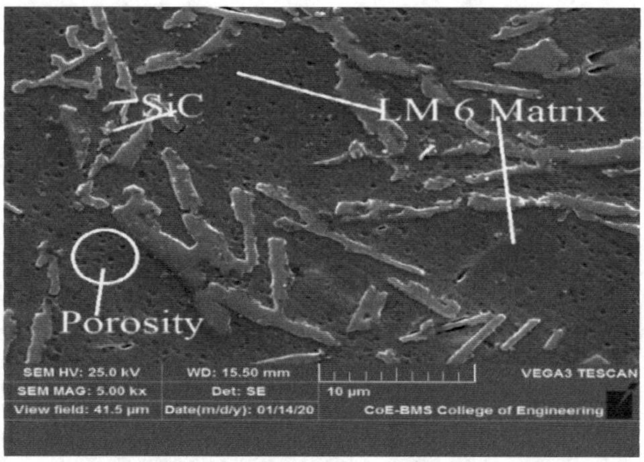

Figure 2. SEM micrograph of LM6+12%SiC

3.2 Effect of process parameters on composite machining

The interaction effect of the process parameters, associated output in terms of graphical representations/3D surface plot derived from DOE using Response Surface Methodology (RSM) approach have been presented below.

3.2.1 Effect of reinforcement on force components

The variation of the force components respectively Feed force, Cutting force and Radial force with the Reinforcement and Flank wear is depicted in Figure 3 (a-c)

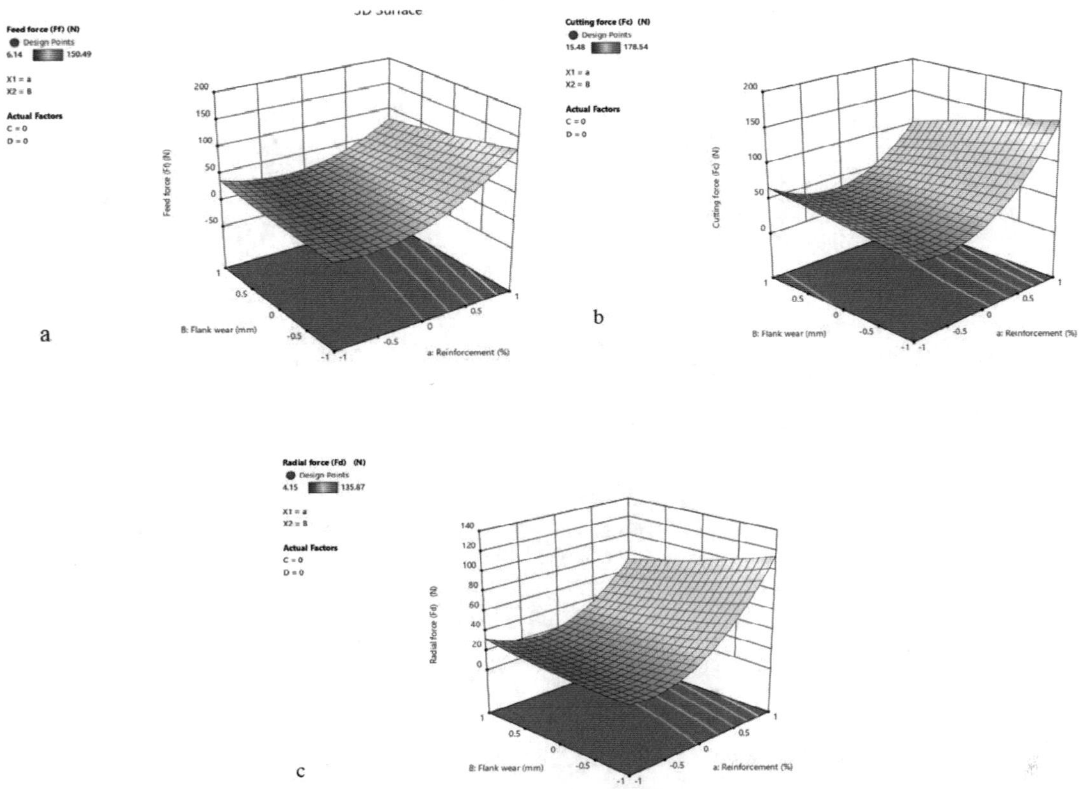

Figure 3 (a-c). Variation of the force component with the reinforcement and flank wear

From the surface plots, it is evidenced that the force components showed a increasing trend with the increase in the addition of reinforcement or in other terms upsurge in the volume fraction of the particulates. When the reinforcing percentage ascends from 0% to 12% the feed force varies from 25-75N, cutting force shifts from 50 N to 100N and radial component of force sways from 25-70 N. The addition of reinforcements/particulate matter to the aluminium matrix can lead to possible increase in yield strength of the composite which is attributable to the entangling/pinning/restrainment of the reinforcement particle in the matrix boundary. These observations are in conformance with the findings of the previous researchers [12,13].

The influence of flank wear on force components is shown in figure 3 (a, b, c). The increasing trend of the flank wear results in significant increase in the force components which could be ascribed to the aggravated ploughing of the tool on the matrix composites with subsequent increase in the size of the wear land. With further progress in the size of the wear land cutting force increases resulting from the effective secondary rubbing action between the wear land and work piece.

3.2.2 Effect of feed on force components

The influence of feed rate on force components is shown in figure 4 (a-c).

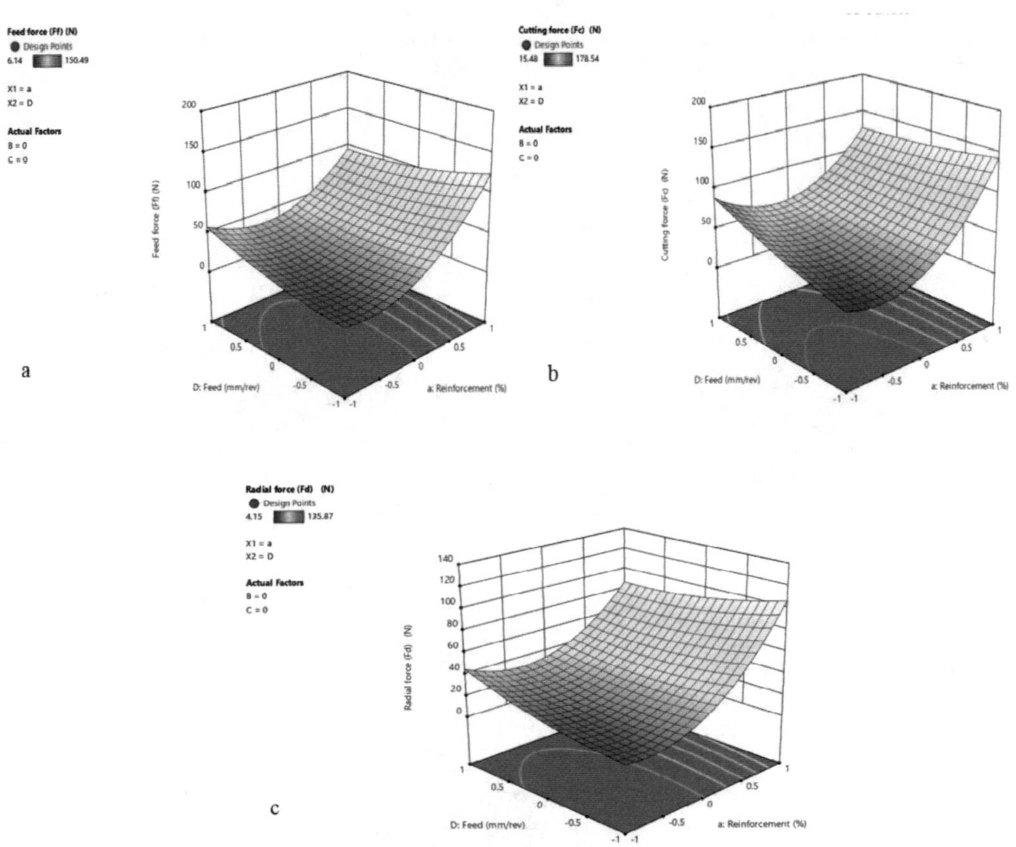

Figure 4. (a-c). Variation of the force component with the reinforcement and feed

It is inferred from the plot that, increase in feed rate is directly associated to enhance metal removal rate within the same time period. As MRR peaks up, the cutting force required to cut the material subsequently increasing the plastic deformation and heat generation. Moreover, increased feed rate promotes the friction between the cutting edge of the tool and workpiece which shoots up the cutting force.

With rise in feed, high forces were essential to deform the material within the short period of time elevating the temperature at tool tip-work material interface. The heat created at the tool interface was dissipated through chips rather than work material retaining the heat. Material removal takes place within a short time requiring high cutting forces

3.2.3 Effect of Speed on force components
The influence of cutting speed on force components is shown in figure 5 (a-c).

The increase in cutting force with increase in cutting speed could be ascribed to the existence of elevated temperature at the tool tip zone/cutting segment leading to increased wear in the flank portion of the tool and thereby peak in cutting force component [12] Further the presence of silicon carbide which can withstand higher/increased thermal gradient does not get softened and so the softening effect is not more pronounced which aids in the reduction of cutting force.

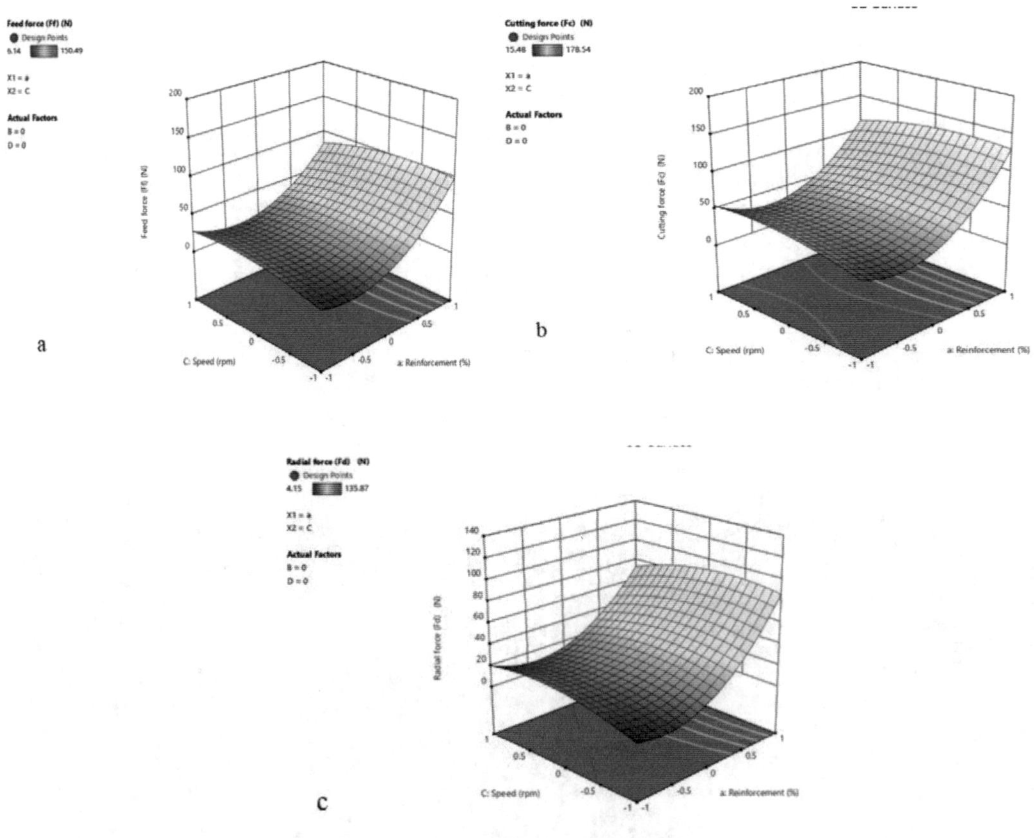

Figure 5 (a-c). Variation of the force component with the reinforcement and speed

3.2.4 Effect of Surface roughness (Ra) on force components

The variation of surface roughness with feed rate is shown in the figure 6

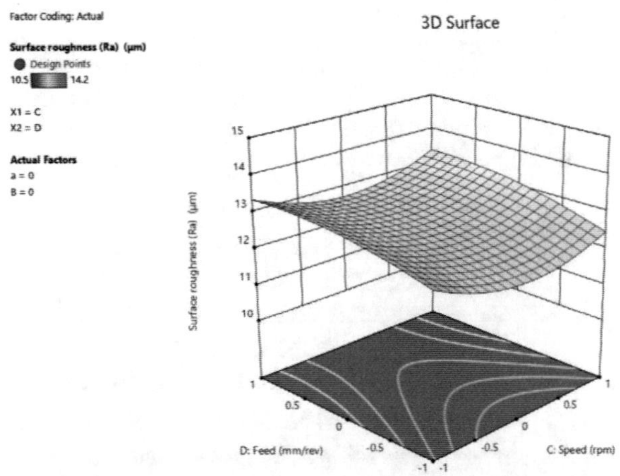

Figure 6. Variation of the surface roughness (Ra) with the feed and speed

With rising trend in feed rate the surface roughness of the machined surface increases. MRR is also high at increased feed rate creating friction at the tool-work interface resulting in pitting marks.With increase in feed rate higher heat is accumulated at the zone of cutting which can be ascribed to higher thermal loading/thermal gradient. Rise in temperature lowers the bonding affinity thereby strength between the matrix and Silicon carbide is reduced. Hence increase in feed rate leads to the generation of pits resulting from pulling out of the SiC particles from MMC that counteracts at the tool-work interface. Similar trend is observed by Dwivedi et al [14]

The variation of surface roughness with cutting speed is shown in figure 6. At lower speed, temperature gradient is more pronounced at the tool-work interfaces promoting the formation of unstable larger Built-Up-Edge (BUE) as shown in the figure 7, accompanied by fracturing of chips creating rough surface and adhesive wear on the tool. As the cutting speed peaks up the time consumed in machining is reduced drastically, decreasing the tool-chip contact length and friction impeding the formation of BUE, occurrence of chip fracture leading to improved surface finish [13,15]

The variation of surface roughness with percentage addition of reinforcement could be attributed to the presence of harder Silicon carbide particles in the matrix. Also presence of pores, shearing /pull out of the surface during machining can deteriorate the surface leading to increased surface roughness.

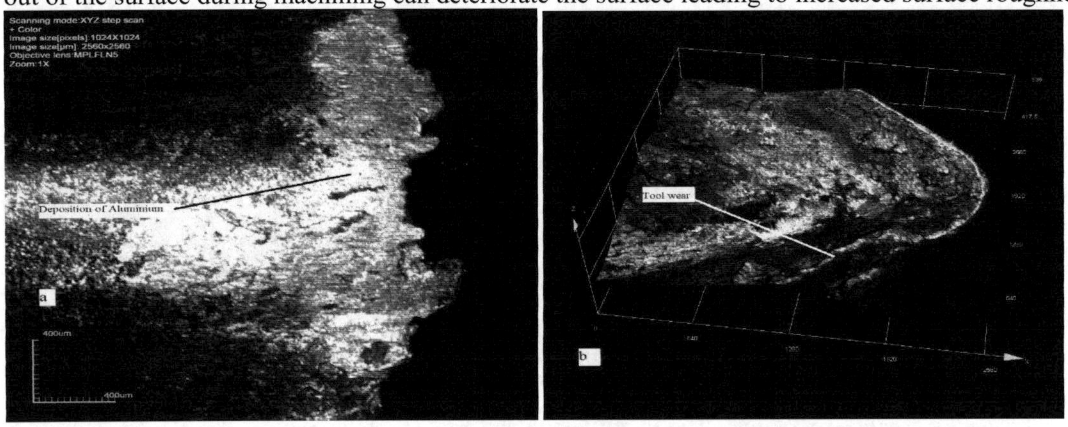

Figure 7 (a-b). Formation of Built up edge and tool wear

3.3 Numerical optimization of machining variables

Desirability chart for solution as shown in the figure 8 is generated thorough numerical optimization, represents the factors and the responses individually in the desirability graph. Overall desirability of the parameters and responses are indicated in the bottom of the histogram bar. The desirability value of responses lies in the range of 0.92 to 1. The value of the complete desirability is 0.970 and it indicates that the regression equation offers better performance under optimal conditions.

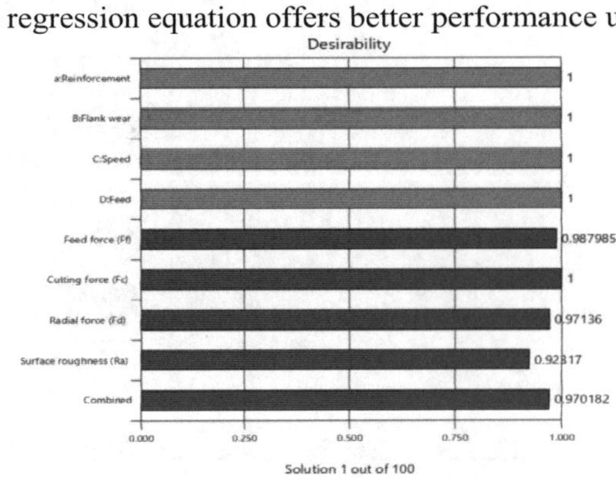

Figure 8. Desirability solution chart

3.4 Validation of model

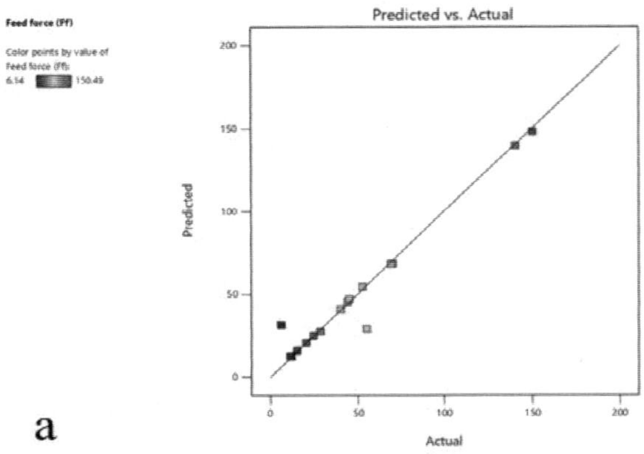

Figure 9. Predicted vs Actual value for Ff

Predicted and actual values shown in the figure 9. The values of predicted and actual experimental data shows the good correlation.

4. Conclusions

The machining characteristics of LM6 aluminum alloy was investigated experimentally. Tungsten Carbide tool tips used in the process were artificially grounded to induce the flank wear. The resulting process responses are optimized using RSM technique for minimizing the component factors. The regression model have been arrived at to predict the output response. The following conclusion are drawn.

- LM6/SiC MMC with varying 0, 4, 8 and 12wt% SiC reinforcement percentage are fabricated for sound casting by using the stir casting method. SEM analysis of the casting shows fairly uniform distribution of SiC Particulates, agglomerates and micro porous sites. EDAX confirms the validation of the fabricated composite. It shows the predominance of aluminium and silicon in the composite.
- Response surface methodology established using the functional relationship between the cutting forces and cutting parameters showed the 95% confident interval for the mathematical model and predicted values of the cutting forces responses. The force components showed an increasing trend with increasing flank wear, feed and addition of reinforcement. Decrement in the feed rate results in better surface finish of the machined component
- Optimized force components for the given set of input attributes are 7.827 N, 11.196N, 7.890 N and 10.786μm

References

[1] Preetkanwal Singh Bains, Sarabjeet Singh Sidhu & H. S. Payal, "Fabrication and Machining of Metal Matrix Composites: A Review," Materials and Manufacturing Processes, 31:5, (2016), p553-573.
[2] Jigar Suthar & K. M. Patel, "Processing issues, machining, and applications of aluminum metal matrix composites," Materials and Manufacturing Processes, 33:5, (2018), p499-527.
[3] Santosh Kumar Tripathy & Senapati A. K. (2020). A review on turning analysis of industrial waste reinforced aluminum metal matrix composite. Materials Today: Proceedings.
[4] Ireneusz zagorski, and Warda T. (2018). Effect of technological parameters on the surface roughness of aluminium alloys after turning. Advances in Science and Technology Research Journal, 12(2), pp.144-149. https://doi.org/10.12913/22998624/87136
[5] Diptikanta das and Vivek Chakraborty (2018). Dry condition machining performance of T6 treated aluminium matrix composites. Materials Today: Proceedings, 5(9), 20145–20151. doi:10.1016/j.matpr.2018.06.383
[6] Joardar H. (2018). Optimization for Turning of Aluminium and Titanium Di Boride Cast Composites Using Response Surface Methodology. Materials Today: Proceedings, 5(11), 23576–23585. doi:10.1016/j.matpr.2018.10.146
[7] Philip Hoier, Surreddi, K B, Klement, U. (2019) Tool wear by dissolution during machining of alloy 718 and Waspaloy: a comparative study using diffusion couples The International Journal of Advanced Manufacturing Technology https://doi.org/10.1007/s00170-019-04805-9

[8] Shubhajit das, Sangeeta Das, & Kakoli Roy. (2020). Modelling and turning investigations of Al2024 based metal matrix composites. Materials Today: Proceedings. doi:10.1016/j.matpr.2020.02.409
[9] Ram Singh, Shadab, M., & Rai, R. N. (2018). Optimization of machining process parameters in conventional turning operation of Al5083/B4C composite under dry condition. Materials Today: Proceedings, 5(9), 19000–19010. doi:10.1016/j.matpr.2018.06.251
[10] D. P. Myriounis , S. T. Hasan & T. E. Matikas, " Microdeformation behaviour of Al–SiC metal matrix composites," Composite Interfaces, 15:5, (2008),495-514.
[11] S. Balasivanandha Prabu , L. Karunamoorthy, S. Kathiresan, B. Mohan, "Influence of stirring speed and stirring time on distribution of particles in cast metal matrix composite," Journal of Materials Processing Technology, 171 (2006) 268-273.
[12] Umesh Khandey, Sudarsan Ghosh & Krishnaswamy Hariharan, "Machining parameters optimization for satisfying the multiple objectives in machining of MMCs," Materials and Manufacturing Processes, 32:10, (2017), p1082-1093.
[13] Balasubramanian K, Nataraj M & Palanisamy Duraisamy, "Machinability analysis and application of response surface approach on CNC turning of LM6/SiCp composites," Materials and Manufacturing Processes, (2019), 34:12, 1389-1400.
[14] Dwivedi, S. P., Kumar, S., & Kumar, A. "Effect of turning parameters on surface roughness of A356/5% SiC composite produced by electromagnetic stir casting," Journal of Mechanical Science and Technology, (2012), 26(12), 3973–3979.
[15] Ciftci, I., Turker, M., & Seker, U. "Evaluation of tool wear when machining SiCp-reinforced Al-2014 alloy matrix composites," Materials & Design, 25(3) (2014), 251–255.
[16] BS 1490:1988, Aluminium and aluminium alloy ingots and castings for general engineering purposes, British Standard Institution
[17] Design-expert trail version 12, Stat-Ease Inc. Minneapolis

Aerodynamic Analysis and Optimization of Airfoils Using Vortex Lattice Method

Chaitanyaa K K[1], Chetan R Patil[2], Dr. Bhaskar Pal[3], Dr. Devendra Dandotiya[4]
1,2,3,4Department of Mechanical Engineering, Presidency University, Bangalore, India.

E-mail: chaitanyaakk45@gmail.com

Abstract: - Air transport is one of the fastest means of transport in the modern world. Aircraft's efficiency depends on the efficiency of its subparts and its subsystems individually. For example, the gas turbine used as main source of thrust in aircrafts has to be largely efficient, which adds to the overall efficiency of the aircraft. The same way, another important part of the aircraft is the wing. Wings are the main reason for an air vehicle to fly. Aerodynamic parameters like lift, drag and the ratio of co-efficient of lift to drag which is C_L/C_D ratio determines the efficiency of the aircraft wing. This C_L/C_D ratio depends on the cross section of the wing, i.e. the airfoil. This paper emphasizes aerodynamic analysis on airfoils NACA 0018, NACA 2412, NACA 4412, USA 45, TSAGI-S12 and B737A used on different planforms of an aircraft wing like Rectangular, Elliptical, Delta and Swept back wing. The results are analyzed and the best airfoil corresponding wing shape is reported. This best airfoil is further refined effectively to get high C_L/C_D ratio. The complete analysis and airfoil optimization were carried on XFLR5 software which is governed by Naiver- Stokes equations. The airfoils were analyzed for a velocity of 50 m/s, at an angle of attack of 7°, up to a maximum Re (Reynolds Number) of 3e+06, the wing chord length at root, C_R=1m and total wing span, L=10m was kept constant throughout the analysis for uniformity. Post analysis, the best three airfoils were further interpolated and refined to obtain higher C_L/C_D ratio.

Keywords: Aerodynamic Analysis; Airfoils; XFLR5; Vortex Lattice Method (VLM); C_L/C_D Ratio, Optimization

1. Introduction

Airfoils refer to the aerodynamic shape of the cross-section of the aircraft wing. The Wing is a prominent structure which suspends the bulky weight of aircraft in air. As the wing glides through the air [1], aerodynamic forces are produced. Component of force perpendicular to the movement is called lift and the component parallel and opposite to the motion of the wing is termed as drag. The aircraft requires more energy to overcome drag and hence airfoils are optimized with an intent to reduce drag force.

Drag force adds on due to the friction between air molecules and the wing surface, this is known as skin friction. A phenomenon called Reynold's effect [2] induces drag as the boundary layer thickens on the surface when the plane flies at slower speeds. The wing's co efficient of lift being a dimensionless number depends on various design parameters like Aspect Ratio (AR), airfoil type and Reynold's number. The pressure difference between the upper surface (suction surface) and the lower surface (suction surface) is the primary reason for lift generation as stated in Bernoulli's Principle. The most important parameter to result in a higher lift is the Angle of Attack (AOA). Angle of attack is the angle between the chord line (straight line connecting leading and trailing edge) and the vector representing the motion of wind. Critical angle of attack is the maximum angle at which maximum lift can be observed. An increase in AOA increases lift up to the critical AOA after which the aircraft stalls. The distance between the upper and lower surfaces of the airfoil measured perpendicular to the camber line (American convention of measuring thickness) is referred to as thickness, which when increased increases the lift.

Another parameter which affects lift is the camber line. Camber line is the reference line equidistant from upper and lower surfaces, joining the leading and the trailing edge as shown in Fig.01. Maximizing camber at specific sections of the airfoil increases lift. In addition, airspeed and area of the airfoil also adds on to the lift whereas lift decreases with higher altitudes.

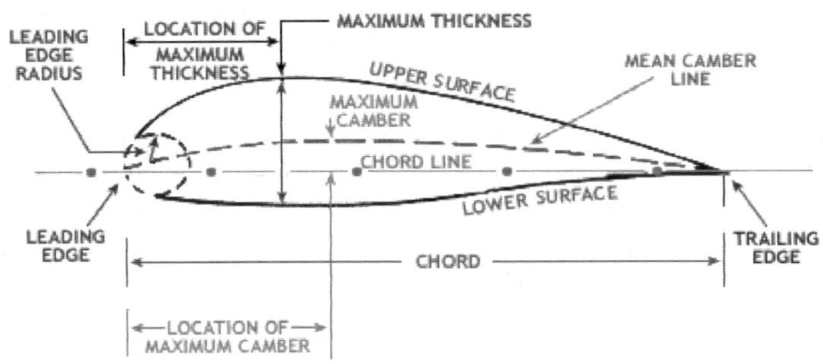

Figure 1. Airfoil Nomenclature

2. Literature review

1. Numerical analysis on NACA 2412 airfoil was performed in ANSYS Fluent with a chord length of C_R = 0.3m and total wing span of L =1.6m. the coefficient of lift C_L =0.16 and coefficient of drag C_D =0.06 was obtained [3], and the same results were replicated in XFLR5 software for validation.
2. Aerodynamic analysis was performed on MH – 60 airfoil with varying airfoil thickness from 10.1% to 26.9% and max camber 1.7% to 36.6%. The analysis was performed using vortex lattice method (VLM) and the total wing span L =1.9m and chord length at root was C_R =0.980m [4].
3. Aerodynamic analysis was performed on 9 different airfoils for sailplane and the analysis was performed using XFLR5 software. The analysis was performed at Reynolds number 2e+05 and from -5° to 20° angle of attack. The results suggested that Wortmann FX S02-196 and FX-Wortmann FX61-184 airfoils were efficient for root section of the sailplane wing and GOE 533 airfoil was efficient for tip section of the wing [5].
4.Aerodynamic analysis was performed on Clark -Y airfoil using XFLR5 software, the analysis was performed at the Reynolds number 2.5e+05 to 5e+05. The chord length at the root is C_R=0.27m, total wing span L =1.8m and angle of attack -5° to 20°. The analysis was carried out using Vortex Lattice Method (VLM) and Maximum coefficient of lift (C_L) was achieved at 12° angle of attack [6].

3. Wing Planforms

A wing planform is the shape of an aircraft wing as viewed from the top. The wing planforms have evolved drastically through generations. Some wing planforms analyzed are

- Elliptical wing, which was used in Spitfire, a famous plane used in World War II.
- Rectangular wing, non-tapered and straight used in small aircrafts like Piper PA 38.
- Delta wing is a triangular wing used in modern fighter jets like Dassault Rafale.
- Swept back wings are angled backwards or forward from the root of the plane, mostly used in modern commercial aircrafts like Boeing 737.

4. XFLR5 Explanation

XFLR5 is an analysis tool for airfoils, wings and planes operating at low Reynolds number. It analyses airfoils and wing design based on Lifting Line theory, Vortex lattice method and 3D panel method [7].

5. Vortex Lattice Method (VLM)

The aerodynamic analysis was performed using VLM on XFLR5 software, VLM is a numerical method used in CFD, where the lifting surfaces such as a wing is modelled as an infinitely thin sheet

of discrete vortices to compute lift and induced drag. The VLM is built on the theory of potential flow (ideal flow). Potential flow is a simplified represented of real flow experience in nature and this method neglects all viscous effects.
Assumptions in Vortex Lattice Method:

- The flow is incompressible, inviscid and irrotational
- The lifting surfaces are thin
- Angle of attack and angle of slide slip are small

6. Methodology

A group of six airfoils were chosen for the aerodynamic analysis. These airfoils were analyzed on four different wing profiles, the aerodynamic characteristics were compared and the best 3 airfoils with the corresponding best wing profile were selected. These selected airfoils were further refined by interpolating to obtain best CL/CD value.

7. Airfoils Chosen

Totally six types of airfoils were chosen [8][9] for the analysis based on their varying camber and thickness. Table 1 contains the data of the airfoils depicted in Figures 2 to 7.

Table 1. Airfoil Name and Data

Airfoil Name	Thickness (%)	Camber (%)
NACA 0018	18	0
NACA 2412	12	2
NACA 4412	12	4
TSAGI S-12	11.9	1.99
USA 45	14.53	4.15
B737A	15.37	1.92

Figure 2. NACA 0018

Figure 3. NACA 2412

Figure 4. NACA 4412

Figure 5. TSAGI S-12

Figure 6. USA 45

Figure 7. B737A

The analysis was performed using vortex lattice method (VLM), the wing root chord and total wing span is kept constant at C_R =1m and L =10m respectively. The density of air ρ = 1.225 kg/m^3, kinematic viscosity of the air ν = 1.789e-05 m^2/s, inlet velocity of air V = 50 m/s (average cruise speed of Cessna 152 aircraft is 180 kmph) and Angle of Attack at 7° are kept constant. The calculations are performed between the Reynolds number (Re) 1e+05 to 3e+06.

7. Four types of wing are chosen for the analysis, namely:

- Rectangular wing
- Elliptical wing
- Swept back wing
- Delta wing

7.1 Rectangular Wing

The chord length remains constant throughout the wing span length as shown in Figure. 8 and this type of wing is commonly used in small planes which are used for training pilots and planes used for recreational purposes. Example: Piper PA-38 Tomahawk, Cessna 152, etc.

Table 2. Initial condition of rectangular wing profile

Chord length (C_R)	1 m
Mean Aerodynamic Chord (MAC)	1 m
Wing area (A)	10 m^2
Aspect ratio	10
Velocity (V)	50 m/s
Re at root	2.79e+06
Re at tip	2.79e+06

Figure 8. Rectangular wing profile

7.1 Elliptical Wing

The wing profile looks like an ellipse when looked from the top, hence the name elliptical wing. Figure. 9 shows an elliptical wing planform. This type of wing is rarely used in aircraft because of its low manufacturability as it is a complex shape. Example: Supermarine Spitfire.

The equation to plot the ellipse is[10]:

$$= C_0\sqrt{1 - 4(\frac{y}{b})^2}$$

y = Distance from the wing root
b = Total wing span
Co = Chord length at root
C = Chord length at 'y'

The equation for offset is:

$$offset = \frac{1}{2}(C_o - C)$$

Co = Chord length at root
C = Chord length at 'y'

Using the above equations, the elliptical wing was designed. Table. 3 contains the co-ordinates for elliptical wing.

Table 3. Plot coordinates for elliptical wing

y	C	offset	y	C	offset
0.00	1.000	0.000	3.70	0.673	0.164
0.50	0.995	0.003	3.80	0.650	0.175
1.00	0.980	0.010	3.90	0.626	0.187
1.25	0.968	0.016	4.00	0.600	0.200
1.50	0.954	0.023	4.05	0.586	0.207
1.75	0.937	0.032	4.10	0.572	0.214
2.00	0.917	0.042	4.15	0.558	0.221
2.10	0.908	0.046	4.20	0.543	0.229
2.20	0.898	0.051	4.25	0.527	0.237
2.30	0.888	0.056	4.30	0.510	0.245
2.40	0.877	0.061	4.35	0.493	0.253
2.50	0.866	0.067	4.40	0.475	0.263
2.60	0.854	0.073	4.45	0.456	0.272
2.70	0.842	0.079	4.50	0.436	0.282
2.80	0.828	0.086	4.55	0.415	0.293
2.90	0.815	0.093	4.60	0.392	0.304
3.00	0.800	0.100	4.65	0.368	0.316
3.10	0.785	0.108	4.70	0.341	0.329
3.20	0.768	0.116	4.75	0.312	0.344
3.30	0.751	0.124	4.80	0.280	0.360
3.40	0.733	0.133	4.85	0.243	0.378
3.50	0.714	0.143	4.90	0.199	0.401
3.60	0.694	0.153	4.95	0.141	0.429
			5.00	0.000	0.500

Table 4. Initial conditions of elliptical wing profile

Chord length (C_R)	1 m
Mean Aerodynamic Chord (MAC)	0.849 m
Wing area (A)	7.842 m^2
Aspect ratio	12.498
Velocity (V)	50 m/s
Re at root	2.79e+06
Re at tip	3.94e+05

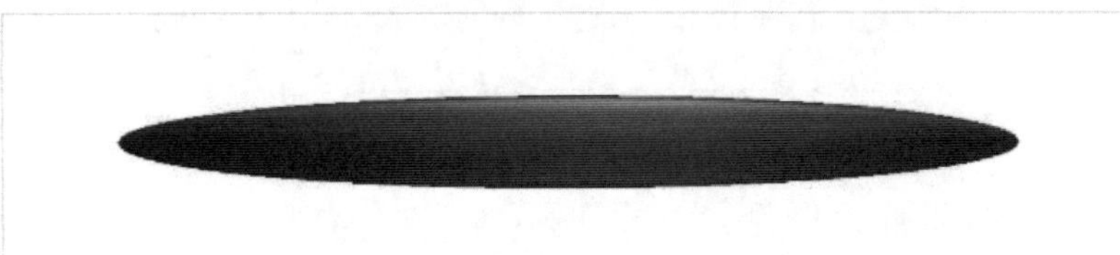

Figure 9. Elliptical wing profile

7.3 Sweptback Wing

As the name suggests, the wings are swept backwards for aerodynamic drag reduction at an angle. for this analysis sweep angle of 37.5° is considered[11]. This type is most commonly used in commercial airplanes and cargo jets. Figure 10 is a swept back wing angled at 37.5 degrees Example: Boeing 737

Table 5. Initial conditions of Sweptback wing

Chord length at root (C_R)	1 m
Chord length at tip (C_T)	0.4 m
Mean Aerodynamic Chord (MAC)	0.743 m
Wing area (A)	7 m^2
Aspect ratio	14.286
Root to tip sweep angle (φ)	37.5°
Velocity (V)	50 m/s
Re at root	2.79e+06
Re at tip	1.11e+06

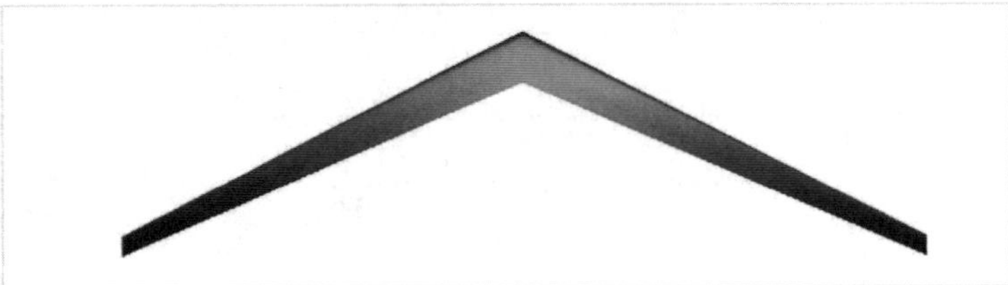

Figure 10. Sweptback wing profile

7.4 Delta Wing

Delta wing is type of trapezoidal shape wing which is commonly used in fighter jets for maneuverability. Delta wing planform is shown in Figure 11. Example: MiG-21

Table 6. Initial conditions of Delta Wing

Chord length (C_R)	1 m
Mean Aerodynamic Chord (MAC)	0.743 m
Wing area (A)	7 m^2
Aspect ratio	14.286
Velocity (V)	50 m/s
Re at root	2.79e+06
Re at tip	1.11e+06

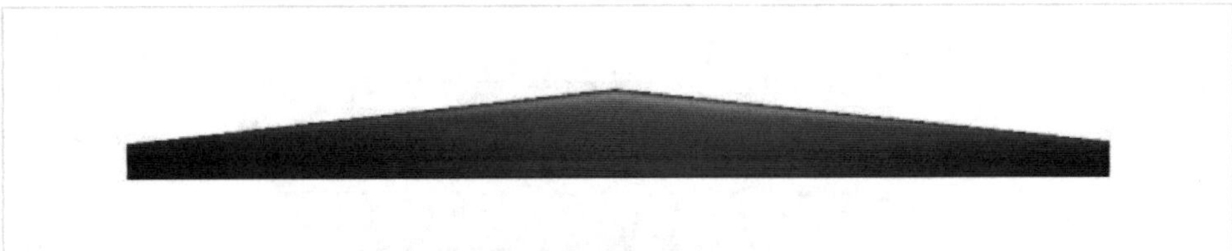

Figure 11. Delta wing profile

8. C_L/C_D Table and Graphs

After analysis, the following results in Table 7 were obtained.

Table 7. C_L/C_D ration of the airfoils with respective to wing planforms

Profile names	Rectangular wing	Elliptical wing	Swept back wing	Delta wing
NACA 0018	31.316	35.833	39.000	38.529
NACA 2412	30.440	34.250	41.000	36.304
NACA 4412	27.235	32.226	38.550	35.000
USA 45	28.033	32.464	38.273	35.500
TSAGI S-12	33.526	36.316	41.500	38.722
B737A	27.920	32.696	35.556	34.864

Graph 1. C_L/C_D ratio of the airfoils with respective to wing planforms

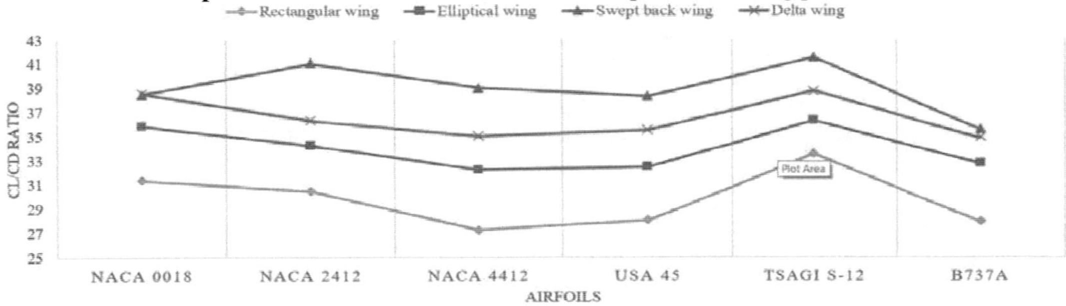

Numerical data tabulated in Table 07 is pictorially represented in Graph 01. From the graph, we can conclude that TSAGI S-12 with swept back wing comparatively has the highest C_L/C_D ratio and NACA 4412 with rectangular wing has the least C_L/C_D ratio.
From the Table 7 it is evident that swept back wing profile coupled with TSAGI S-12 has the highest C_L/C_D ratio among the airfoils and wing profiles chosen. The NACA 2412 and NACA 4412 results were also promising, so these 3 airfoils are further refined by interpolation.

9 Refinement

Refinement of airfoils or wing designs is undertaken to enhance lift and reduce drag or to obtain a higher CL / CD ratio. This can be done either by increasing the pressure difference across the top and bottom surfaces of airfoil/wing or by distributing the pressure over larger length of airfoil. But parameters like camber, thickness and angle of attack also induces higher lift[12]. Certain parameters like camber and thickness of airfoils was modified to derive higher performances.
The three following airfoils were interpolated to obtain 3 test foils as we discussed above:

- TSAGI S-12
- NACA 2412
- NACA 4412

All the following test foils were used on swept back wing profile for analysis.

9.1 Test foil 1

This airfoil was obtained by interpolating NACA 2412 and NACA 4412 airfoils. The thickness (%) and camber (%) was 10.2 and 3.01 respectively.

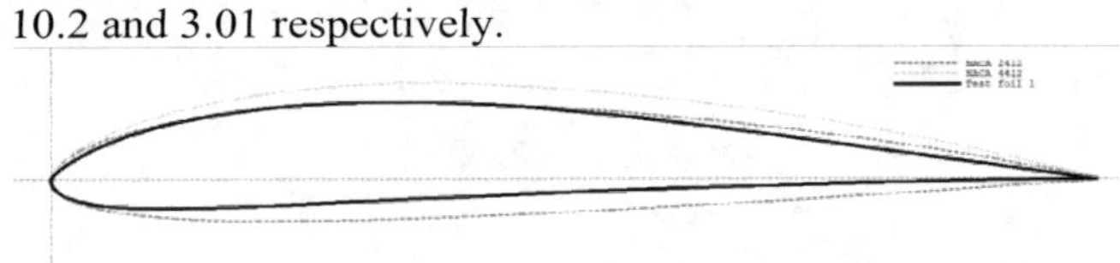

Figure 12. Test foil 1

9.2 Test foil 2

This airfoil was obtained by interpolating NACA 4412 and TSAGI S-12 airfoils. The thickness (%) and camber (%) was 12.3 and 2.89 respectively.

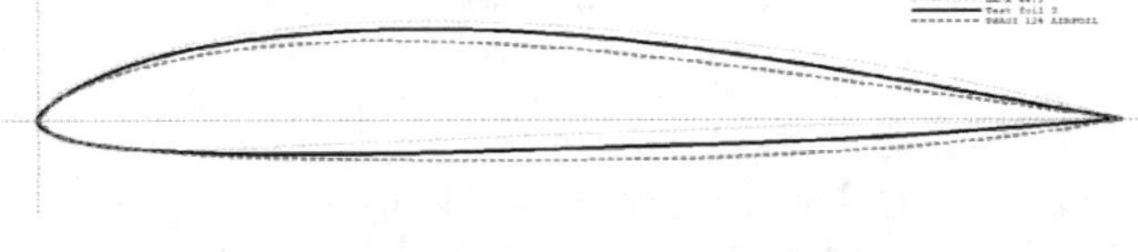

Figure 13. Test foil 2

IOP Conf. Series: Materials Science and Engineering 1013 (2021) 012023 doi:10.1088/1757-899X/1013/1/012023

9.3 Test foil 3

This airfoil was obtained by interpolating NACA 2412 and TSAGI S-12 airfoils. The thickness (%) and camber (%) was 11.95 and 1.93 respectively.

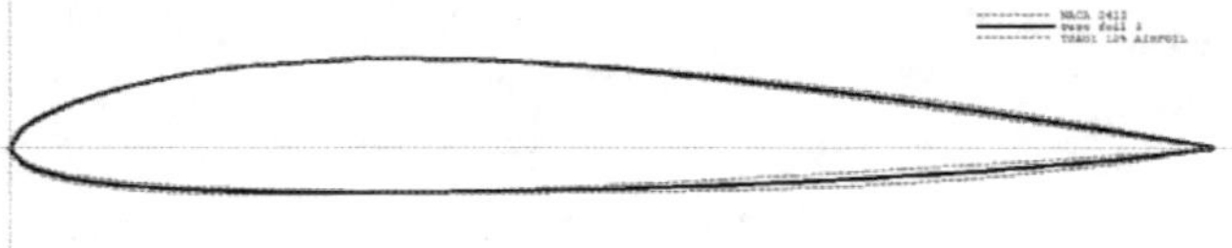

Figure 14. Test foil 3

After aerodynamic analysis of the above test foils yielded the following results:

Table 8. C_L/C_D ratio of testfoils with sweptback wing planforms

Airfoil name	C_L / C_D ratio
Test foil 1	38.316
Test foil 2	41.765
Test foil 3	42.600

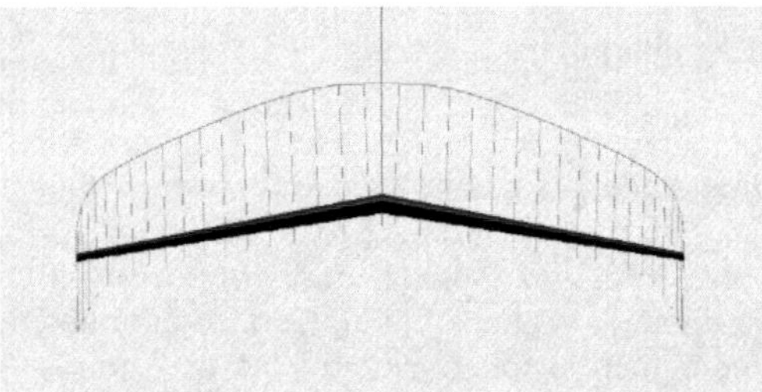

Figure 15. Lift and Downwash of test foil 3

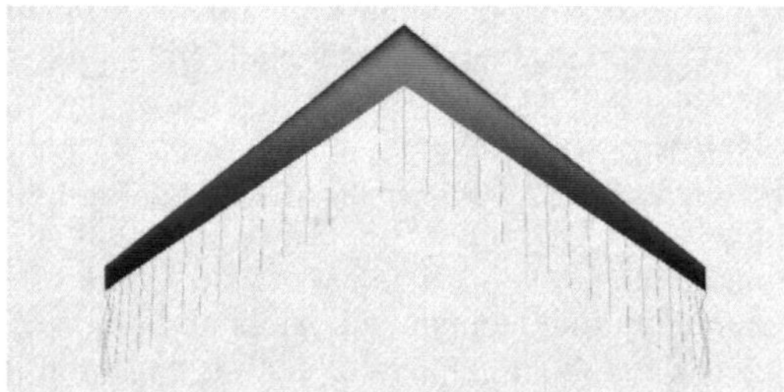

Figure 16. Streamlines of test foil 3

Graph 2. C_L/C_D ratio of test foils with sweptback wing planform

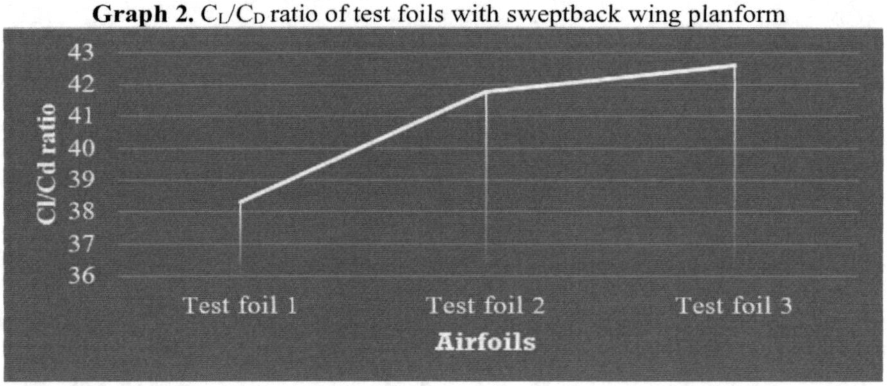

10 Results and Conclusion

After analysis, it was found that Test foil 3 has the maximum C_L / C_D ratio when paired with swept back wing. In straight wing design such as rectangular wing the air flow is parallel to the wing, this is called the chord wise flow. The air above the wing accelerates and generate lift, but this type of wing design is not suitable for high speed flights as the wing plan form and the straight wing design generate more drag in high speed flight. This type of airflow reduces the critical Mach number of the plane. A swept will introduce a new type of flow along with chord wise flow, called span wise flow. The airflow is diverted into span wise flow which does not accelerate over the wing like the chord wise flow which allows the aircraft to fly at high speeds as the critical Mach number is further increased. This effect is more prominent when the sweep angle is higher.

Figure. 16 shows a distortion at the wing tips of Swept back wing. Due to the pressure imbalance between upper and lower surfaces, the air tends to spiral at the wingtips. This is referred as wingtip vortices. A vortex is a flow pattern in which the elements of fluid follow circular paths about a point [13]. As the angle of attack increases during higher lift, there is higher pressure difference, as a result the effect of fast spinning vortices increases. These violent vortices increase turbulence in the flow and induce drag. Optimizing the wing design by introducing winglets at the wingtip reduces vortices. Winglets are small upright tails which avoid formation of vortices by reducing vortex interference with laminar airflow near wingtips.

References

[1] A. Kulshreshtha, S. K. Gupta, and P. Singhal, "FEM/CFD analysis of wings at different angle of attack," Mater. Today Proc., no. xxxx, 2020, doi: 10.1016/j.matpr.2020.02.342.
[2] K. Kumar, O. Prakash Mishra, and S. Kumar, "Simulation of Airfoil Shape for Optimum Wing Characteristics," Mater. Today Proc., vol. 24, pp. 2231–2237, 2020, doi: 10.1016/j.matpr.2020.03.749.
[3] N. M. Triet, N. N. Viet, and P. M. Thang, "Aerodynamic Analysis of Aircraft Wing," vol. 31, no. 2, pp. 68–75, 2015.
[4] S. V. Sandapeta, S. K. Parre, Y. Dedeepya, H. Jaffar, M. Vamshi, and K. Reddy, "Design and Analysis of Flying Wing UAV using," vol. 6, no. 1, 2019.
[5] K. Fen and B. Dergisi, "Karadeniz Fen Bilimleri Dergisi The Black Sea Journal of Sciences," vol. 8, no. April, pp. 47–65, 2018.
[6] V. Prisacariu and L. Siminonescu, "FLIGHT PERFORMANCES ANALYSIS OF UAV MFD NIMBUS," pp. 225–232, 2018, doi: 10.19062/2247-3173.2018.20.30.
[7] A. Deperrois, "Guidelines for XFLR5: Analysis of foils and wings operating at low Reynolds numbers," Availiable from http//xflr5. sourceforge. net/xflr5. htm, no. February, pp. 1–72, 2013, [Online].
[8] "Airfoil Tools." http://www.airfoiltools.com/ (accessed Jun. 30, 2020).
[9] "UIUC Airfoil Data Site." https://m-selig.ae.illinois.edu/ads/coord_database.html (accessed Jun. 30, 2020).
[10] D. Ignjatovic, L. Cole, and T. Spitfire, "Decoding the spitfire," pp. 1–31, 1934.
[11] "Sweep Angle and Supersonic Flight | AeroToolbox." https://aerotoolbox.com/intro-sweep-angle/ (accessed Jul. 01, 2020).
[12] T. Srinivasa Rao, T. Mahapatra, and S. Chaitanya Mangavelli, "Enhancement of Lift-Drag characteristics of NACA 0012," Mater. Today Proc., vol. 5, no. 2, pp. 5328–5337, 2018, doi: 10.1016/j.matpr.2017.12.117.
[13] Ira H. Abbott, Albert E. Von Doenhoff "Theory of wing sections," J. Franklin Inst., vol. 249, no. 3, p. 254, 1950, doi: 10.1016/0016 0032(50)905

STRUCTURAL ANALYSIS OF A TWO-WHEELER DISC BRAKE

Mr. Adarsh Bhat, Dr. Bhaskar Pal, Dr. Devendra Dandotiya

Department of Mechanical Engineering, Presidency University, Bengaluru

E-mail: adarshb365@gmail.com

Abstract: Automotive brakes have come a long way, from drum brakes being widely used for a long time until disc brakes came along and began slowly wiping out the usage of drum brakes. This is primarily due to the number of advantages that the disc-pad setup brings - better stopping distance being one of the most important aspects of any braking system. Apart from this, key advantages of disc brakes over drum brakes include better heat dissipation, fade resistance, self-adjusting ability, compact packaging and so on. Although drum brakes too have certain advantages over disc brakes with lesser cost of manufacturing being one of them, the pure superiority of disc brakes outweighs the few benefits of the drum brakes and this has also reflected in the market. However, there is always scope for improvement when it comes to the structural and thermal characteristics of the brake disc. Researches have been carried out to discover better suited materials and brake disc designs to improve the overall structural and thermal characteristics and reduce uneven wear, squealing etc. The aim of this research is to analyze a new design and material combination based on earlier researches to decrease the deformation and provide a base for improving thermal characteristics.

Keywords: Disc brake, brake pads, silicon carbide single crystal, Solidworks, ANSYS, static structural analysis.

1. Introduction

A brake is a device which makes use of frictional resistance to stop the motion of any machine or vehicle. It absorbs the kinetic energy and dissipates it in the form of heat. Brake systems must be able to fulfill certain key requirements such as:

- Minimum stopping distance - Vehicle must be able to come to a complete halt within a regulated distance in case of emergency braking.
- Prolonged application of the brake should not lead to brake fade.
- Must have anti-wear properties [1].

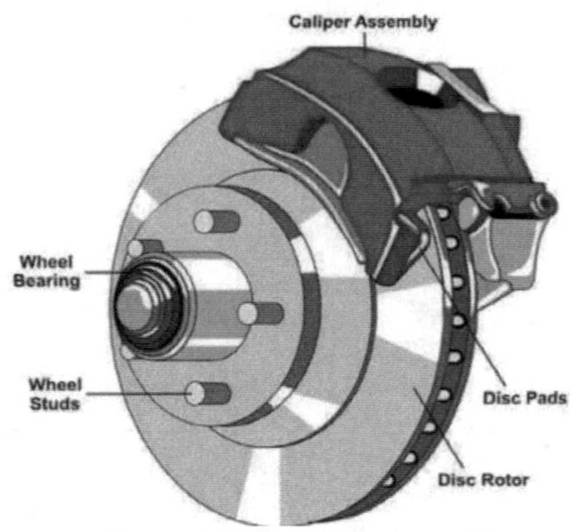

Figure 2. Disc Brake Assembly

Figure 1. Motorcycle Disc Brake (Duke 200)

A disc brake assembly, be it in the case of any vehicle, consists mainly of calipers which house the brake pads and the disc/rotor as shown in fig. 1.

Brakes are important components of any vehicle's retardation system. The disc brake uses a metallic rotating disc which is sandwiched between two pads that are activated by a cylinder backed in a caliper mounted on the stud shaft. When the brake pedal is depressed, the brake pads are pressed either mechanically or pneumatically on to the rotating disc. The kinetic energy of the vehicle is converted into heat due to the friction between the disc-pad interface and is absorbed mainly by the rotor and pads in the form of heat [2][3]. This heat is then dissipated into the surrounding atmosphere. As there is generation of frictional heat on the disc-pad interface, there is a rise in temperature. If this temperature exceeds the critical value of the given material it leads to disastrous events such as brake failure, premature wear, failure of bearing and thermal cracks or vaporization of brake fluid [4]. In addition, due to the heat generation at the disc-pad interface, global deformation occurs in the disc and pad. Some of the usual deformations are coning and buckling [5-7]. Apart from all this, macro cracks could appear on the disc brake in radial direction after a number of braking cycles in turn affecting the life and performance of the disc brake [8-9]. Lee and Yee [4] found out that the unequal distribution of temperature on the surface of frictional pad causes thermal distortion popularly known as coning and this is responsible for judder effect and variation in the thickness of the rotor.
Researches have been carried out to analyze various designs of the rotor by including different types of ventilations such as circular, oval, square holes and so on. On the other hand, different metals and composites have also been analyzed over the years to find structurally and thermally superior solutions especially for higher end vehicles such as sports cars and bikes which require high performance brakes. This paper focuses on improving the structural behavior of the disc brake by designing and analyzing a new design keeping in mind the need for better thermal characteristics as both the parameters are closely linked.

2. Literature Review

Belhocine studied the thermo-mechanical behaviour between brake disc and pad to identify the factors affecting the design of ventilated disc based on transient temperature conditions. He used grey cast iron due to its good thermo-physical properties and concluded that the variations in temperature observed were less compared to solid conventional disc. Further, his work proves that ventilation plays a key role in the disc as it provides good high temperature thermal resistance. He carried out static structural analysis on a composite disc and analysed its performance by comparing it with that of a standard steel disc. The deformation of composite structure was found to be considerably less than steel [10]. C. Radhakrishnan, Yokeswaran, Naveen Kumar, et al. focused on use of Titanium alloy 550 for the brake disc due to its thermal stability, high strength and forgeable qualities. They concluded that the deformation of Titanium alloy 550 was very less compared to cast iron and stresses induced in Titanium alloy 550 were six times less than that of grey cast iron [11]. Shinde, Sagar, & Baskar worked on enhancing the cooling performance of disc brake by varying the cut patterns on the disc. They concluded that elliptical type disc has better heat transfer than circular ones. However, the strength to withstand the braking force was more in case of circular type [12]. Huang & Chen calculated heat transfer coefficient of disc brake surface and its effect on design parameters. Using Reynold's equation, they calculated heat transfer coefficient and amount of flux generated. After tabulating the results of heat transfer coefficients depending on operating speeds and temperature, it was observed that temperature distribution patterns were different for different models. They concluded that the model can be used as a design tool for designing brake discs [13].
Radhika worked on circular and oval hole profiles and proved that oval/elliptical holes not only provide better heat transfer but also lesser deformation as significantly lesser stresses are induced [14]. Amit and Sachin further worked with similar hole profiles and showed the effect of spacing between the cut patterns on thermal characteristics of the rotor. They were also able to show that oval cut patterns perform better [15].

IOP Conf. Series: Materials Science and Engineering 1013 (2021) 012024 doi:10.1088/1757-899X/1013/1/012024

3. Experimental Details

3.1 Material Selection

Based on literature review, it was decided that three models would be prepared based on a single design. The following are the materials selected:

- Model 1 – Magnesium Alloy
- Model 2 – Aluminum Alloy
- Model 3 – Cast Iron + Aluminum Alloy + Silicon Carbide Single Crystal

Carbon Ceramic was selected for brake pads, which was common across all the three models. Aluminum alloy has been widely used due to its less affinity towards galling, lower production costs and superior thermal characteristics compared to conventional materials. Magnesium alloy is commonly used in the casting of automotive parts. These are among the lightest alloys and aluminum alloy also happens to have a significantly better performance index when compared to titanium alloy at a relatively lower cost. Further, to try to improve structural characteristics, a combination of cast iron at the center of the disc due to less stresses induced in the region, aluminum for majority of the contact patch with the brake pads and silicon carbide tips with high strength and thermal conductivity was considered.

3.2 Material Properties

Table 1. Material Properties of Brake Pad

Property	Units	Carbon Ceramics
Density	Kg/m^3	1600
Elastic Modulus	Pa	7e+10
Poisson's Ratio	–	0.1
Ultimate Compressive Strength	Pa	1.9e+8
Maximum Temperature Use	°C	1750
Thermal Conductivity	W/mK	8
Specific Heat	J/gK	1.123

Table 2. Material Properties of Disc

Property	Units	Al Alloy	Mg Alloy	Cast Iron	SiC
Density	Kg/m^3	2770	1800	7200	3200
Elastic Modulus	Pa	7.1e+10	4.5e+10	1.1e+11	7e+11
Yield Strength	Pa	2.8e+8	1.93e+8	–	2.1e+10
Poisson's Ratio	–	0.33	0.35	0.28	0.192
Thermal Conductivity	W/mK	171	156	52	350
Specific Heat (Cp)	J/KgK	875	1024	447	1090

Carbon ceramics possess high thermal resistance as well as strength- weight ratio. They produce significantly less noise during braking and due to their composition wear rate is lower than that of conventional pad materials. One of the most important reasons behind this is the ideal operating temperature of the carbon ceramics, which is much higher compared to the that of conventional brake pad materials. The ability to maintain their functionality at high temperatures is their biggest advantage. In high-performance vehicles, as much as six tons of actuating force is generally required to stop the vehicles which produces a lot of heat, over 800°C, whereas the boiling point for brake fluid is between 600 to 750°F. However, the carbon ceramic brakes can diffuse the heat before it affects the brake fluid. Repeated heavy braking at high speeds is easily achieved for longer. On the other hand, there is also less brake dust accumulation from carbon ceramic brakes. Carbon ceramic brake pads are durable and do not wear out as fast as the regular brake pads in comparable driving conditions.

3.3 Calculations

Given: Based on a standard 200-250cc motorcycle
Mass of the vehicle (M): 230 kg
Speed of the vehicle (u): 27.77 m/s
Pad Area (A): 5021.16 mm^2
Radius of the disc (R): 140 mm
Braking Time (T): 5 sec
Now, let us consider the extreme condition of hard braking, wherein the vehicle is supposed to be stopped within a distance of 22m from a speed of 100km/hr to 0 km/hr.

According to Newton´s law of motion,

$$v^2 = u^2 + 2as$$

Therefore, we get,

$$a = 17.526 \text{ m/s}^2$$

Calculation of Braking force,
In two wheelers, 70% of the weight acts on the front-end during braking, hence we have,

$$F_b = 0.7 * M * a$$

$$F_b = 2821.79 \text{ N}$$

Calculation of Brake Torque,

$$T_b = F_b * R$$

$$T_b = 397.31 \text{ N-m}$$

Calculation of Pad Pressure,

$$P_b = \frac{Brake\ force}{Pad\ Area}$$

$$P_b = 0.5619 \text{ Mpa}$$

Table 3. Disc Specifications and Automotive Parameters

Specifications	Units	Values
Outer diameter of disc	mm	280
Inner diameter of disc	mm	100
Thickness of disc	mm	3
Thickness of pad	mm	6
No. of clamp holes	–	6
Pad area	mm^2	5021.16
Weight of the vehicle	kg	230
Speed of the vehicle	km/h	100
Stopping distance	m	22

The dimensions in the table above are similar to those found on 200-250cc bikes in the market.

3.4 Finite Element Analysis

The steps followed to conduct the finite element analysis of the disc brake were:

- Modelling the geometry
- Meshing
- Applying the boundary conditions
- Evaluating results

IOP Conf. Series: Materials Science and Engineering 1013 (2021) 012024 doi:10.1088/1757-899X/1013/1/012024

Geometry

As mentioned earlier, Solidworks was used to design the disc brake setup as per dimensions in table 3. Three different models were considered with different materials but the same design. As oval cut patterns have been proven to be better than circular, a disc was designed with oval cuts of gradually increasing area in the contact region of the disc and pads. The disc and two pads were assembled and saved as three separate models in iges format, namely model 1, model 2 and model 3.

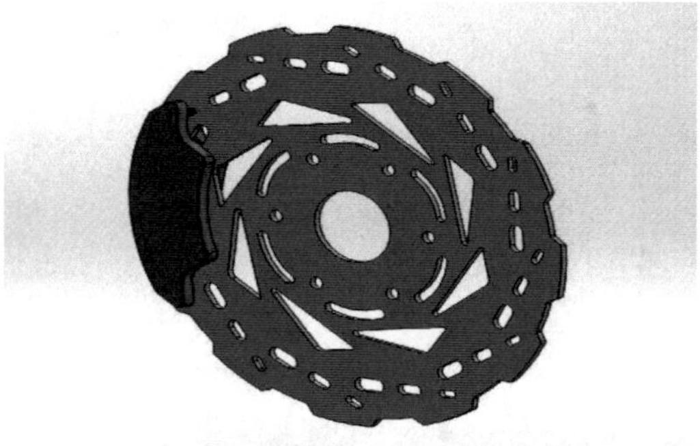

Figure 3. Disc Brake CAD model (Solidworks)

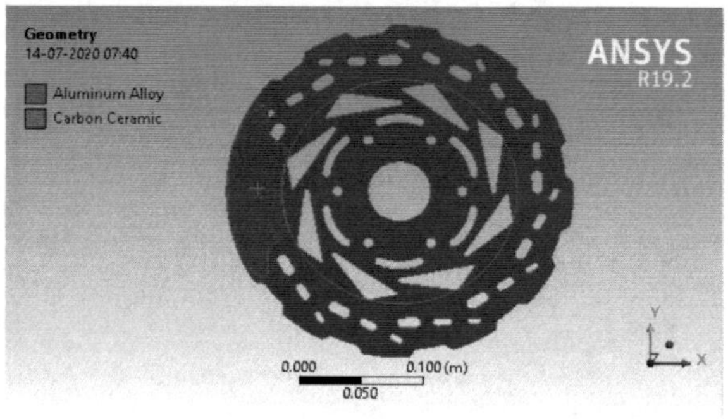

Figure 4. Model 1 CAD

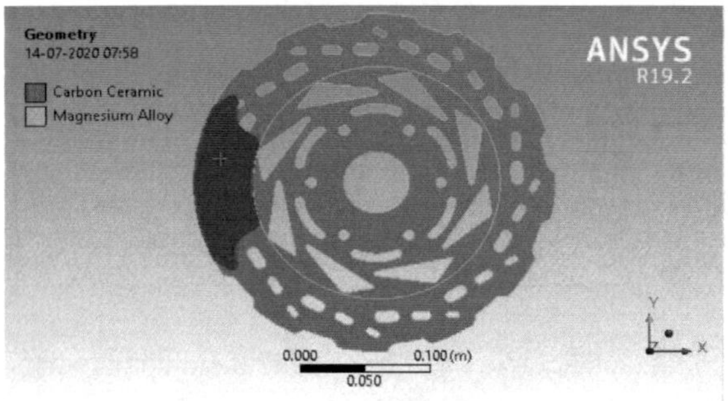

Figure 4. Model 2 CAD

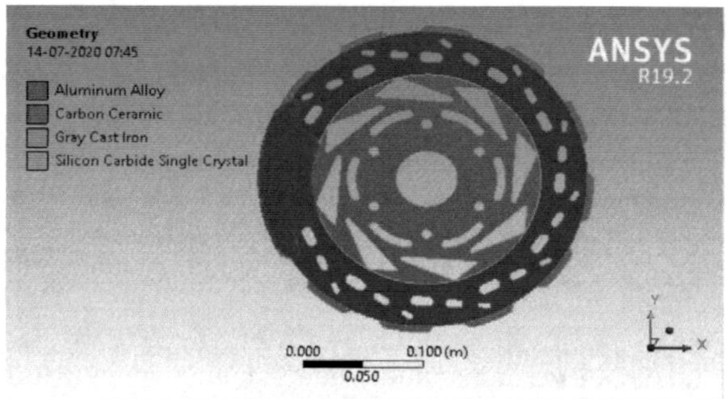

Figure 3. Model 3 CAD

The three models were imported to ANSYS 19.2 and material properties were assigned as shown above.

Meshing

Meshing is the most important step in FEA as it decides numerical convergence to the solution. Finer meshing offers more accurate results. In this research, fine meshing was done using triangulated surface mesher with a minimum element size of 4 mm.

Table 4. Mesh Details

	Model 1	Model 2	Model 3
No. of elements	73703	72481	87189
No. of nodes	141730	139655	229368

Boundary Conditions

To conduct the structural analysis, disc was fixed at the six hub mounting points and at calliper. Moment was then applied on the disc about the axis in a plane perpendicular to the plane of disc. Calculated pad pressure was applied on both the brake disc pads. Geometry for the structural analysis with boundary conditions is as shown in figure.

This simulates the real-world scenario wherein the brake disc is rotating with the wheel it is connected to at the wheel hub and the callipers being actuated as the driver slams the brakes, resulting in pad pressure being applied on the spinning disc.

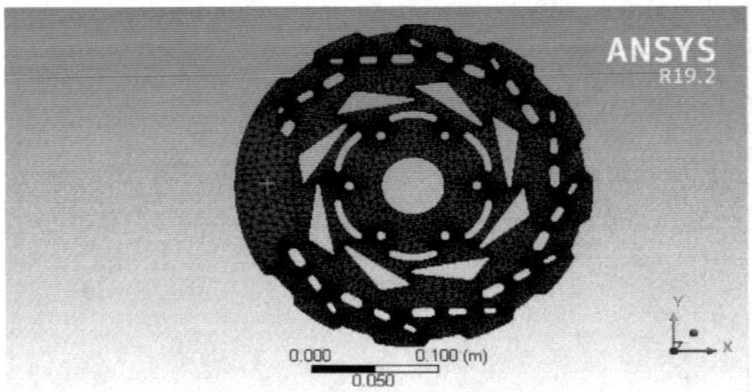

Figure 6. Model 1 Mesh

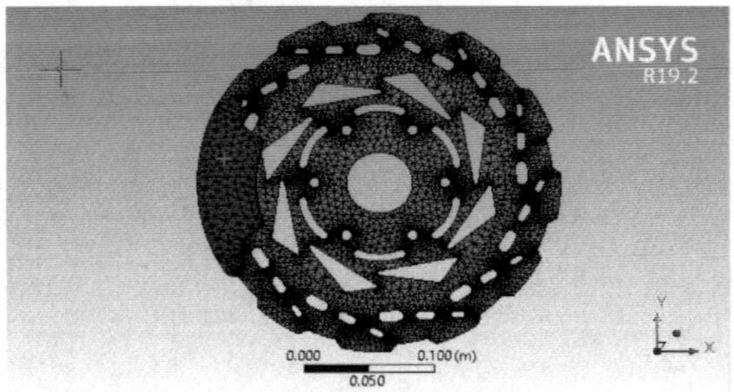

Figure 8. Model 2 Mesh

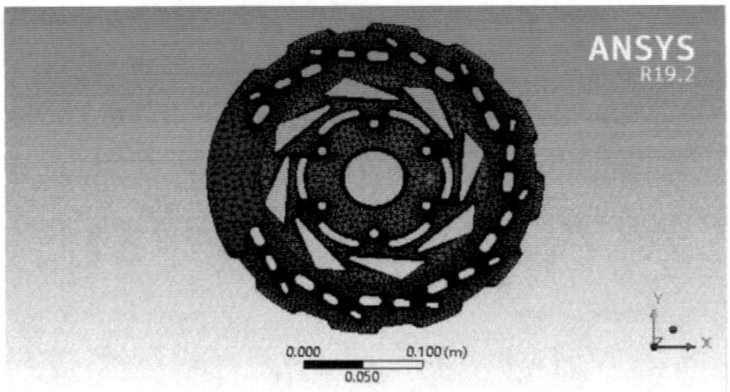

Figure 5 Model 3 Mesh

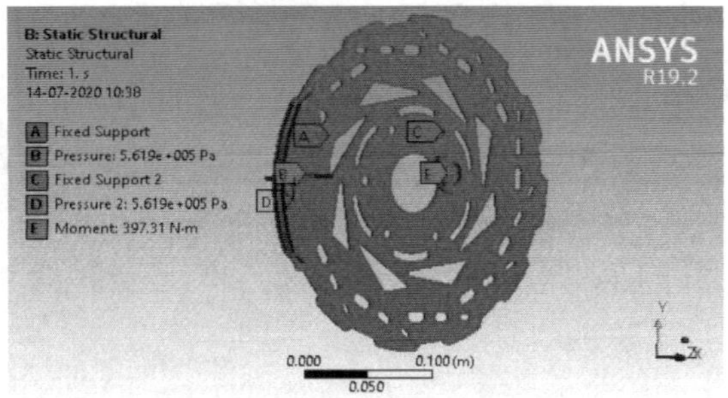

Figure 10. Model 1 Boundary Conditions

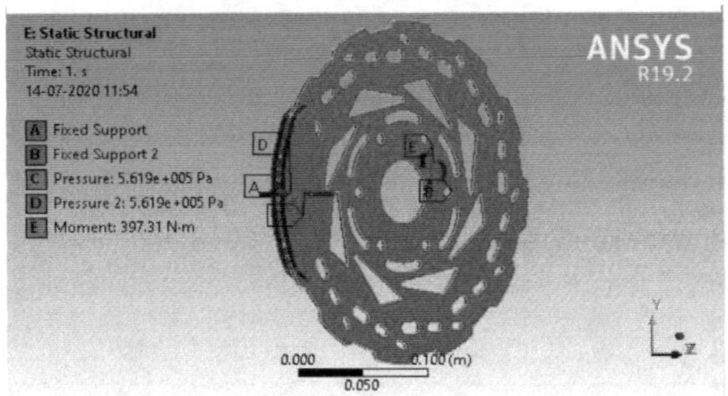

Figure 11. Model 2 Boundary Conditions

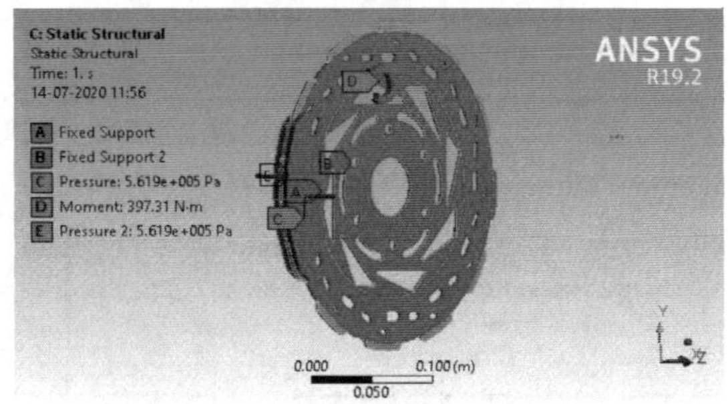

Figure 12. Model 3 Boundary Conditions

Structural Analysis

As moment is provided about the hub mounting points of the brake disc and clamping forces are generated at the contact patch between the disc and the pads, the region in contact is held firmly by pads whereas the outer region i.e. the region diametrically opposite to that of contact tends to follow the rotation of disc thereby resulting in greater deformation at the periphery, whatever be the material selected, as shown below. The results in case of aluminium alloy as well as magnesium show minimal deformation for a large chunk of the disc around its centre. Based on this understanding, model 3 has been conceptualized. Cast iron has been used at the central region where stresses are minimal to help cut costs.

Aluminium alloy was retained for majority of the contact region. Silicon carbide single crystal was chosen for the tips at the periphery to enhance both structural and thermal characteristics which is undoubted due to its high thermal conductivity, while balancing the possible increase in cost by using cast iron around the hub region. Fig. 15 shows the deformation of model 3.

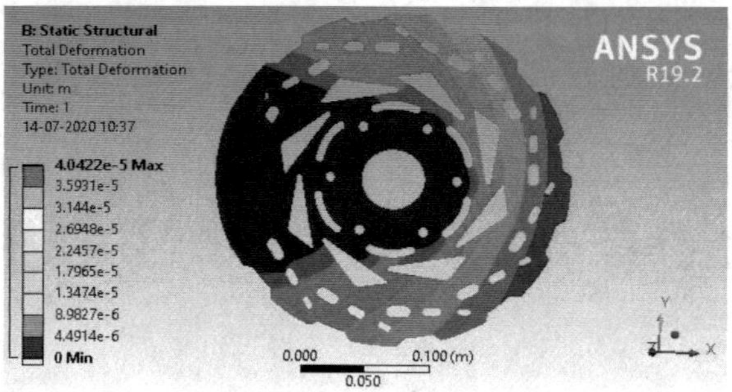

Figure 7 Model 1 Deformation

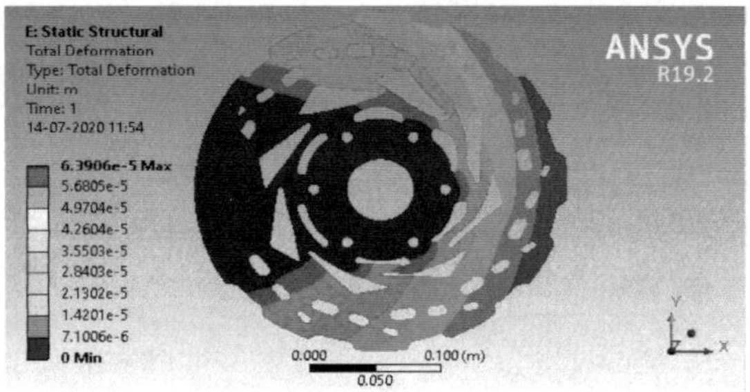

Figure 14. Model 2 Deformation

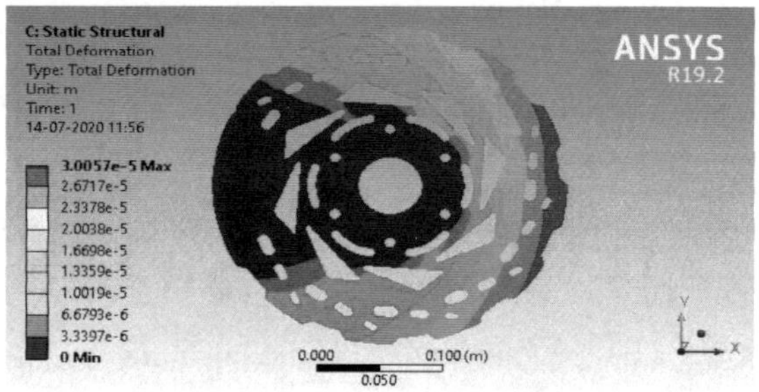

Figure 15. Model 3 Deformation

4. Results and Discussions

	Unit	Model 1	Model 2	Model 3
Maximum Deformation	mm	0.040908	0.063906	0.030057
Minimum Deformation	mm	0	0	0
Average Deformation	mm	0.014645	0.023531	0.0095401

As earlier researches have shown, aluminium alloy undergoes lesser deformation than magnesium alloy. The newly conceptualized model 3 performs even better as it has the least deformation compared to the other two models. When compared to [16] from which the disc and pad parameters as well as material properties have been obtained, the design approach of using gradually increasing area of oval holes seems to have paid off well.

5. Conclusions

Model 1 and 2 that have been replicated in terms of materials previously worked with have shown results that are in line with the researches carried out earlier. In terms of cut patterns, the oval hole approach also performs similar to what has been shown earlier. The gradually increasing area of the slots on the contact region could aid heat dissipation. Model 3 which consists of a combination of three materials and performed better than model 1 and 2 structurally, appears to a very promising concept that could perform better than the other 2 thermally. This will be explored next in an extensive thermal analysis carried out using the same parameters.

References

[1] V. Chengal Reddy, M. Gunasekhar Reddy, Dr. G. Harinath Gowd, Modeling and Analysis of FSAE Car Disc Brake Using FEM, ISSN 2250-2459, 3(9), September 2013.
[2] Praveena S, Lava Kumar M, Sreekanth Reddy S, Modeling and Structural Analysis of Disc Brake, ISSN: 2319-8753, 3(10) October 2014.
[3] Jung, S. P., Park, T. W., Kim, Y. G., A study on thermal analysis and shape optimization of a ventilated disc, International Journal of Precision Engineering and Manufacturing 13 (1), 57-63, 2012
[4] S. Lee and T. Yeo, Temperature and coning analysis of brake rotor using an axisymmetric finite element technique, Proc. 4th Korea-Russia Int. Symp. on Science & Technology, Vol.3, pp.17-22, 2000.
[5] G. Ranjith Kumar, S. Thriveni, M. Rajasekhar Reddy, Dr. G. Harinath Gowd, Design Analysis & Optimization of an Automotive Disc Brake, International Journal of Advanced Engineering Research and Science.
[6] Haider Hussain and Dr. A.I. Khandwawala. Application of Transient Thermal Analysis For Three Feeder Design Optimization For Sand Casting, International Journal of Mechanical Engineering and Technology, 4(6), 2013, pp. 241-248.
[7] P. Dufr´enoy and D. Weichert. A thermomechanical model for the analysis of disc brake fracture mechanisms. Journal of Thermal Stresses, 26(8):815–828, 2003.
[8] S. Panier, P. Dufr´enoy, and D. Weichert. An experimental investigation of hot spots in railway disc brakes. Wear, 256(7-8):764–773, 2004.
[9] T.K. Kao, J. W. Richmond, and A. Douarre. Brake disc hot spotting and thermal judder: an experimental and finite element study. International Journal of Vehicle Design, 23(3):276–296, 2000.
[10] Belhocine, A., & Bouchetara, M. (2012). Thermal analysis of a solid brake disc. Applied Thermal Engineering, 32, 59-67.
[11] Radhakrishnan, C, Yokeswaran, K., Vengadeshprasadh, M., Vishnuhasan, A., Vimalraj, T., & Velusamy, M. (2015). Design and analysis of disc brake with titanium alloy. International Journal of Innovative Science, Engineering & Technology, 2(5), 1044-1050.
[12] Shinde, V. V., Sagar, C. D., & Baskar, P. (2014). Thermal and structural analysis of disc brake for different cut patterns. International Journal of Engineering Trends and Technology, 11(2), 84-87.
[13] Huang, Y. M., & Chen, S. H. (2006). Analytical study of design parameters on cooling performance of a brake disk (No. 2006-01-0692). SAE Technical Paper.
[14] Radhika S, Sathiyamurthy S (2017) Analysis of structural and thermal characteristics of disc brake with circular/oval shaped holes for two wheeler. IJEI 6:20–27
[15] Amit M. Patil, Sachin S. Kudte Springer Nature Switzerland AG 2020 P. M. Pawar et al. (eds.), Techno-Societal 2018.
[16] Chetan Babasaheb et al. Static structural and thermal analysis of brake disc with different cut patterns. **J. appl. res. technol**, México, v. 16, n. 1, p. 41-52, 2018.

FAULT DIAGNOSIS IN BELT TRANSMISSION USING WAVELET ENVELOPED SPECTRUM

Sujesh Kumar[1], Dr. Lokesha M[2], Kiran Kumar M V[3] and Dr. Ramachandra C G[4]
1,2,3Department of Mechanical Engineering, Mangalore Institute of Technology and Engineering, Mangaluru, India.
4Department of Mechanical Engineering, Presidency University, Bengaluru, India.

E-mail: sujesh@mite.ac.in

Abstract: - In the recent years, there is a need for the evolution of methodologies for vibrational analysis by condition monitoring. In this regard, wavelet analysis has become a tool to efficiently detect the features of any vibration signal. The application of wavelet analysis for finding fault in the belt drive is presented in this paper. Vibration signal has been obtained from a healthy and faulty belt drive using a belt drive testing apparatus for experimental studies. The fault diagnostic capability of wavelet analysis has been compared by processing an experimental data. The experiment is done for various working conditions of belt drive. The advantages of wavelet analysis is shown in comparison with Fast Fourier Transform.

Keywords: Fault Diagnosis, Vibration signature; Fast Fourier transform; Continuous wavelet transform; Vibration Measuring Techniques; Envelope power spectrum; Wavelet;

1. Introduction

The machines have made the life easy. It has become the most important part of any activity. Machines reduces the effort that is required to achieve any task. It is very important to have good longevity for the machines. It can be achieved by regular maintenance. The regular maintenance also gives reliability and safety. The maintenance period depends on the behaviour and health of the machine. Scientists are trying to maximize the machine work. The condition monitoring is the technique, which helps to make the machine to work for maximum time by predicting its time of failure. There are many techniques in condition monitoring. Vibration monitoring is one of them. It analyses the running condition of the machine and identifies the faults in it. The belt conveyors are used mainly in coalmines. It consists of many idlers. The daily production of the industry is affected by the faults in belt conveyors. Diagnosis methods are used to determine the faults in the idlers. The belt drive test setup is used for experimental investigation and the report is presented in this paper.

2. Literature review

The machineries in the industries may develop fault in various parts and these faults reduces the productivity and increases the maintenance cost. The production can be increased and the maintenance cost can be decreased by using the predictive maintenance technique. The malfunctioning and breakage of the system during operation can be reduced by developing a suitable condition monitoring technique. Early detection of fault can help in avoiding serious damage to the complete system. The vibration based condition monitoring techniques are broadly used in many mechanical components like gears, bearings [1], [2]. There are different techniques like time; frequency and time-frequency domain used in the analysis of vibration signal [3], [4], [5]. The statistical characteristics such as impulse factor, shape factor, crest factor, standard deviation, peak level, Kurtosis are used in the time domain techniques. Fast Fourier Transform of time domain, cepstrum, power spectrum etc. is used in frequency domain techniques. These techniques assume that there is a stationary signal of vibration. The FFT is used in analyzing harmonic signals because it represents a signal with finite time duration.

It is poor in analyzing transient signal components since it has a constant time and frequency resolution [6]. The non-stationary signal are produced due to the fault. The information about the machine faults can be obtained from these non-stationary components. Hence, non stationary signal needs to be analyzed. The time frequency technique helps in the time evolution of frequency content [7], [8]. The non-stationary signals can be analyzed by using new time variant techniques. The energy distribution in can be described by the time and frequency distribution. The STFT can be used to produce the energy density spectrum known as spectrogram. The fast variations obtained from narrow time windows results in poor resolution of frequency. Analysis of the signal becomes complicated while using signal with low and high frequency components [3]. The WVD can provide better resolution as compared to the STFT. The main deficiency of the WVD is cross term interference [3], [9]. The transient signal characteristic features of non-stationary vibration signals produced by the faults can be extracted by the wavelet transform (WT) in time and frequency domain. The wavelet analysis includes the identification of similarity in the wavelet base function. The fault related features are extracted by analyzing the wavelet coefficients. There are many wavelet-based functions with high sensitivity for the mechanical fault detection. The fault features in rolling elements, such as ball bearing and gear are extracted by using Morlet wavelet and Impulse wavelet. The fault detection process is then enhanced by optimizing the wavelet parameters based on maximum Kurtosis. [14], [15]. It is also used for the diagnosis of intake valve of an internal combustion engine [17]. At present there is no online condition monitoring system. The Morlet wavelet is not flexible in signal analysis, but it provides smoothing. The Haar wavelet does not provide smoothing, but it is very easy to implement.

3. Experimental details

3.1 Continuous Wavelet Transfer

The enlarged or compressed versions of the basic function and translation are a family. A continuous wavelet transform is given by the equation (1) as follows.

$$W(a, b) = \int_{-\infty}^{\infty} x(t)\frac{1}{\sqrt{|a|}}\Psi * \left(\frac{t-b}{a}\right)dt \quad \text{-----(1)}$$

Where, x(t) is the finite energy signal, Ψ is the analyzing signal and 'b' is the dilation parameter. The wavelet function Ψ is stretched, when the value of 'a' is greater than one. For the value of 'a' less than one, the function Ψ is contracted along the time axis. The negative value of 'a' flips the probing function on the time axis [22]. The value of wavelet function Ψ is in the oscillatory form. Hence, it is called as wavelet. When the value b = 0 and a = 1, the base wavelet is generated. This is the natural form of the wavelet and it is termed as mother wavelet. At a given combination of scale and position 'a, b', the similarity between the waveform and the wavelet are given by the wavelet coefficients W (a, b). In wavelet analysis, a most clear approximation of the shape of the wavelet is searched over the waveform. The search may be carried out over a range of wavelet sizes. The shape of the wavelet remains the same, but the time span varies. The wavelet is designed to have net area as zero. If the wavelet has constant waveform throughout its length, then it gives zero coefficients. When the wavelet coefficients have same scale as the wavelet, they respond more strongly. However, when the wavelet coefficients resemble the wavelet, they respond most strongly.

3.2 Enveloped wavelet power spectrum

The modulating signal can be extracted from an amplitude-modulated signal by the envelope detection or amplitude demodulation technique. Hence, the time history of the modulating signal is obtained. The modulating signal has envelope analysis as the FFT frequency spectrum. The vibration signal obtained from a faulty machine part may be considered as a carrier signal. This signal has a resonant frequency of that part, which is modulated by a decaying envelope. The oscillation caused by the impact needs to be replaced with a single pulse over the complete period of the impact.
The Morlet wavelet is given by the following equation.

$$\Psi(t) = e^{-t^2}\cos\left(\pi\sqrt{\frac{2}{ln2}}\,t\right) \quad \text{-----(2)}$$

Where x (t) is the finite energy signal, Ψ (t) is the mother wavelet and Ψ*ab is the scaled and conjugate wavelet. For calculating the wavelet transform, both real and complex wavelet are employed. Hence, the result of the wavelet transform is also an analytical signal. This is shown in equation (3) and (4).

$$WT\{x(t),a,b\} \leq x(t), \Psi_{a,b}(t) \geq \frac{1}{\sqrt{a}} \int_{-\infty}^{\infty} x(t)\Psi^{*}_{a,b}(t)dt \quad \text{-----(3)}$$

$$R_e[WT(a,b)]+iIm[WT(a,b)] = A(a,b)e^{i\theta(a,b)} \quad \text{-----(4)}$$

Where Ψa, b is the family of daughter wavelet. It is defined by the dilation parameter 'a' and translation parameter b. The factor 1/√a ensures the energy preservation A (a, b) is the instantaneous envelope of the resulting wavelet transform (EWT). It extracts the slow time variation of the signal. It is given by the following equation.

$$A(a,b)=EWT(a,b) = \sqrt{\{Re[WT(a,b)]\}^2 + \{Im[WT(a,b)]\}^2} \quad \text{-----(5)}$$

A series of coefficients are produced as an inner result for each wavelet. These coefficients indicate the closeness of the signal to that particular wavelet. The wavelet scale power spectrum (energy/unit scale) may be used for the extraction of the frequency content of the enveloped correlation coefficients. It is given by the equation (6) as,

$$SWPS\ (a, \omega) = \int_{-\infty}^{\infty} |SEWT(a,\omega)|^2 d\omega \quad \text{-----(6)}$$

The Fourier transform of EWT (a, b) is given by SEWT (a, ω). The total energy of the signal x (t) is given by the following equation (7),

$$TWPS = \int |x(t)|^2 dt = \frac{1}{2\pi} \int_{-\infty}^{\infty} SWPS(a,\omega)da \quad \text{-----(7)}$$

3.3 Experimental Setup and Procedure

The vibration data is collected from the experimental setup, which is designed and fabricated as per the requirements as shown in figure 1. The driver pulley is connected to a motor of 0.5hp power. The power rating capacity is 0.37kW. Bearings are used to support shaft. Pulleys are used for mounting the belts. The outer diameter of larger pulley is 100mm. The outer diameter of smaller pulley is 50mm. Two pulleys are mounted on the motor and two pulleys are mounted on the shaft. The driver and driven pulleys are connected by a V-Belt. The entire setup is mounted on a robust wooden platform.

Figure1. Belt Drive Experimental Setup

Accelerometer (Model 621B40, IMI sensors, sensitivity is 1.02 mV/m/s2 and frequency range up to 18 kHz) is used for collecting the vibration data from the belt drive by using NI Data Acquisition Device.

Table 1. Stages of Induced Fault

Stage	Condition of the Belt	Fault description
Stage 0	Healthy Belt	No fault is induced.
Stage 1	Faulty Belt	5mm side cut in
Stage 2	Faulty Belt	2mm side cut out
Stage 3	Faulty Belt	2mm loose

Healthy condition: The experiment was conducted for a speed of driver pulley at 1451 RPM and 755RPM. The speed is measured by tachometer. The vibration signals are collected in three axes using

LabView7.1 software. The signals collected in healthy condition of the system are used for comparing the signals at faulty condition.

Faulty condition: The belt was induced with three kinds of faults for the study and analysis under different speeds. One fault was induced at inner surface of the belt and is known as Side-Cut-In. The second fault was induced at the outer periphery of the belt and is known as Side-Cut-Out fault. The third fault is loose fault, which is induced at the outer side of the belt with slight loose condition.

Belt frequency calculation

The Belt speed can be obtained by the formula,

$$\text{Belt speed} = \frac{\pi\ X\ PS\ X\ PD}{BL} \quad (8)$$

Where PS = Pulley speed = 1451RPM and 755RPM
PD = Pulley diameter = 0.1m
BL = Belt Length = 36inch = 0.9144m
The higher speed of belt is 498RPM. The belt frequency is 8.3Hz.
The lower speed of belt is 259RPM. The belt frequency is 4.3Hz.

3.4 Wavelet Power Spectrum Implementation

The localized fault in the belt drive is detected by using several applications that helps to visualize the performance of the proposed approach. The fault condition is indicated by the increment in the number and amplitude of sidebands. The three dimensional accelerometer is used for obtaining the time domain signal from the experimental setup. The signal hence obtained is processed using various signal-processing techniques.

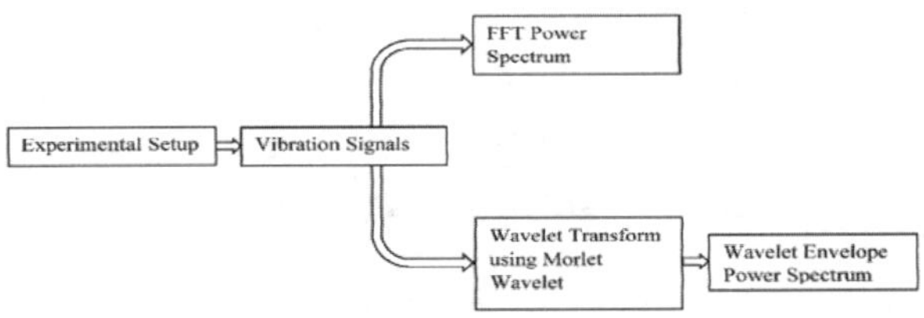

Figure 2: Methodology of envelope power spectrum implementation

4. Results and Discussions

A. FFT Power spectrum for high speed in X direction

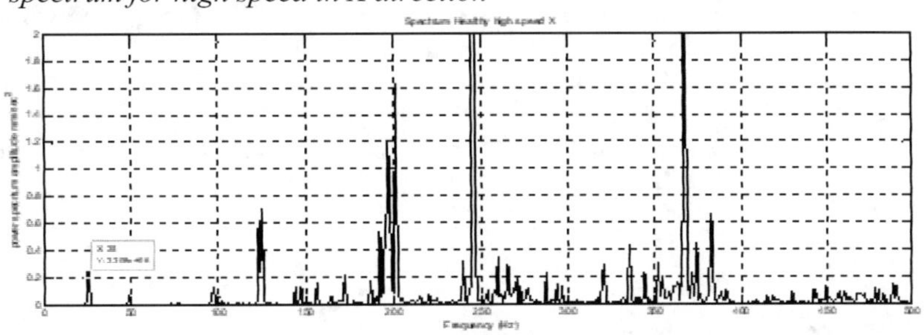

(a)

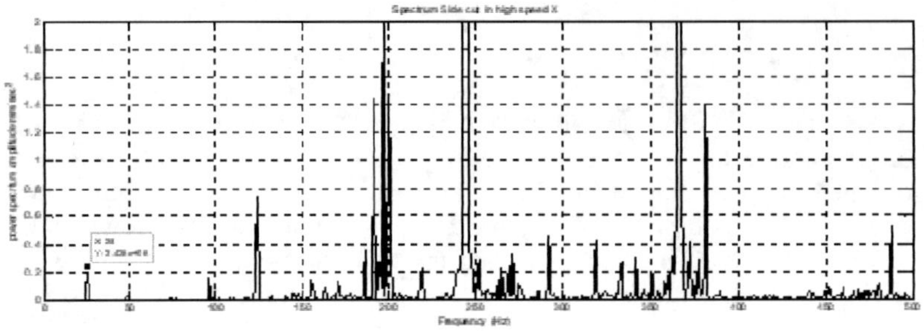

(b)

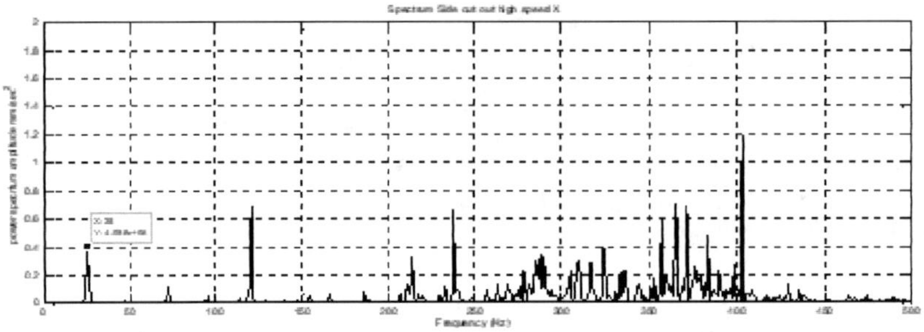

(c)

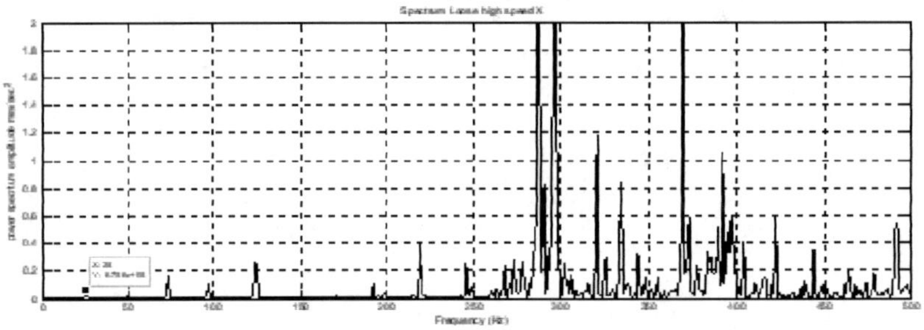

(d)

Figure 3: FFT Power spectrum for high speed in X direction. (a) Healthy, (b) Side cut in, (c) Side cut out, (d) Loose condition.

The above figure 3 shows the FFT Power spectrum, which helps to identify the fault in the belt by comparing the vibration signal with the healthy condition. The FFT power spectrum gives a poor reading of the amplitude and it cannot recognize the small amplitude.

B. Morlet Wavelet Based Enveloped Power Spectrum for high speed in X direction

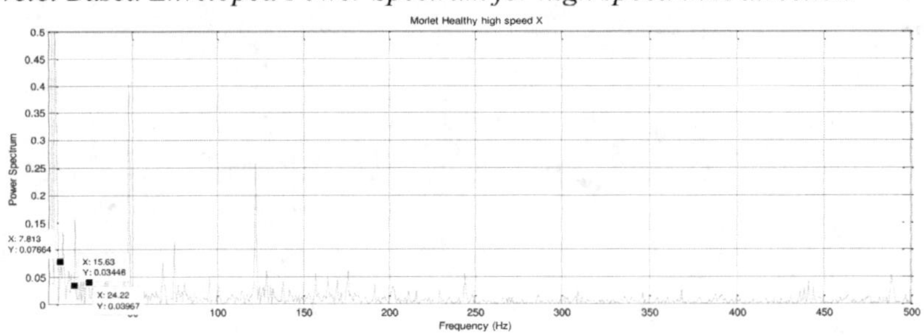

(a)

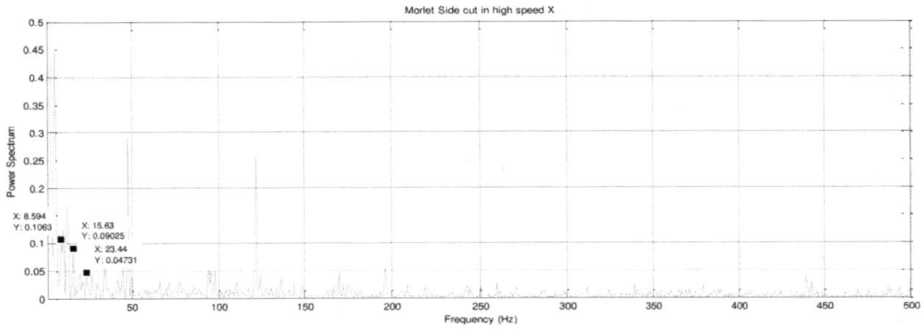

(b)

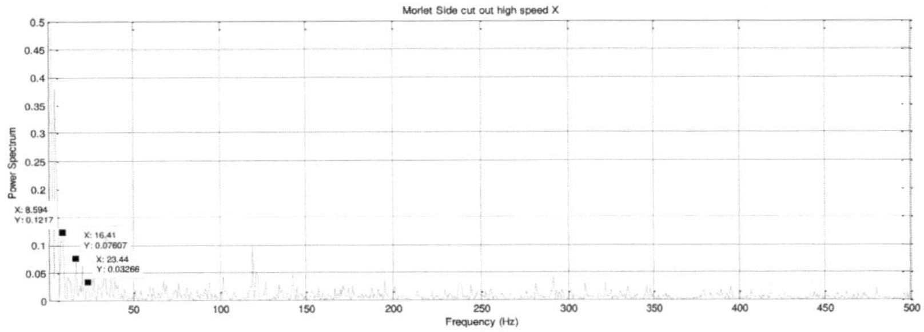

(c)

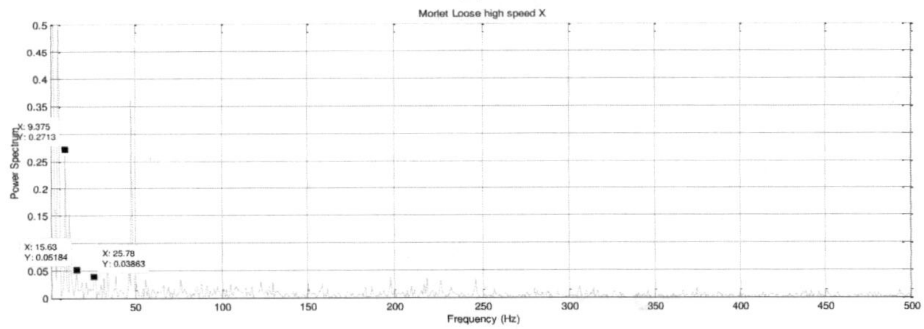

(d)

Figure 4: Morlet Wavelet Based Enveloped Power spectrum. (a) Healthy, (b) Side cut in, (c) Side cut out, (d) Loose condition

Figure 4 shows Morlet Wavelet Based Enveloped Power spectrum for healthy and faulty belt conditions. The Morlet Wavelet Based Enveloped Power Spectrum gives a better reading of the vibration amplitude and hence helps to compare the faulty condition with the healthy condition.

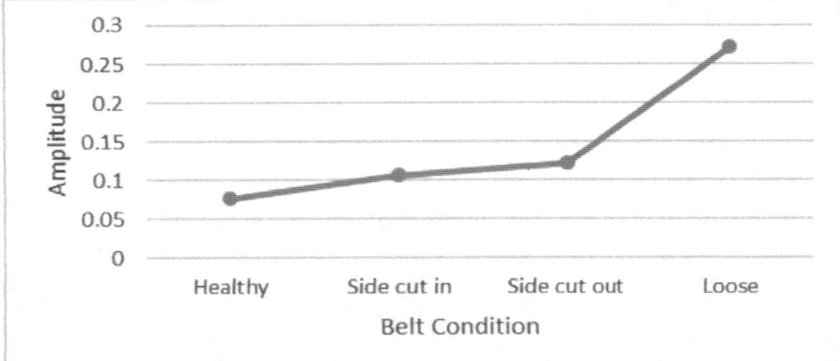

Figure 5: Comparison of amplitude of belt frequency for different belt conditions.

IOP Conf. Series: Materials Science and Engineering 1013 (2021) 012025 doi:10.1088/1757-899X/1013/1/012025

The above figure 5 shows the comparison of amplitude of vibration under different faults in the belt. The amplitude was minimum for the healthy condition. The amplitude increases while inducing fault in it. The loose belt was found to be having maximum vibration.

C. FFT Power spectrum for low speed in X direction

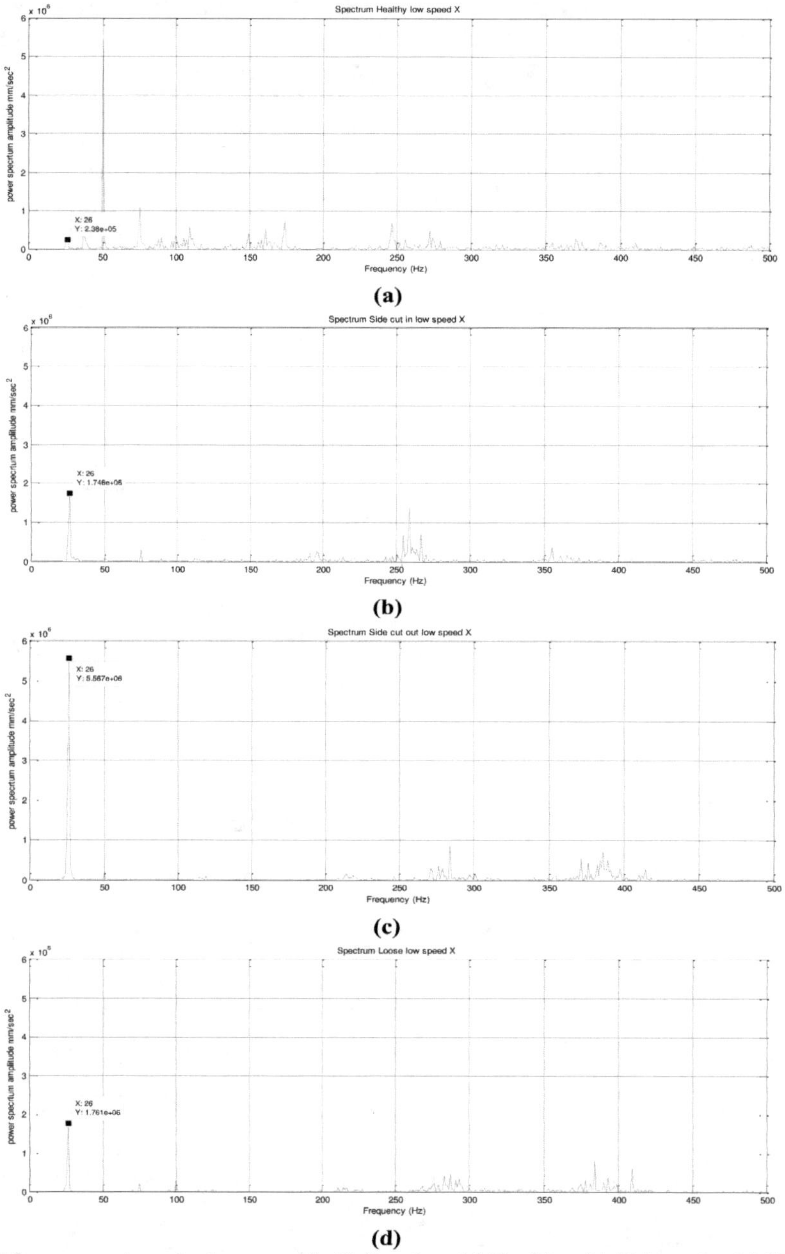

Figure 6: FFT Power spectrum for low speed in X direction. (a) Healthy, (b) Side cut in, (c) Side cut out, (d) Loose condition.

The above figure 6 shows the FFT Power spectrum. It can be used to identify the fault in the belt by comparing the vibration signal with the healthy condition. The amplitude of vibration was found low at lower speed.

D. *Morlet Wavelet Based Enveloped Power Spectrum for low speed in X direction*

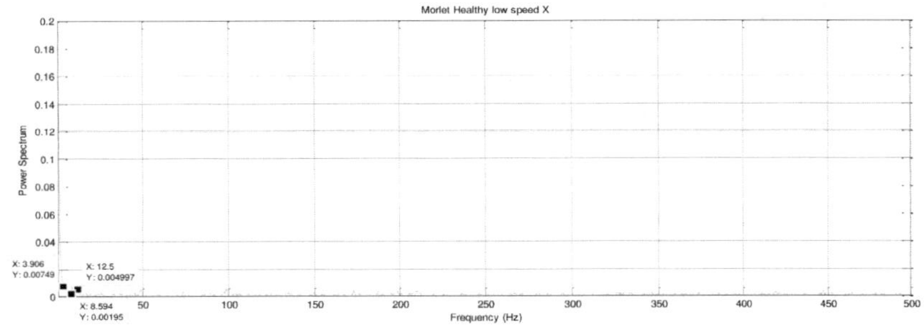

(a)

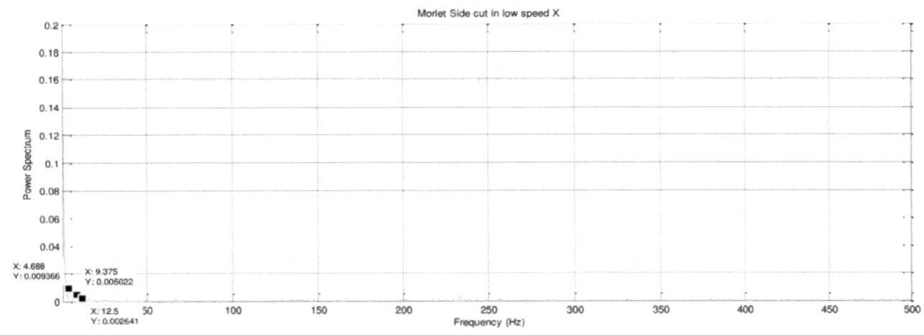

(b)

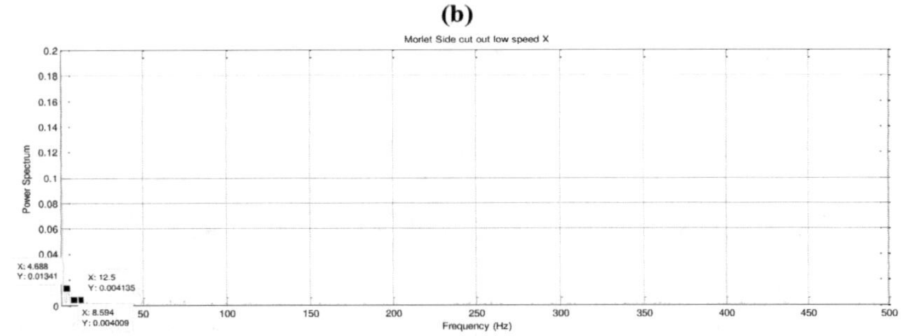

(c)

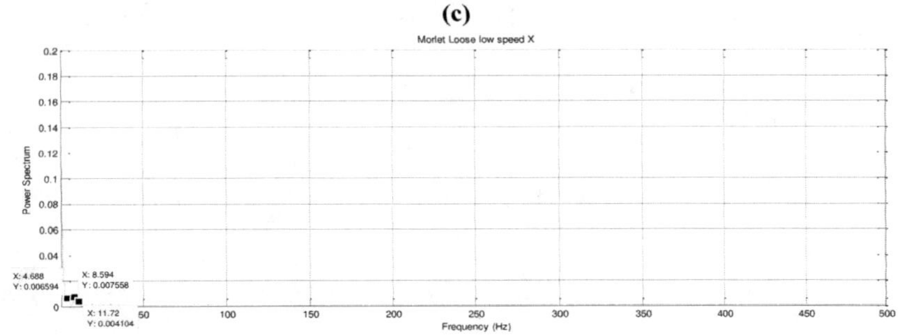

(d)

Figure 7: Morlet Wavelet Based Enveloped Power spectrum for low speed in X direction. (a) Healthy, (b) Side cut in, (c) Side cut out, (d) Loose condition

Figure 7 shows Morlet Wavelet Based Enveloped Power spectrum for healthy and faulty belt conditions at low speed. The Morlet Wavelet Based Enveloped Power Spectrum gives a better reading of the vibration amplitude and hence helps to compare the faulty condition with the healthy condition.

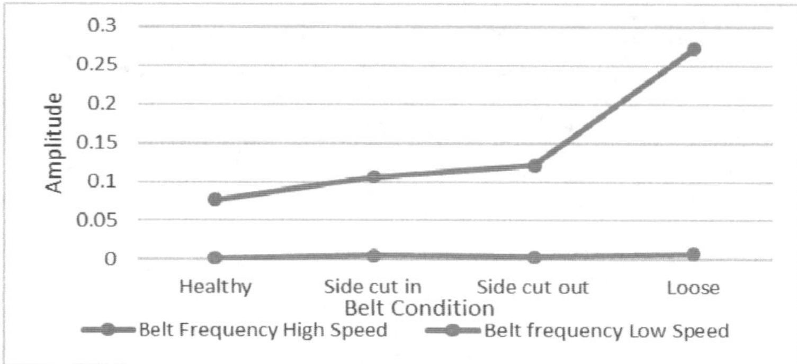

Figure 8: Comparison of amplitude of belt frequency for different belt speed.

In the above figure 8, a comparison has been made for 2 different speed of the belt. The amplitude of the belt was small for the lower belt speed and it increases with increase in the belt speed.

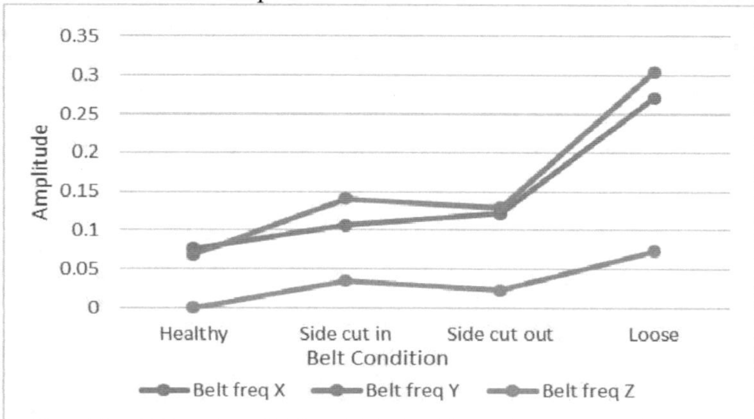

Figure 9: Comparison of amplitude of belt frequency measured in X, Y, Z directions.

The amplitude of vibration, which is directly proportional to the degree of fault, also depends on the direction of measurement. The experiment shows that the horizontal X direction and vertical Y direction gives better reading of the vibration as compared to the axial Z direction.

5. Conclusions

The vibration signal is processed by using signal-processing techniques, such as Fast Fourier Transform and Morlet wavelet enveloped power spectrum. The results are compared for three different faults induced in the belt. The result shows that the Morlet wavelet enveloped power spectrum is more sensitive to the fault as compared to the Fast Fourier Transform. The amplitude of the vibration was found to be maximum for the loose condition. The vibration level was less during the low speed of the belt. The vibration level increases with increase in the belt speed. The vibration signals were recorded in all the 3 directions and the quality of the signal was found to be good in the horizontal and vertical directions.

References

[1] Omar, Farag K., and A. M. Gaouda. "Gearbox diagnostics using wavelet-based windowing technique." In *Journal of Physics: Conference Series*, vol. 181, no. 1, p. 012089. IOP Publishing, 2009.

[2] Abu-Mahfouz, Issam A. "Gearbox health monitoring experiment using vibration signals." *Journal of Engineering Technology* 24, no. 1 (2007): 26.

[3] Wändell, Johan. "Multistage gearboxes: Vibration based quality control." PhD diss., KTH, 2006.

[4] Choy, F. K., S. Huang, J. J. Zakrajsek, R. F. Handschuh, and D. P. Townsend. "Vibration signature analysis of a faulted gear transmission system." (1994).

[5] Saravanan, N., and K. I. Ramachandran. "Incipient gear box fault diagnosis using discrete wavelet transform (DWT) for feature extraction and classification using artificial neural network (ANN)." *Expert Systems with Applications* 37, no. 6 (2010): 4168-4181.

[6] Peter, W. Tse, Wen-xian Yang, and H. Y. Tam. "Machine fault diagnosis through an effective exact wavelet analysis." *Journal of sound and vibration* 277, no. 4 (2004): 1005-1024.

[7] Dalpiaz, G., A. Rivola, and R. Rubini. "Gear fault monitoring: Comparison of vibration analysis techniques." In *Proceedings of the 3rd International Conference on Acoustical and Vibratory Surveillance Methods and Diagnostic Techniques*, vol. 13, pp. 623-637. 1998.

[8] Saravanan, N., VNS Kumar Siddabattuni, and K. I. Ramachandran. "Fault diagnosis of spur bevel gear box using artificial neural network (ANN), and proximal support vector machine (PSVM)." *Applied Soft Computing* 10, no. 1 (2010): 344-360.

[9] Peng, Z. K., and F. L. Chu. "Application of the wavelet transform in machine condition monitoring and fault diagnostics: a review with bibliography." *Mechanical systems and signal processing* 18, no. 2 (2004): 199-221.

[10] Purushotham, V., S. Narayanan, and Suryanarayana Prasad. "Multi-fault diagnosis of rolling bearing elements using wavelet analysis and hidden Markov model based fault recognition." *Ndt & E International* 38, no. 8 (2005): 654-664.

[11] Shi, D. F., W. J. Wang, and L. S. Qu. "Defect detection for bearings using envelope spectra of wavelet transform." *Journal of vibration and acoustics* 126, no. 4 (2004): 567-573.

[12] Nikolaou, N. G., and I. A. Antoniadis. "Demodulation of vibration signals generated by defects in rolling element bearings using complex shifted Morlet wavelets." *Mechanical systems and signal processing* 16, no. 4 (2002): 677-694.

[13] Junsheng, Cheng, Yu Dejie, and Yang Yu. "Application of an impulse response wavelet to fault diagnosis of rolling bearings." *Mechanical Systems and Signal Processing* 21, no. 2 (2007): 920-929.

[14] Lin J, Zuo M.J., 2003. Gear box fault diagnosis using adaptive wavelet filter, *Mechanical Systems And Signal Processing*, Vol. 17, No. 6, pp. 1259-1269.

[15] Al-Raheem K.F, Roy. A, Ramachandran.K.P. 2006. Detection of rolling element bearing faults using optimized – wavelet denoising technique. *International Conference on Signal Processing*, Beijing, China.

[16] Freudinger L.C., Lind R., and Brenner M.J., 1998. Correlation filtering of modal dynamics using the Laplace wavelet, *Proceeding of the 16th International Modal Analysis Conferences*, Vol. 2, pp. 868-877.

[17] Yanyang Z., Xuefeng C. Zhengila H. and Peng C., 2005. Vibration based modal parameters identification and wear fault diagnostics using Laplace wavelet. *Key Engineering Materials*, Vol. 293-294, pp. 183-190.

[18] Smith C., Akujuobi C.M., Hamory P., Kloesel K., 2006. An approach to vibration analysis using wavelets in an application of aircraft health monitoring, *Mechanical Systems and Signal Processing*, Vol. 21, pp. 1255–1272.

[19] Randall R.B. 1982. A new method of modeling gear faults. *Journal of Mechanical Design*, Vol. 104, pp. 259-267.

[20] vanKhang N., Cau T.M., Dien N.P., 2004. Modeling parametric vibration of gear pair systems for a tool for aiding gear fault diagnosis, *Technische Mechanic*, Band 24, Heft 3-4, pp. 198-205.

[21] Prokop M., Mohr F. 2003. Gear box vibration analysis, *University of Applied Sciences Pforzheim*, Germany online.

[22] Semmlow J.L. 2004. Bio signal and biomedical image processing, *Marcel Dekker*, Inc. New York, USA, ISBN0-8247-4803-4

[23] Al-Raheem K.F., Roy A., Ramachandran K.P, Harrison D.H., Grainger, S. 2007. Rolling element gearing fault diagnosis using Laplace-wavelet envelop power spectrum. *EURASIP Journal on Advances in Signal Processing*, Vol. 2007, article id 73629, 14pages

[24] Al-Raheem K.F., Kareem W.A., 2011...Rolling element bearing fault diagnostics using Laplace wavelet kurtosis, *IJMSE, World Academic Publishing*, Vol.1 No.1, pp. 17-25.

VERTICAL AXIS WIND TURBINE PERFORMANCE ENHANCEMENT VIA UTILIZING THE REPULSIVE CHARACTERISTICS OF PERMANENT MAGNETS

Sandesh Hegde[1] Dr. Ramachandra C G[2]

[1]Mechanical Department, Srinivas Institute of Technology Mangluru, India
[2]Mechanical Department, Presidency University Bangalore, India

Email: sandeshh.hegde92@gmail.com

Abstract: Humans have been using wind power for past thousands of years to propel sailboats and ships. But from calendar year 1980s, in the energy sector we can see dramatic development in the form of wind energy. Today in renewable power, other than the hydropower, together with solar power, the wind is also most important source energy. Here the wind energy conversion system; consist of turbine which transforms the kinetic energy available in the wind into the mechanical energy, then the mechanical energy is boosted up with the help of gearbox and then generators are used to transform the rotational mechanical energy into the electrical energy. Now this paper mainly concentrates on the implementation of "magnetic propelling phenomenon" to improve the efficiency of the VAWT system and technology will be implemented between the rotating plane and fixed plane below the turbine, in such way that like polarities facing each other to create magnetic repulsion. The natural property of magnetic repulsion which is created between like poles of magnets can be used as a source of energy. Thus this system will operate with an additional feature like, magnetic repulsion characteristics. The force which is created as the result of the magnetic repulsion will also adds kinetic energy to the turbine while converting the wind kinetic energy into the rotational mechanical energy by the turbines, Hence here our PMP-VAWT system has the ability to operate in the both high & low wind speed environment condition with better efficiency due to the implementation of magnetic repulsive characteristics, and here our choice is to demonstrate the efficiency of PMP-VAWT system as compared to the Traditional VAWT system.

Keywords: Wind Energy, Vertical Axis, Efficiency, Turbine, Magnets, Repulsion.

1. Introduction

In wind energy conversion system, wind energy is harnessed by transforming the wind kinetic energy available in the wind into the rotational mechanical energy. Then this rotational mechanical energy boosted up by using gearbox and then used to turn generators shafts to generate electricity [2]. Here in our paper we mainly concentrated on the implementation of the "permanent magnet's propelled VAWT system which has potential to run under both the high & low wind environment. Here our course of action is to compare efficiency of both PMP-VAWT system & Traditional VAWT system in the varying wind speed conditions.

2. Wind Sail Outline Composition

Normally the wind turbine system mainly consists of a set of blades, rotor shaft, and generator with gear box. When the air with kinetic energy pass through the turbine blades will result in change in pressure & this causes the rotational force at the turbine rotor shaft. When blade rotates, the kinetic energy available in wind is transformed in terms of rotational mechanical energy, later which boosted up in gear box and used to rotate generator shaft to produce electrical energy. Here the horizontal axis wind turbine & vertical axis wind turbine are the two main types of wind turbines. The horizontal axis wind turbine is like pedestal fan, Rotor shaft is arranged in vertical manner in VAWT, as its plane of rotation is vertical and will spins in a cylindrical in motion, which is older and lesser known in family of wind machines family [4]. The VAWT system provides diverse benefits over the standard HAWT system which is common in world due to its efficiency. VAWT can be classified in to two types: Darrieus & Savonius.

The savonius WT system is a drag based VAWT System, and will operate on principle of paddles. S-Shaped cross section rotors are fabricated in the savonius type rotor system. And the turbine which are based its operation on drag principle will have a much higher torque & little less rotation speed at its shaft [3]. Hence here after studying both kind of VAWT's rotors arrangements, finally we marked to support the substructure of our outline on savonius model and the design is visualized in Figure 1. As design was attained with a triangular shape cut out shape from aluminium metal sheet, we implemented a bit different outline in turbine blades, by adapting a curvature design from top of the sail to bottom of the sail at its outer line while resembling the standard outline of savounis models.

Figure 1: Modified Savonius Turbine

The scoop which was created due to implementation of magnetic propelling section will be eliminated by implementing curvature outline in design and to allow it for smoother torque during turbine rotation.

3. Magnetic Propelling Section

Here the magnetic propelling phenomenon implemented based on the repulsive characteristics of permanent magnet. The Nd-Fe-B (neyodymium iron born) magnet and substantial supports are used to experience the magnetic repulsion, array type settings is provided in such way that the like polarities of magnets face each other [6]. Hence the magnetic repulsion created and will adds some kinetic force to turbine while transforming kinetic energy available in wind into the mechanical energy.

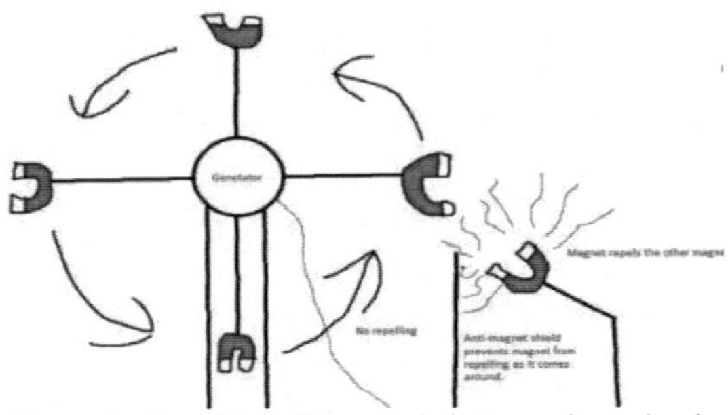

Figure 2: Magnetic Repulsion Characteristic to produce circular motion.

In our work, magnetic repulsion technology concept is enacted to cause the higher efficiency of turbine system [8]. The magnets are arranged in array like structure in between the base of the turbine i.e on rotating plane and fixed plane on supper structure when the same polarities face each other in arrangement, the magnetic repulsion will be experienced and this repulsion will give some kinetic energy to the rotary system. This force which was created due to the magnetic repulsion will also adds up some kinetic energy to the turbine while transforming kinetic energy available in wind into rotary mechanical energy and then in to electrical power [5].

4. Fact-Finding Methodology

Total power in the main wind stream is corresponding to the air density and the cross sectional area which is nothing but swept area of turbine which is perpendicular to the air flow direction & the wind speed of main wind stream [8]. That which can be written as follows,

$$Pw = 0.5\rho AV^3 \quad (1)$$

Power in main Wind Stream (W/m^2) = P_w
Air Density of main wind stream ρ = 1.2 Kg/m^3
Turbine swept area /cross sectional area which is perpendicular to wind stream = 0.174m^2
Main wind stream average velocity (m/s) = V

Table I: Wind power at various wind velocity

Sl. No	V= Average Wind Velocity (m/s)	Pw=0.5ρAV3 (w/m²)
1	6	22.55
2	4.5	9.51
3	3.2	3.42
4	2.8	2.29

Now, the turbine mechanical output power (P_T) can been written by considering the function of tangential force (F) on wind turbine and Turbine shaft rotational speed per Minute (N_r) in RPM.

$$P_T = 2\pi NrF/60 \quad (1)$$

Tangential force (F) = Mass of the turbine x turbine angular acceleration
Now,
Angular Acceleration = $\frac{acceleration\ (a)}{radius\ of\ the\ urbine\ (r)}$

The change in velocity of the turbine = v-u (rev/sec)
Angular acceleration = $\frac{a}{r}$ (rev/m^2)
Mass of the turbine = 3.1kg
Nr = Turbine Revolution per Minute

Now the efficiency of the turbine can be considered as power coefficient (C_p) which is given by the ratio of mechanical output power of the turbine to the total power which can be accessed from the main wind stream at the cross section area of the wind turbine [6]. Hence Power Coefficient C_{P} can be written as follows

$$Cp = \frac{P_T}{Pw} \quad (3)$$

The positioning of permanent magnets is experimented with various orientation combinations. The practical experiments are carried out at various wind speed condition.

Table II: Efficiency of PM- Propelled VTWT at different orientation of magnets

Sl. No	V (m/s)	Magnets Orientation	P_T	Efficiency = P_T/ P_W (%)
Traditional - Vertical Axis wind turbines				
1	6	Traditional Vertical Axis Wind Turbine without Magnets	38.52	1.71
2	4.5		13.70	1.44
3	3.2		3.71	1.08
4	2.8		1.75	0.76
Permanent Magnet Propelled VAWT				
5	6	6 Sets of Magnets are	46.48	2.06

6	4.5	Placed in 180° angle Orientation (Parallel to Each other)	16.48	1.73
7	3.2		3.61	1.05
8	2.8		1.57	0.69
9	6	6 No's of Magnets are Placed in fixed Plane & 1 No in Rotary Plane in 180° angle Orientation	38.52	1.71
10	4.5		13.70	1.44
11	3.2		3.71	1.08
12	2.8		1.75	0.76
13	6	6 No's of Magnets are Placed in fixed Plane and 3 No in Rotary Plane in 180° angle Orientation	42.64	1.89
14	4.5		13.70	1.44
15	3.2		1.80	0.53
16	2.8		1.11	0.48
17	6	6 No's of Magnets are Placed in fixed Plane in 180° angle Orientation and 1 No in Rotary Plane 45° angle orientation	39.40	1.75
18	4.5		13.70	1.44
19	3.2		3.66	1.07
20	2.8		1.67	0.73
21	6	6 No's of Magnets are Placed in fixed Plane in 180° angle Orientation and 3 No in Rotary Plane 45° angle orientation	40.28	1.79
22	4.5		16.69	1.75
23	3.2		3.76	1.10
24	2.8		1.80	0.79
25	6	6 No's of Magnets are Placed in fixed Plane in 45° angle Orientation and 3 No in Rotary Plane 45° angle orientation	45.43	2.01
26	4.5		17.51	1.84
27	3.2		5.64	1.65
28	2.8		3.30	1.44
29	6	6 Sets of Magnets are Placed in 45° angle Orientation	67.42	2.99
30	4.5		24.21	2.54
31	3.2		6.72	1.96
32	2.8		3.66	1.60

4. Simulation

Team Mathwork has been developed an application called MATLAB simulink with additional feature like Simulation & Linking, which offers modeling of a system, and simulation & analyzing the same in a multi domain of a dynamic system under GUI Interface environment. Here the models of multi domain dynamic system are easily constructed by just a click on selection box with suitable customized set of libraries block and these library block will allows matrix operation, creating data table for analysis and plotting of those functional data's on graph by implementing the customized algorithm. Hence we have selected MATLAB Simulink software for creating experimental model.

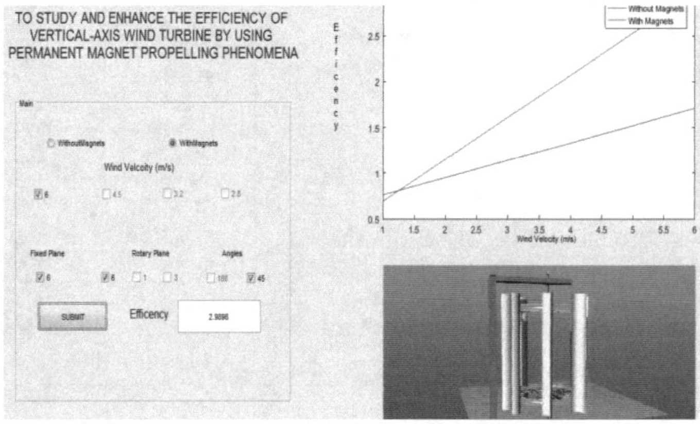

Figure 3: Simulation counterpart via Simulink MATLAB application

4. Results and Discussions

The attraction and repulsion characteristics of permanent magnets can be considered as a source of energy. The repulsion characteristic of permanent magnets will also adds up some sort of kinetic energy to the turbine while transforming the kinetic energy available in main wind stream to rotation mechanical energy [7].

For performance analysis comparison in consider Table 1 & Table 2, and its observed that the Permanente Magnet Propelled- VAWT have an efficiency of 2.99% which is improved when compared to Traditional- VAWT having an efficiency of 1.77%. So now efficiency got increased due to the implementation of magnetic propelling section by using permanent magnets.

5. Conclusions

In the present situation there is no doubt that the energy production in the world must be increased drastically on immediately on continual basis. Un-even heating of the earth surface which caused by earth rotation will create change in air pressure which creats wind energy and in 20th century we are facing a global warming means temperature on earth is increasing rapidly, hence there will be a change in air pressure which will results in air in motion . Hence there won't be a scarcity in wind energy in future. Here additional option of free energy concept from permanent magnets looks like cutting edge technology, but it's feasible at given appropriate condition and in future it will become more economical

References

[1] N. Talijan, J. Staji-Tros, T. Zak, V. Menushenkov, A. Milutinovi- Nikoli, Nd-Fe-B Alloys Synthesis-Structure-Properties, Proc. European Congress & Exhibition on Powder Metallurgy PM 2001, Nice, France, 2 (2001) 280-284
[2] World Wind Energy Association (WWEA), "World wind energy report 2010", (April 2011).
[3] N.Vaughn, "Renewable Energy & the Environment ", CRC Press, Ed. 2009 pp 63 -101
[4] Gary L. Johnson "Wind Energy Systems", Electronic Edition Manhattan, October 10, 2006, pp. 61-70-157
[5] J. P. Hennessey, Jr., "Some aspects of wind power statistics, and performance analysis of a 6MWwind turbine-generator". J. Appl. Meteorol., vol. 16, no. 2, pp. 119–28, Feb. 1997.
[6] Er. Girt, K. M. Krishnan, G. Thomas, E. Girt, Z. Altounian, Coercivity limits and mechanism in nanocomposite Nd-Fe-B alloys, Journal of Magnetism and Magnetic Materials, 231 (2001) 219 – 230
[7] A.D. Wright and L.J. Fingersh, Advanced Control Design for Wind Turbines: Part I, Design, Implementation, and Initial Tests. NREL /TP-500-42437, Golden, CO, National Renewable Energy Laboratory (2008).
[8] Pallabazzer R. "Parametric analysis of wind siting efficiency", J. Wind Eng Ind Aerodyn 2003; 91:1329-1352.

NUMERICAL INVESTIGATION OF HEAT TRANSFER IN AIRCRAFT ENGINE BLADE USING $k-\epsilon$ AND SST $k-\omega$ MODEL

Chethan R Patil1, Devendra Dandotiya2, Bhaskar Pal3, Nitin Banker4
1,2,3 Department of Mechanical Engineering, Presidency University, Bangalore, 560064, India
4 School of Engineering and Applied Sciences, Ahmedabad University, Gujarat, 380009, India
E-mail: devendradandotiya@presidencyuniversity.in

Abstract - Even the smallest part engineered in gas turbines adds on hugely to its output performance. The blades form a major part which extracts energy from high temperature and high-pressure gases produced from the combustion chamber. This high temperature may cause the blades to undergo creep or fatigue failure eventually. The blade material would not withstand such high temperatures. With an intention to overcome this thermal barrier, various cooling technologies like jet impingement, film cooling, mist assisted film cooling, pin fin cooling, transpiration cooling have been adapted in the gas turbines. Using these cooling technologies, the temperature on the blades can be reduced by 200 to 300 K and higher performance may be achieved without much of a compromise. This study mainly focuses on how much the temperature can be reduced if only convective cavity cooling was used, wherein a hollow blade is used and cold fluid passes through it. Computational Fluid Dynamics (CFD) in ANSYS was used to compute the temperature difference on the blade and high temperature region. The computations were carried out on both CFD and CFX. Two different turbulence models were considered, k-epsilon and k-omega SST (Shear Stress Transport). Analysis were carried out with nearly 3 million elements. Velocity, temperature flows at selected regions are reported.

Keywords - Gas turbines, blades, combustion chamber, high temperature, cooling technology, jet impingement, film cooling, pin fin cooling, convective cooling, CFD, Ansys, CFX, K epsilon, K omega SST (Shear Stress Transport)

1.Introduction

Gas turbines also called combustion turbine, is an internal combustion engine which works on the principle of Brayton cycle. These gas turbines are used in many processing plants and industries for power generation, widely used for aircraft propulsion to provide thrust [1]. Safety and efficiency have to be of prime importance as it finds application in major areas. The air which enters the engine, is first compressed in the compressor, which adds on to its pressure. Later fuel is allowed to combine with this pressurized air and ignited, which adds on to its energy and temperature. This pressurized high temperature air is fed onto the series of blades in turbine which extracts energy and this extracted energy is converted to useful power depending on the application .The energy extracted by the blades is partially used to run the compressor and the remaining is used to do external work .The external work maybe thrust in case of aircrafts ,a propeller in ships or a generator used in industrial areas [2].

In order to survive the harsh environment in turbines like extreme temperature at high rpm, blades are manufactured by superalloys, which have high melting points. One of the earliest materials ever introduced was Nimonic which was used in British Whittle engines. In the modern era, blades are manufactured using Nickel superalloys with some percentage of chromium, cobalt and rhenium. Inconel is another widely used material to manufacture blades .IN-738 was first used by GE specifically for land-based turbine for the second series or stage of turbine blades [5].

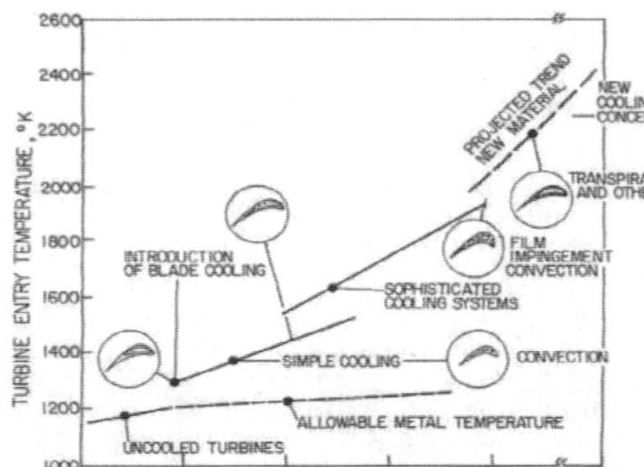

Fig.1 Evolution of Cooling Methods [6]

Cooling methods like jet impingement and film cooling use air or water as primary fluid to achieve heat transfer between the blade and the coolant, whereas mist assisted cooling which is a new research uses air along with minute droplets of water which achieves more cooling than using air alone. Employment of water in cooling may lead to corrosion of the parts, leakage, choking and other problems. Using air at high pressure and higher mass flow rates may bring about turbulence in the main stream bringing down efficiency. About 1-3% of main flow is generally used to avoid such problems [7].

There are cooling techniques which use either water or air as cooling medium. Nano fluids with higher heat capacity or any other mixture of fluids can be used as coolant. Such synthesized coolants may possess density higher than air and water, so passing it through holes or multiple passes maybe difficult. In order to reach a higher level of research and interaction of these fluids, a basic research with a simple blade design and traditional method of cooling i.e. convective cooling with air as coolant is employed. The coolant is passed through the hollow blade where heat transfer happens between itself and the blade surface. Heat flows through the blade (conduction) and later this heat conducted heat is lost to the coolant which acts a sink through convection. An improvised model of convection cooling is that the coolant can be allowed to stay for a longer time within the blade by creating multiple passes in the hollow region of the blade. This improved method brings about more heat transfer, hence more cooling can be achieved.

2. Geometric Model and Working Conditions:

The blade profile used is the Rolls Royce Conway blade profile but with a different cooling technique. It is a widely and commonly used blade profile in modern gas turbines. The blade has a cavity in its centre so as to facilitate the flow of coolant through it. The blade geometry was modelled in Solid works. The material used for the blade is nickel and the cooling medium used is Air. The surrounding hot domain is also air, relatively at a higher temperature. The blade material as mentioned above is Nickel (Ni) bearing properties thermal conductivity of 91.74w/m-k, specific heat (C_p) of 460.6 J/kg-K, density of 8900 kg/m^3.The air in the hot domain representing air from the combustion chamber has temperature of 1560 K and air depicting the coolant flowing through the cavity of blade has temperature of 750 K, with density 1.225 kg/m^3 ,thermal conductivity 0.0242 W/m-K and specific heat(Cp)1006.43J/kg-K [8] .

Table 1. Properties of hot domain

Temperature	1560K
Thermal Conductivity(k)	0.0242w/m-K
Specific heat (Cp)	1006.43J/kg-K
Density(ρ)	1.225kg/m^3

Table 2. Properties of cold domain

Temperature	750K
Thermal Conductivity(k)	0.0242w/m-K
Specific heat (Cp)	1006.43J/kg-K
Density(ρ)	1.225kg/m^3

Table 3. Properties of blade material

Thermal Conductivity(k)	91.74w/m-K
Specific heat (Cp)	460.6J/kg-K
Density(ρ)	8900kg/m3

3. Meshing:

Meshing for the model was carried on in Ansys Model-Meshing component systems of Ansys software. The model consists of 2363493 (2.3million) elements and 620843 (0.6million) nodes. An image of the meshing has been shown below in Fig.2. The number of elements was increased to 3.5 million, with an intention to witness any variation in results. But there was no appreciable change, so the model with 2.3 million elements was used for further analysis.

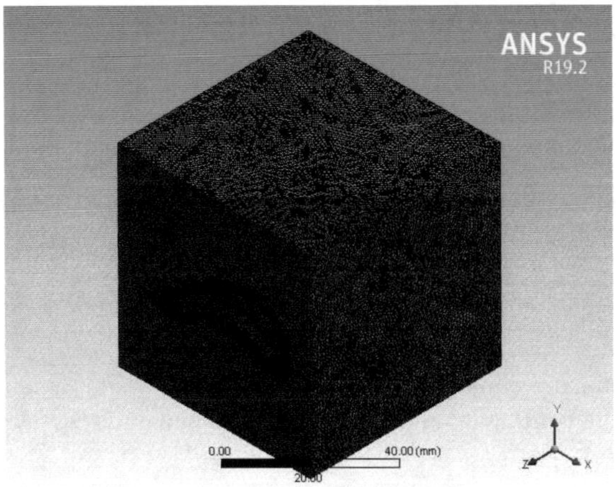

Fig.2 Meshed Component

4. Computational Model and Governing Equations:

As mentioned, the analysis is carried out in CFD and Ansys Fluent. The analysis mainly focuses on comparing the results obtained from K-Epsilon model (CFX) and K-Omega SST turbulence model (Ansys Fluent). Analysis can be carried out theoretically or computationally which abides by certain governing equation for different types of turbulence model. The total energy equation which comprises of the conservation of mass equation, the conservation of momentum equation and the conservation of energy equation is solved internally to yield results in any kind of fluid dynamics problem. Corresponding governing equations for CFX are [8],

$$\frac{D\rho}{Dt} + \rho\frac{\partial u_i}{\partial x_i} = 0 \tag{1}$$

$$\rho\frac{Du_i}{Dt} = -\frac{\partial P}{\partial x_i} + \frac{\partial \tau_{ij}}{\partial x_j} + \rho F_i \tag{2}$$

$$\rho\frac{De}{Dt} = -p\frac{\partial u_i}{\partial x_i} + \emptyset + \frac{\partial}{\partial x_i}(K\frac{\partial T}{\partial x_i}) \tag{3}$$

where

$$\emptyset = \tau_{ij}\frac{\partial u_i}{\partial x_j} \tag{4}$$

$$\tau_{ij} = \mu(\frac{\partial u_i}{\partial x_j} + \frac{\partial u_j}{\partial x_i} - \frac{2}{3}\frac{\partial u_r}{\partial x_r}\delta_{ij}) \tag{5}$$

The turbulence model SST (Shear Stress transport) is based on two equations. This model manages calculations of both k-ε and k-ω thus tackling adverse pressure gradients and turbulence properties.

$$\frac{\partial k}{\partial t} + U_j\frac{\partial k}{\partial x_j} = P_k - \beta^* k\omega + \frac{\partial}{\partial x_j}[v + \sigma_k v_T)\frac{\partial k}{\partial x_j}] \tag{6}$$

$$\frac{\partial \omega}{\partial t} + U_j\frac{\partial \omega}{\partial x_j} = \alpha S^2 - \beta\omega^2 + \frac{\partial}{\partial x_j}\left[(v + \sigma_k v_T)\frac{\partial k}{\partial x_j}\right] + 2(1 - F_1)\sigma_{\omega 2}\frac{1}{\omega}\frac{\partial k}{\partial x_i}\frac{\partial \omega}{\partial \omega_i} \tag{7}$$

The transport equation in transitional SST model is governed by the following equation

$$\frac{\partial(\rho k)}{\partial t} + \frac{\partial(\rho u_j k)}{\partial x_j} = \tilde{P}_k - \tilde{D}_k + \frac{\partial y}{\partial x}[(\mu + \sigma_k\mu_t)\frac{\partial k}{\partial x_j}] \tag{8}$$

$$\tilde{P}_k = \gamma_{eff}P_k \tag{9}$$

$$\tilde{D}_k = \min(\max(\gamma_{eff}, 0.1), 1.0)D_k \tag{10}$$

The concept of thermal conductivity plays a major role in the analysis. The high temperature of hot gases from the combustion chamber is absorbed by the solid blade material through thermal convection. The convective heat transfer as per Newton's law of cooling:

$$q = h(T - T_\infty) \tag{11}$$

Thereafter the heat gets conducted from the topmost layer of the blade to the underlying layers through thermal conduction. The interaction of high temperature gas molecules with the blade molecules creates a temperature gradient between the layers of the blade. Heat transfers along the gradient of descending temperature, also sometimes referred to as thermal diffusion. Thermal conduction is best explained by Fourier's equation:

$$q = -\lambda \frac{dT}{dx} \tag{12}$$

A layer of TBC (Thermal barrier coating) on the blade helps to reduce the effect of heat on the blade. The heat from the high temperature gas is transferred to the coating first, later onto the solid blade material. Then the heat gets transferred to the coolant flowing through the cavity. Such TBCs help to increase the life of the blade to a certain extent, but levies high maintenance charges as it undergoes wear and tear very often.

5. Cooling Effectiveness:

In order to know how effective the employed cooling method is, a parameter known as cooling effectiveness is determined. Cooling effectiveness is the ability of the coolant to reduce the blade temperature. The value scales from 0 to 1. The value reaches 1 when the blade temperature reaches the coolant temperature, indicating efficient cooling and is 0 when the blade temperature reaches the main stream temperature. Cooling effectiveness can be calculated using,

$$\varepsilon = \frac{T_g - T_m}{T_g - T_{ci}} \tag{13}$$

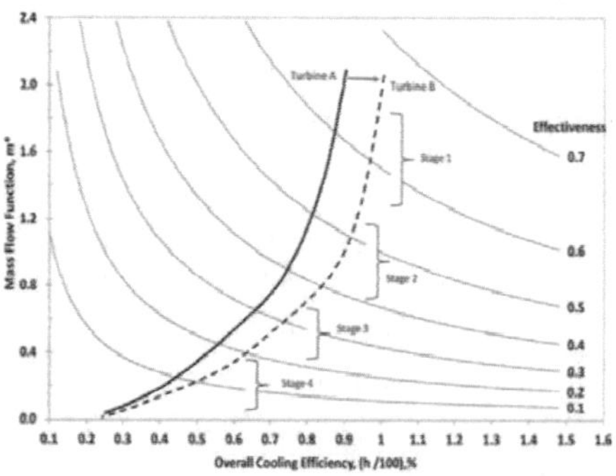

Fig.3 Cooling Efficiency and Effectiveness at Different Stages

6. Results and discussions:

The figure 4 shows comparison of the results generated in Ansys Fluent and CFX wherein the same model with the same number of elements was used. No significant variation was noticed out of these results. The iso-surface generated in CFX gave out relatively more accurate visuals around the blade.

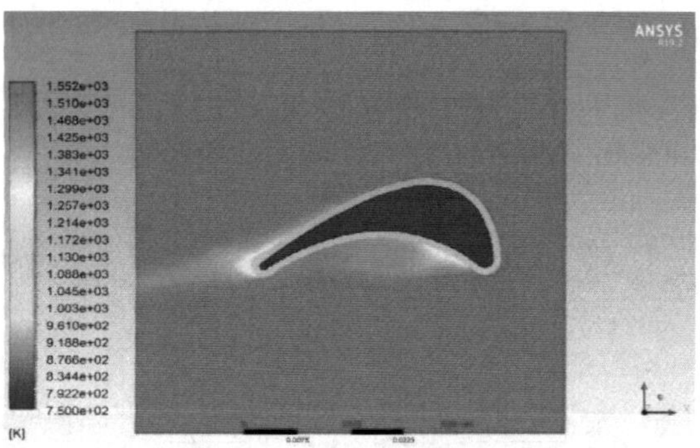

Fig.4(A) Iso- surface generated in Ansys fluent

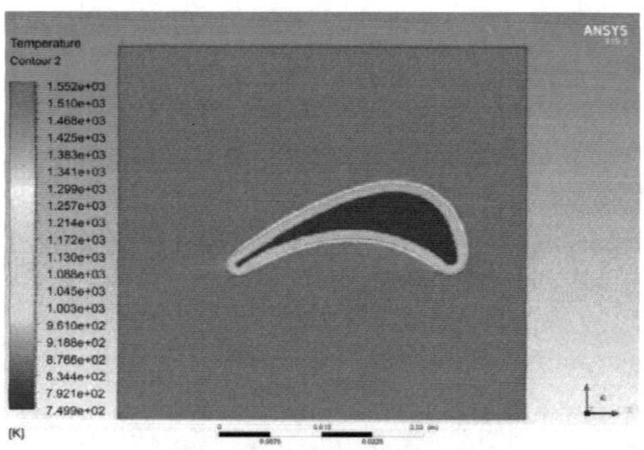

Fig.4(B) Iso- surface generated in Ansys CFX

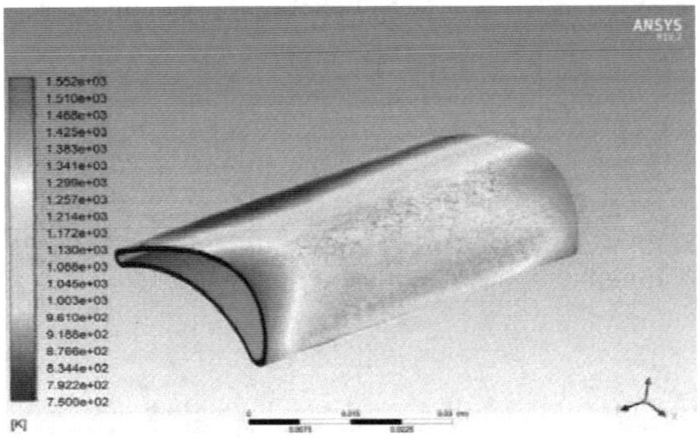

Fig.5(A) Blade Contour Generated in Ansys Fluent

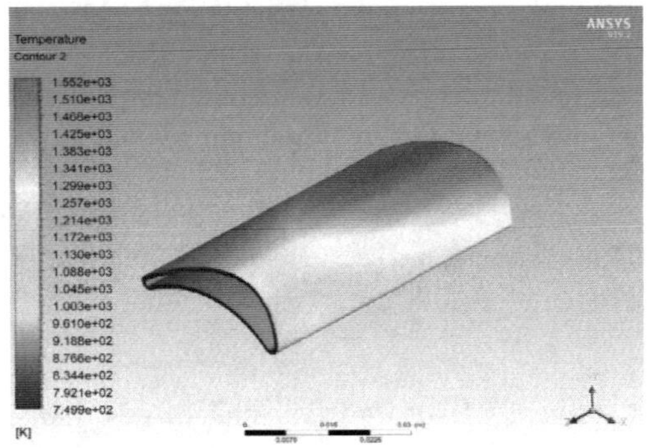

Fig.5(B) Blade contour generated in CFX

The figure 5 show cases the temperature contours of blade, These contours show region of high and low temperatures. Regions of high and low temperatures can be analyzed through these contours and further research can be carried out to increase effectiveness at high temperature region.

Fig.4 shows one of the possible hot spots at the pressure surface near the trailing edge of the blade where the temperature is comparatively higher than other points. It is also known as the stagnation point where the gas flow halts for a moment and thereby this area experiences combined effect of kinetic and static energy that the gas carries. A temperature of 948K was observed at the leading-edge hot spot whereas the other points where lying at a temperature ranging from 850K to 920K. Another most probable point of hot spot is at the trailing edge.

7. Conclusion and Future Research:

The present analysis on gas turbine blade with convective method of cooling generated cooling effectiveness of 0.5. This level of effectiveness is not sufficient to have high efficiency and power output of the gas turbine on the whole. The current model can be made efficient by using different coolant medium with a higher heat capacity. But this may require refabrication of the turbine. So, an upgraded model of convective cooling with multiple passes helps increase the efficiency. An updated model of cooling technology is Film cooling where the coolant gushes out of the blade through small holes creating a layer, thereby reducing the blade temperature.

Gas turbines are majorly used in Airplanes. Commercial flights fly typically at a range of 6 to 8 miles above the Earth surface. The temperature ranges from +15 °C to -50 °C. The idea here is to somehow direct this cold atmospheric air onto the blades or through the blades so that the temperature can be reduced. The air has to be directed towards the blade in such a way that the air doesn't come in contact with the main stream air or the combustion chamber. This may yield more efficiency than the existing cooling method technologies.

8. References

[1] Zohuri, B., and Zohuri, B., 2015, "Gas Turbine Working Principles," Combined Cycle Driven Efficiency for Next Generation Nuclear Power Plants, Springer International Publishing, pp. 147–171.

[2] Çengel, Y. A., and Boles., M. A., 2011, 9-8. Thermodynamics: An Engineering Approach, McGraw-Hill, New York.

[3] Dambrine, B., and Mascarell, J. P., 1988, "Designing Turbine Blades for Fatigue and Creep," Def. Sci. J., 38(4), pp. 413–429.

[4] "NPTEL :: Aerospace Engineering - Turbomachinery Aerodynamics" [Online]. Available: https://nptel.ac.in/courses/101101058/. [Accessed: 03-May-2020].

[5] Schilke, P. W., 2004, "Advanced Gas Turbine Materials and Coatings, General Electric Company," GE Energy, p. 30.

[6] Rice, T., 2017, "Numerical Analysis Of The Film Cooling Effectiveness On A Highly Loaded Low Pressure Turbine Blade In Conjunction With Endwall Effects."

[7] Yahya, S. M., 2010, Turbines Compressors and Fans, Tata McGraw-Hill.

[8] ANSYS, 2015, "ANSYS Fluent Theory Guide", ANSYS 16.2

CFD ANALYSIS ON HEAT TRANSFER ENHANCEMENT IN A PIPE IN PIPE HEAT EXCHANGER WITH TANGENTIAL INJECTION

Ketan Vinayak Warghat [1], Sathish Kumar D [1] and Dr. S Jayavel [1]
[1] Department of Mechanical Engineering, Indian Institute of Information Technology Design and Manufacturing, Kancheepuram Chennai, India, 600127
E-mail: ketan.v.warghat@gmail.com

Abstract: - Heat transfer enhancement is a significant requirement in process industries. There are various passive and active methods available to reinforce the heat transfer. In the present study, tangential injection and its effect on the advancement in heat transfer rate are analysed. This is an active technique for enhancing heat transfer accompanied by negligible increment in pressure drop. Design and operating parameters like geometry, pipe material, working fluid, and flow rate play a vital role in heat transfer characteristics. Taking into account the design and operating parameters, the present study is aimed to optimize the orientation of the injecting nozzle. FLUENT software is employed to simulate the heat transfer and flow characteristics. A thorough grid independence study has been administered, and the present results are validated using the author's previous experimental work. The tangential orientation considered is 30°, 45°, and 60°. The results show 60° inclination of the nozzle for parallel flow gives maximum overall heat transfer coefficient.

Keywords: Heat Exchanger, Tangential Injection, Heat Transfer Enhancement

1. INTRODUCTION

The heat exchanger is a device used for efficient heat transfer between multiple mediums which are at different temperatures. Heat exchanger employs the fact that energy flow occurs whenever there is a temperature difference between two mediums. So, the thermal energy flows from high energy medium to a low energy medium. They are widely employed in refrigerators, air conditioners, power plants, chemical process industry, food processing industry, and ice making plants. The general sample of a heat exchanger is car radiator which works as across flow heat exchanger which takes away the heat energy from the coolants and gives away the heat to the environment, it works as a cooling device in cars.

A. Classifications of heat exchanger

Classification according to construction:

1. Tubular-Pipe in Pipe, Shell & tube, Spiral tube
2. Plate type
3. Extended surface
4. Regenerative

B. Pipe in pipe heat exchanger

A simplest heat exchanger is a circular duct inside another larger circular duct as shown in Figure 1, where one fluid flows through the inner duct and the other fluid flows over the annulus space between the two ducts. The wall between the two fluids acts as a heat transfer surface. It can be operated as parallel flow as well as counter flow, but the counter flow gives more efficiency.

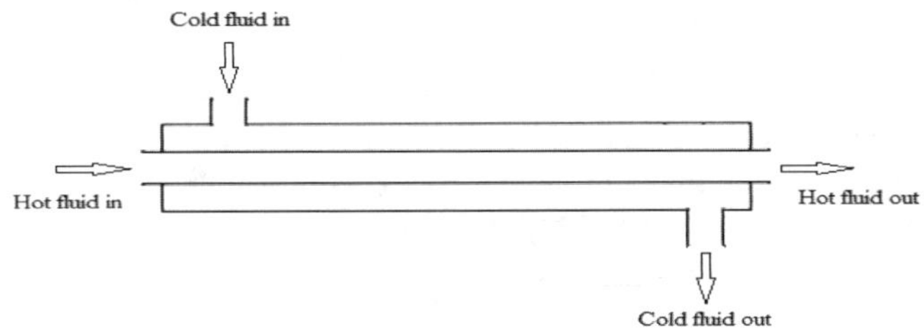

Figure 1:Pipe in Pipe Heat Exchanger.

C. Heat transfer improvement techniques

Heat transfer improvement techniques are very crucial for laminar flow as the heat transfer rates are low for laminar flow in a plane duct. The heat transfer rate can be typically enhanced by instigating turmoil in the flow which can be attained by including twisted tape or turbulator inserts in the duct. The Enhancement techniques can be divided into three different categories:

1. Passive Techniques.
2. Active Techniques.
3. Compound Techniques.

Passive technique uses internal design modification by introducing inserts or other obstacles. The inserts or obstacles introduce disturbance in the fluid flow results in increase in the heat transfer coefficient with the consequence of pressure drop. Due to the extended surface the heat transfer area increases. In the Passive Technique no external power is required. The heat transfer improvement by this technique can be attained by using:

- Inserts
- Treated Surfaces
- Rough surfaces
- Extended surfaces
- Additives for liquids

The active technique is much compound when it comes to designing the heat exchanger it as it requires external power for fluid input to give improved flow patter to enhance the heat transfer rate. It has limited applications due to the requirement of external sources. Enhanced heat transfer rate by active technique can be attained in the following ways:

- Mechanical Aids
- Surface vibration
- Fluid vibration
- Electrostatic fields
- Injection
- Jet impingement

The compound technique represents the amalgamation of active or passive technique with the sole objective to provide an improved thermohydraulic effect on heat exchanger.

Different works have been done by numerous specialists on heat transfer augmentation in heat exchanger. Akpinkar et al. [1] experimentally studied the pipe in pipe heat exchanger with a swirl generator at the inlet. The heat transfer rate in the heat exchanger with a swirl generator is 30 % more compared to the normal pipe. He concluded the pipe gives a higher heat transfer rate with a number of holes with a smaller diameter. Baig M et. al. [2] experimentally studied a concentric tube heat exchanger with twisted tape, twisted tape with a hole, and baffles to increase the heat transfer rate. It can be concluded that, there is an 8.9 % more heat transfer rate with twisted tape inserts, while inserts with hole and baffles gives a slighter rate compared to simple twisted tape. [2] Jawarneh [3] experimentally studied the system by giving swirl motion at the entrance of the pipe in the annulus space, from the experiment it is concluded that the swirls parameter at the entrance of the pipe have positive effect on the heat transfer characteristics. Swirl motion give better mixing of the fluid. The thermal and fluid property in the annular space was studied by finite volume method. Vijay K. Dhir et. al. "experimentally investigated a process for improving the efficiency of heat exchanger in which multiple tubes are arranged in one tube and incorporating injectors along the circumference to give tangential flow at the inlet. The injectors are provided to each of the inner tubes along the circumference to give tangential flow to the fluid. A notable improvement in the heat transfer rate was noticed due to modification in the design." [4]. Kreith et.al. [5] experimentally investigated the heat transfer enhancement in double tube heat exchanger using inserts which increase the Nusselt Number by 25%. Abdullah [6] experimentally studied the double pipe heat exchanger with convergent divergent with twisted tape insert which and concludes that the concentric convergent and divergent profile and twisted tape was 52% to 280% greater than the plain heat exchanger. Chang et al. [7] experimentally investigated the heat transfer in pipe having a tangential injection of air through the nozzle having a uniform the temperature at the outer pipe, he concluded that tangential injection gives swirl motion to the fluid results in an increase in performance.

From the above literature it can be perceived that turbulence in fluid flow gives more heat transfer rate. The method with insert and system with tangential injection was studied from which twisted tape gives high heat transfer rate with an increase in pressure drop. The working fluid can be water, air, or any nanofluids. The rate of increase of heat transfer rate depends on Nusselt number and the mass flow rate or Reynolds number, it also depends on the material used, geometric structure, twist ratios in inserts, etc.In the present scenario, the productivity is a key strategy's for the manufacturing industries to compete in the competitive world or in the global markets. The main aim to achieve increase the productivity with the availability of the raw materials and some of the very good high precession technologies and maintenance system used in the world-class industries.

2. PROBLEM STATEMENT

Heat transfer improvement in a pipe in pipe heat exchanger with multiple tangential entries through the nozzle at different angles along the length of the pipe with water as a working fluid.

3. COMPUTATIONAL DOMAIN/GEOMETRY

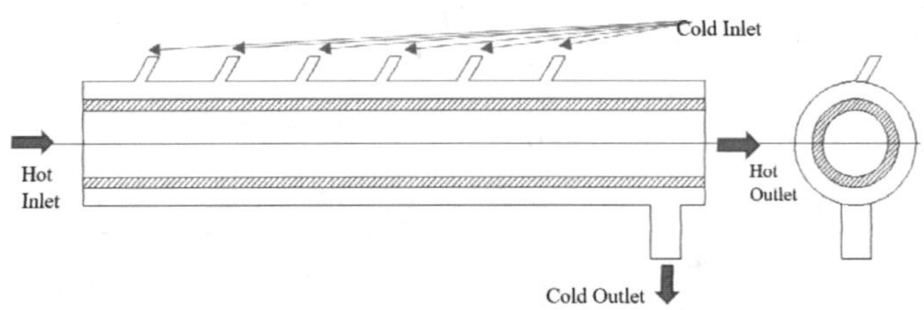

Figure 2: Computational Domain.

Cylindrical concentric pipe heat exchanger with pipe made of aluminium is used as an inner pipe, PVC as an outer pipe having insulation at the top with water used as a working fluid. The fluid is injected in the annulus space with the nozzles on the outer pipe with cold fluid as inlet and inner pipe having hot fluid as working a medium. Six nozzles are used as a cold fluid inlet.

Length-1000 mm Outer radius-21.08 mm
Inner radius-10.75 mm Nozzle Diameter- 4.5 mm No. of Nozzles- 6

4. MATHEMATICAL MODEL

The flow in the pipe is considered as incompressible, turbulent, and steady. The heat transfer between the hot water to the cold region is modelled as conjugated heat conduction with fluid flow. The wall material and working fluid properties were fixed depending on the working temperature, and the vapor density of the working fluid followed the perfect-gas law.

The elliptic class of the conservation equations are solved as a single domain conjugate problem. Body force and gravity terms were neglected.

It follows mass, momentum and energy conservation equations.

Mass Conservation Equation

$$\nabla.\vec{V} = 0 \qquad \text{(1)}$$

Momentum Equation

$$\rho(\vec{V}.\nabla\vec{V}) = -\nabla p + \nabla.(\mu\nabla\vec{V}) \qquad \text{(2)}$$

Energy Equation

$$\rho C_p(\bar{V}, \nabla T) = k\nabla^2 T \qquad \text{(3)}$$

Heat Transfer in hot and cold fluid section

$$Q_h = m_h \times C_{ph} \times (T_{hi} - T_{ho}) \qquad \text{(4)}$$

$$Q_c = m_c \times C_{pc} \times (T_{co} - T_{ci}) \qquad \text{(5)}$$

Log mean temperature difference

$$\Delta T_m = \frac{\Delta T_1 - \Delta T_2}{\ln\left(\frac{\Delta T_1}{\Delta T_2}\right)} \qquad \text{(6)}$$

Where, $\Delta T_1 = T_{hi} - T_{ci}$ $\Delta T_2 = T_{ho} - T_{co}$

Overall heat transfer coefficient

$$U_o = \frac{Q_{avg}}{A \times \Delta T_m} \qquad \text{(7)}$$

Where, $A = \pi \times d_o \times L$

V= Volume (m^3)	k= Thermal Conductivity	C_p= Specific Heat (j/K)
ρ= Density (kg/m^3)	m_h, m_c= Mass flow rate	T_{hi}= Hot inlet temperature(K)
T= Temperature (K)	T_{ho}=Hot outlet temperature(K)	T_{co}, T_{ci}= Cold fluid inlet and outlet Temperature
A= Surface Area (m^2)	ΔT_m= LMTD	U_o= Overall Heat Transfer Coefficient

5. SIMULATION SETUP AND BOUNDARY CONDITION

Flow and turbulent equations are solved simultaneously to capture the flow field. The energy equation is solved to obtain the temperature field. Appropriate thermophysical properties are assigned for the fluids. Hot fluid flows through the inner pipe with a constant 300 LPH flow rate whereas cold fluid flows through the annular space with flow rate varying from 300 LPH to 600 LPH. The pressure outlet condition is given at the outlet for both cold and hot fluids. The outer pipe is insulated to prevent heat loss.

6. RESULTS AND DISCUSSION

Temperature distribution along the length of the pipe. With hot fluid inlet 0.1111 kg/s and outlet fluid inlet as 0.125 kg/s. Thermal and Material properties were defined for each material from property chart.

The thermal energy from the hot fluid through aluminium pipe transfers to cold fluid which results in an increase in the temperature of the cold fluid. LMTD is calculated by the outlet and inlet temperatures of cold and hot fluids respectively. Averaged heat transfer coefficient values are calculated at the surface using CFD post and related temperature variation graphs were plotted.

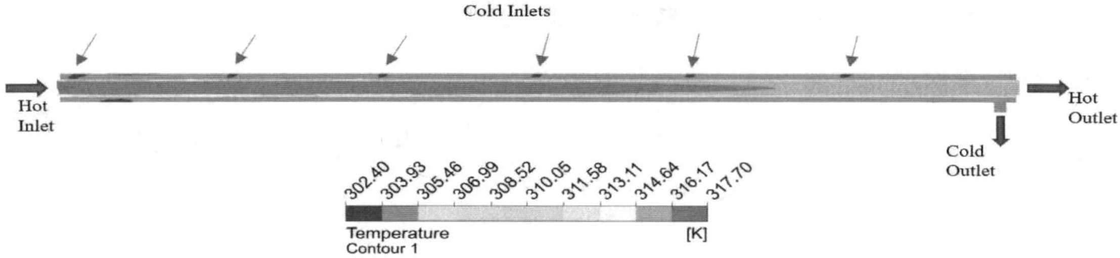

Figure 3:Temperature Contour.

Figure 3 shows the temperature distribution on the mid plain of the pipe the inner tube temperature decreases along the length and the temperature of cold water increases through the length of the pipe in annulus space. **Figure 4** shows the streamline through the annulus space showing a swirl motion around the hot pipe to extract more heat when compared to the normal fluid flow.

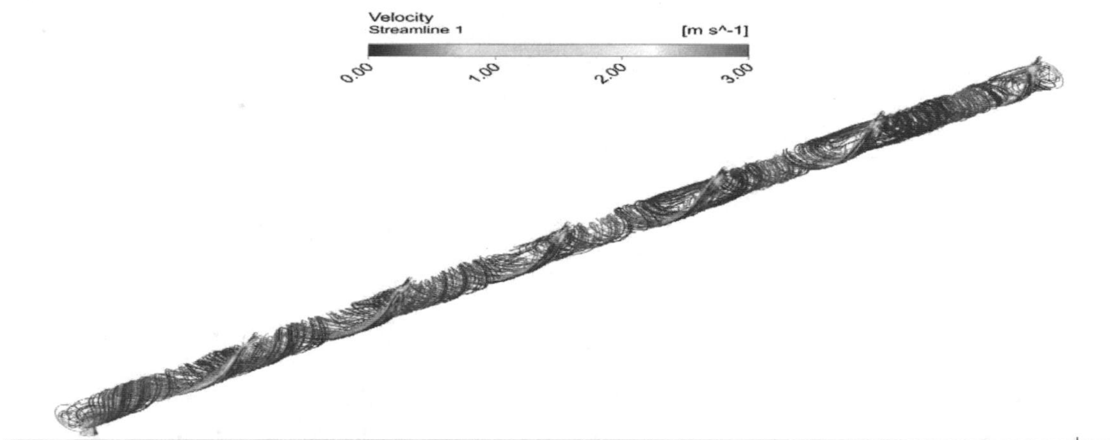

Figure 4: Streamline in annulus space.

Figure 5 display the validation of numerical results using experimental results taken at nozzle with an inclination of 60° with a parallel flow configuration which gives a maximum deviation of 2.17% for 600 LPH cold water flow rate.

Figure 6 shows the comparison of overall heat transfer coefficient with flow rate for tangential nozzle inclined at 60° configuration. It is noticed that, the average overall heat transfer coefficient increases, with the increment in mass flow rate. For 60° angle tangential entry for parallel flow give more overall heat transfer coefficient compared to counter flow, for higher Reynolds number the overall heat transfer coefficient is almost the same with a small deviation.

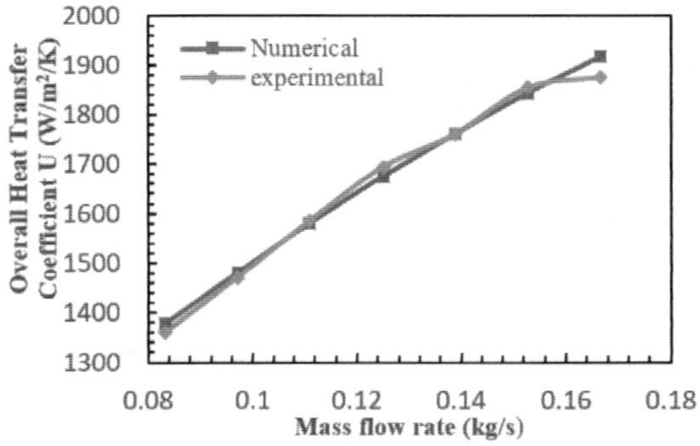

Figure 5: Validation of numerical results with experimental results at 60° inclination parallel flow.

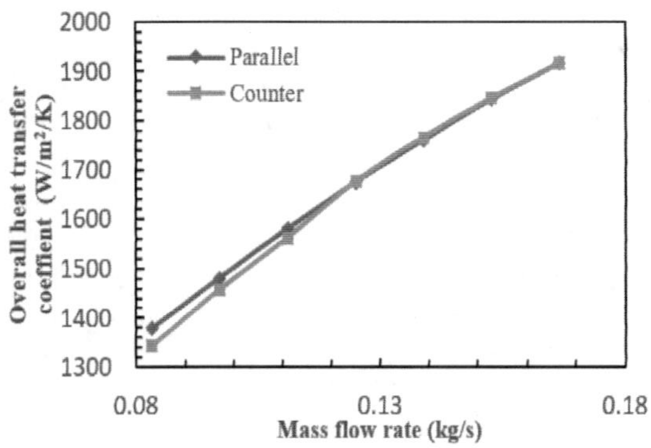

Figure 6: Correlation of overall heat transfer coefficient with flow rate at 60° inclination at cold inlet.

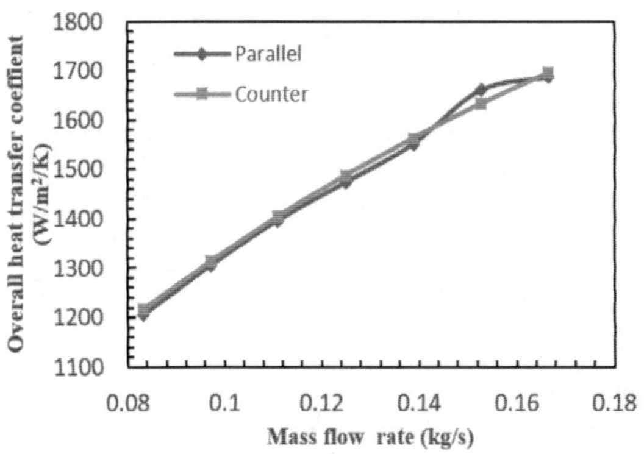

Figure 7: Correlation of overall heat transfer coefficient with flow rate at 30° inclination at cold inlet.

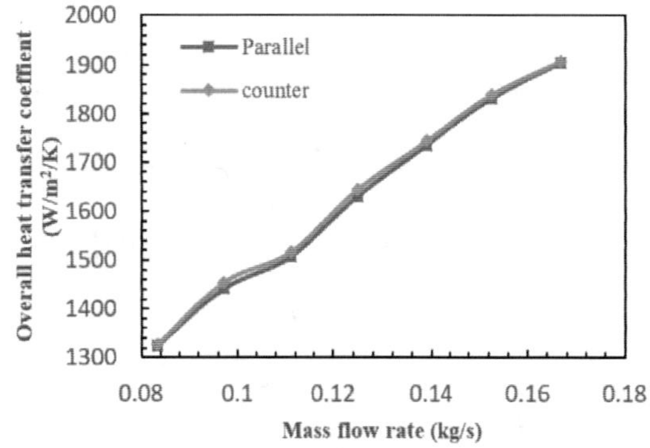

Figure 8: Correlation of overall heat transfer coefficient with flow rate for 45° inclination at cold inlet.

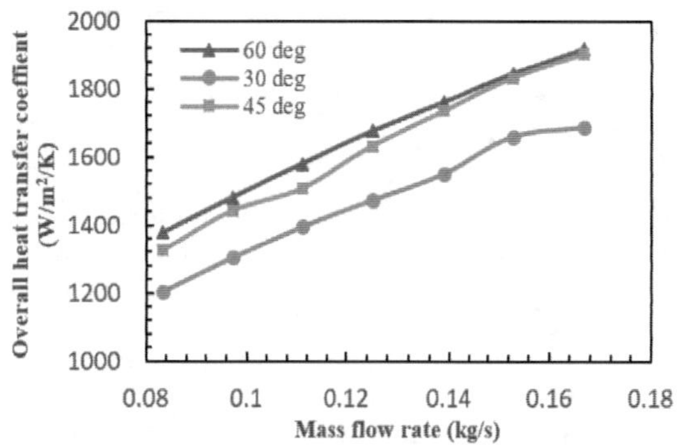

Figure 9: Correlation of overall heat transfer coefficient with flow rate for parallel flow at different inclination angle at cold inlet.

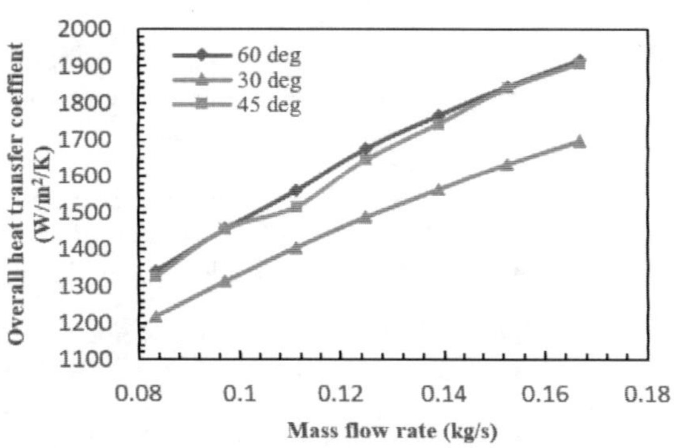

Figure 10: Correlation of overall heat transfer coefficient with flow rate for counter flow at different inclination angle at cold inlet.

Figure 7 shows the correlation of overall heat transfer coefficient with the flow rate for the tangential nozzle inclined at 30° configuration. It is noticed that, the overall heat transfer coefficient increase with the increment in mass flow rate. For 30° angle tangential entry, counterflow gives a more overall heat transfer coefficient to that for parallel flow at lower flow rate. Overall heat transfer coefficient is almost the same with a small deviation with the increment in mass flow rate. A little hike in the overall heat transfer coefficient for 550 LPH for counter flow is observed.

Figure 8 shows the correlation of overall heat transfer coefficient with the flow rate for tangential nozzle inclined at 45° configuration. It is noticed that, the average overall heat transfer coefficient increases with the increment in mass flow rate.

Figure 9 shows the correlation of overall heat transfer coefficient with flow rate for tangential nozzle inclined for parallel flow configuration. It is noticed that, the average overall heat transfer coefficient increases with the increment in mass flow rate. While comparing with different inclination angles, parallel flow at 60º gives 2.7 % more overall heat transfer rate than 45º inclination and 11.66 % more than 30º inclination.

Figure 10 shows the correlation of overall heat transfer coefficient with flow rate for tangential nozzle inclined for counterflow configuration. It is noticed that, the average overall heat transfer coefficient also increases with the increment in mass flow rate. For counterflow the efficiency is more for 60° angle of injection compared to 45° and 30°, it gives 10.8 % more overall heat transfer coefficient matched with 30° inclination.

7. CONCLUSION

Overall heat transfer rate can be augmented by giving turbulence inflow which gives better mixing and gives more convective heat transfer coefficient. From the present and previous studies, it can be concluded that heat exchanger with tangential injection gives 35.95 % more heat transfer coefficient than simple circular duct heat exchanger for parallel flow configuration and gives 33.58 % for counter flow configuration. Whereas, 60° parallel flow configuration gives more heat transfer coefficient compared to other computed cases like 30° and 45° (both parallel and counter flow configuration) and also this 60° configuration gives a maximum of 10.8% more overall heat transfer coefficient than 300 configuration.

REFERENCES

[1] E. K. Akpinar, Y. Bicer, C. Yildiz and D. Pehlivan, "Heat transfer enhancements in a concentric double pipe exchanger equipped with swirl elements," *International Comm. Heat Mass Tranfer,* vol. 31, no. 6, pp. 857-868, 2004.

[2] A. A. M. Baig, I. Nouzil and S. Azzam, "Heat Transfer Augmentation in Concentric Tube Heat Exchanger Using Twisted Tapes," *International Journal of Engineering Research and Applications (IJERA)* , vol. 3, no. 3, pp. 1491-1496, May-June,2013.

[3] A. M. Jawarneh, "Heat Transfer Enhancement in Swirl Annulus Flows," *5th WSEAS Int. Conf. on environment, ecosystems and development, Tenerife, Spain,* pp. 56-61, 2007.

[4] V. K. Dhir, M. Santa and Calif, "Heat transfer enhancement using tangential injection". California, Oakland Patent 5,291,943, 8 March 1994.

[5] F. Kreith and D. Margolis, "Heat transfer and friction in turbulent vortex flow," *Appl. sci. Res.,* vol. 8, pp. 457-473.

[6] D. M. N. Abdullah , "Heat Transfer and Pressure Drop in Turbulent Flow through an Eccentric Converging-Diverging Tube with Twisted Tape Inserts," *Journal of Engineering and Development,* vol. 16, no. 2, pp.

178-192, 2012.

[7] F. Chang and V. K. Dhir, "Mechanism of heat tranfer enhancement and slow decay of swirl in tubes using tangential injection," *International Journal Heat and Fluid Flow,* vol. 16, no. 2, pp. 78-87, 1995.

[8] R. K. Shah and D. P. Sekulic, Fundamentals of heat exchanger design, New Delhi: Wiley India Pvt. Ltd., 2013, pp. 1-73.

[9] I. Sing and D. N. Diwakar, "Heat transfer enhancement using different geometry of twist tape turbulators," *International Journal of Applied Research in Mechanical Engineering (IJARME),* vol. 3, no. 2, pp. 16-29, 2013.

[10] A. B. V. B. C. Ketan Warghat, "Heat Transfer Enhancement in a Concentric Tube Heat Exchanger with Tangential Injection and Twisted Tape Insert.," PUNE, 2016.

DESIGN ANALYSIS AND STUDY OF HEAT TRANSFER FOR DIFFERENT FINNED QUAD TUBED HEAT EXCHANGER

Akshath K R[1], Chaitanyaa K K[2], Basavaraj Devakki[3]
1,2,3 Department of Mechanical Engineering, presidency University, Bangalore, India

E-mail: akshathkr3@gmail.com

Abstract: - The finned quad tubed heat exchanger has been modelled, and analyzed numerically. On the surface of each inner tube, the different shapes of fins have been modelled i.e., square, triangular and Y shape throughout the entire length. The analysis has been carried for Reynold's number 8748, 10935 and 13123 and cold fluid temperature kept constant at 25 °C and hot fluid in annulus shape varied at 40 °C, 50 °C and 60 °C as inlet temperatures. It was found that triangular and Y shaped fins show higher heat transfer enhancement as compared to square fins. An increase of 152% and 145% in heat transfer found for Y and triangular shaped fins respectively as compared to square fin for same inlet conditions. It is the reason being the surface area for triangular and Y shapes fins is much higher. Hence the Y shaped fins can be used in most of the applications which involves combination of conduction and convection heat transfer i.e., in evaporator, condensers, radiators tubes etc. which results in higher heat transfer and increases the efficiency.

Keywords: Quad tubed heat exchanger; Y- shape fins; heat transfer enhancement; triangular fins;

1. Introduction

Double Pipe Heat Exchanger (DPHE) is commonly used in oil, petrochemical, nutrition industries and heat pumps. Double pipe heat exchangers are simple in geometry and construction, it can be used in high pressure and temperature environments. DPHE can also be used in harsh conditions since it can be easily cleaned and good parallel or counterflow combinations can be achieved. Thermal engineers in the HVAC industry are using DPHE to save energy consumption as they are commonly used in condensers, evaporators or in air conditioning systems. Small scale industries prefer DPHE because of its low cost in design and maintenance. The literature contains many efforts to increase the performance of DPHE. The transfer of heat between two or more fluids in a constrained chamber is known as heat exchanger. At present scenario, it plays a very important role in industrial applications in transferring heat in between fluids, here Quad Tubed Heat Exchanger is one of the basic formats of heat exchanger that transfers heat from hot fluid to cold fluid through a cylindrical wall that separates both hot and cold fluid. Its efficiency can be increased by increasing the performance of heat exchanger this can be done by increasing the contact area of hot fluid with cold fluid, the efficient configuration of the heat exchanger in turn results with decreased overall size of the heat exchanger, decreased manufacturing cost of the heat exchanger etc. So, this phenomenon encourages even small-scale industries to use heat exchangers. In heat exchangers initial heat transfer occurs between liquid to solid which known as Convection and then from one side of the wall to other side of the wall i.e. from solid to solid known as Conduction and then again from solid to liquid that is convection.
In this paper, heat transfer analysis has been performed numerically for Quad Tubed Heat Exchanger in ANSYS Fluent software under steady state heat transfer condition. The K-epsilon method on various fins i.e. inner tube - without fins, with triangular fins, with squared fins and with Y- shaped fins used. Parallel flow type heat exchanger is considered and finally its heat transfer rate and Reynolds number are determined, and graphs plotted. For every model totally 9 trials are done by assuming the cold-water temperature at 25 °C and by varying hot water temperature at 40 °C, 50 °C, and 60 °C. The study is carried out assuming that cold water is made to flow through inner tubes and

hot water is made to flow through annulus space. And it is observed that efficiency is increased for Y shaped Quad Tubed Heat Exchanger.

2. Literature review

Dynamic characteristics of DPHE can be predicted with numerical and experimental methods. Experiment on the SCO2 water heat transfer was conducted on the DPHE and correlation between the experimental data and the genetic algorithm is obtained. Numerical studies on DPHE with longitudinal fins of triangular and rectangular cross-sections with variable thickness of the tip is conducted [2], [5] and [7]. By up to 178% increase in the Nusselt number (Nu) relative to the rectangular cross-sections has been obtained. Numerical analysis of DPHE was made with twisted pipe cross-section. It was observed that an increase in the pressure drop increases the heat transfer rate, which was considered as a friction coefficient. The experimental study was made by using MgO nanofluid in a DPHE. 27% increase in heat transfer was achieved for 0.3 wt% as compared with the base fluid [9]. It has been done an experimental study on the DPHE with helical wire in the inner pipe. An increase of up to 2.64 times in Nusselt number (Nu) was observed in comparison to empty pipe [10]. The numerical studies of heat transfer in the DPHE with exponential fins are conducted. With the increase in the thermal conductivity the Nusselt number (Nu) increased significantly. The performance of the exponential fins was better than the triangular fins by 0.02%–15.09% [11]. The viscosity of the fluid had a little effect on the Nu number and significant effect on the Newtonian fluids [12], [8]. The optimal design for the DPHE has been performed [13]. The solution gives the optimum values for the inner and outer diameter of the DPHE. Many engineering techniques have been performed to increase the efficiency of the DPHE, one such effective method is to impart the helical fins around the pipe to allow the helical flow of the fluid. The numerical analysis of thermal performance of the DPHE with helical fins on the annulus will increase as the increase in the baffle spacing and Re number. The thermal and fluid flow properties of the helically twisted DPHE were studied [17]. It was observed 3 times enhancement in the heat transfer in compare to smooth double pipe and a factor of 2–4 increase in the corresponding pressure drop. The experimental study of air bubbles injection in the fluid has shown an overall increase in heat transfer by 10.3% [20].

3. Modelling and Meshing

The complete geometries of the Quad tubed Heat exchanger of all the models are constructed using SolidWorks and then it is imported to ANSYS to study the heat transfers of different constructed models. Geometry of different Quad tubed Heat Exchanger with dimensions of its fins are shown in Table.1. The parallel flow typed Quad Tubed Heat Exchanger with different models are shown with the dimensions of the fins. The dimensions of the complete model such as length and diameter of the tubes in the heat exchanger are kept constant for all the models and the same is shown below.

Figure 01. Modelled heat exchanger

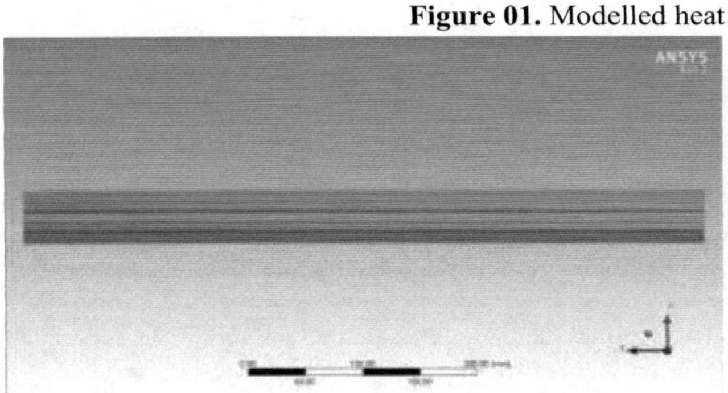

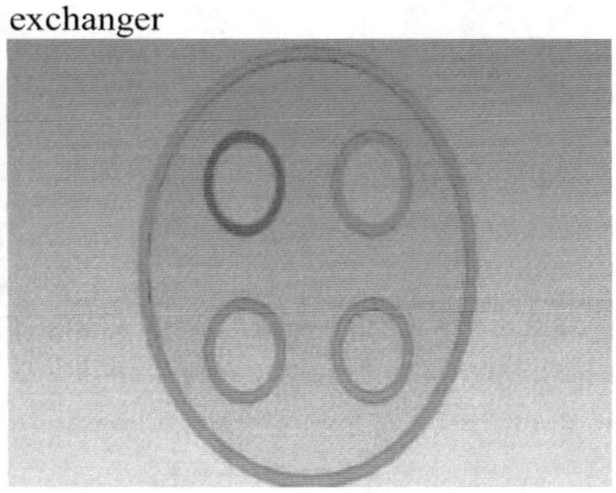

IOP Conf. Series: Materials Science and Engineering 1013 (2021) 012029 doi:10.1088/1757-899X/1013/1/012029

Table 1. Details of heat exchanger

Particulars	Dimensions (mm)
Length of the tube	600
Inner diameter of the inner tube	12.5
Inner diameter of the outer tube	53
Thickness of the inner tube	2.5
Thickness of the outer tube	3

3.1 Model 1 - Quad Tubed Heat Exchanger without fins

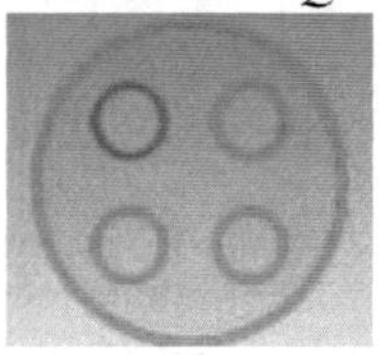

Nos. of Elements	244648
Nos. of Nodes	346280
Avg. Skewness	0.14306

3.2 Model 2 - Quad Tubed Heat Exchanger with Triangular fins

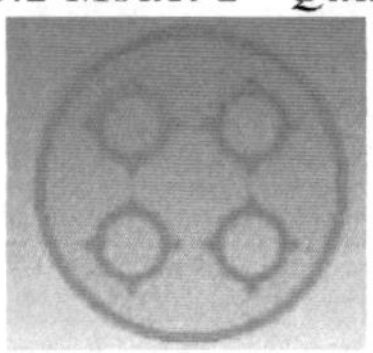

Nos. of Elements	233269
Nos. of Nodes	301408
Avg. Skewness	0.2407

3.3 Model 3 - Quad Tubed Heat Exchanger with Square fins

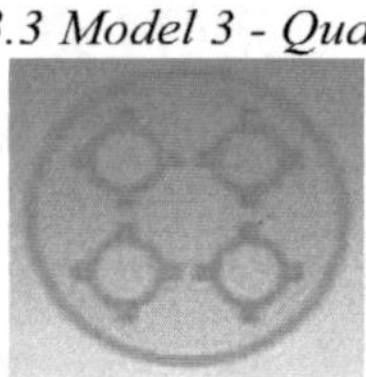

Nos. of Elements	219921
Nos. of Nodes	288126
Avg. Skewness	0.2457

3.4 Model 4 - Model 3 - Quad Tubed Heat Exchanger Y-shaped fins

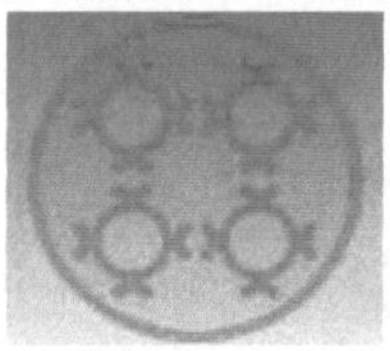

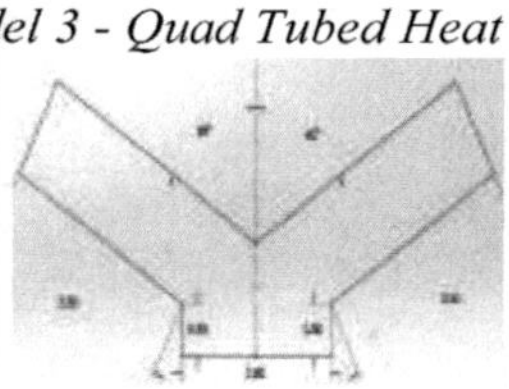

Nos. of Elements	249961
Nos. of Nodes	341691
Avg. Skewness	0.2929

4. Simulation

As discussed in the previous sections, totally of 9 trials have been analyzed by K-epsilon method with standard wall function and considering parallel flow of water with constant cold-water inlet temperature at 25 °C and hot water inlet temperatures at 40 °C, 50 °C, 60 °C. The cold water is made to flow in the inner tubes at velocity of 0.1060895 m/s and the hot fluid water is made to flow in the annulus space at different velocities i.e., 0.0303113 m/s, 0.0252594m/s, and 0.0202075m/s and the Reynolds number equal to 13123, 10935 and 8748 respectively corresponding to these velocities. The pressure outlet boundary is enforced for all the trials. To plot graphs to compare heat transfer of various models, the dimensionless Reynolds number is being calculated using

$$\mathrm{Re} = \frac{\rho v d}{\mu}$$

IOP Conf. Series: Materials Science and Engineering 1013 (2021) 012029 doi:10.1088/1757-899X/1013/1/012029

5. Governing Equations

- Fourier law of conduction

$$Q = -k * A * \frac{dT}{dx}$$

Where, k = Thermal conductivity of the metal
A = Area of cross-section
$\frac{dT}{dx}$ = Temperature gradient

- Newton law of cooling

$$Q = h * A * \Delta T$$

Where, h = Heat transfer coefficient
A = Area of the object in contact with the medium
ΔT = Change in temperature

6. Data Reduction

$$T_{LMTD} = \frac{((T_{hi} - T_{ci}) - (T_{ho} - T_{co}))}{\ln\left(\frac{(T_{hi} - T_{ci})}{(T_{ho} - T_{co})}\right)}$$

Heat transfer rate is computed using Equations

$$Q_{avg} = \frac{Q_h + Q_c}{2}$$

$$Q_h = m_h \times C_{ph} \times (T_{hi} - T_{h0})$$

$$Q_c = m_c \times C_{pc} \times (T_{co} - T_{ci})$$

Where, T_{hi} is the inlet temperature of hot fluid, T_{ho} is the outlet temperature of hot fluid, T_{ci} is the inlet temperature of cold fluid, T_{co} is the outlet temperature of cold fluid, Q_{avg} is the total heat transfer, Q_h is the heat transfer by hot fluid, Q_c is the heat gained by cold fluid, m_h is the mass flow rate of hot fluid, m_c is the flow rate of cold fluid, C_{ph} and C_{pc} is the specific heat of hot and cold fluid respectively.

7. Results and Discussion

7.1 without fins

velocity of cold fluid (m/s)	velocity of hot fluid (m/s)	cold fluid inlet (°C)	cold fluid outlet (°C)	hot fluid inlet (°C)	hot fluid outlet (°C)	NET. (°C)
0.1060895	0.0303113	25	27.861101	40	37.617976	31.892284
0.1060895	0.0303113	25	29.768502	50	46.02996	36.487141
0.1060895	0.0303113	25	31.675903	60	54.44194	41.081997
0.1060895	0.0252594	25	27.79738	40	37.23134	31.78918
0.1060895	0.0252594	25	29.662301	50	45.38556	36.315307
0.1060895	0.0252594	25	31.52722	60	53.53979	40.84143
0.1060895	0.0202075	25	27.70883	40	36.68869	31.64473
0.1060895	0.0202075	25	29.51472	50	44.48115	36.07455
0.1060895	0.0202075	25	31.320619	60	52.27362	40.50437

7.2 Triangular fins

velocity of cold fluid (m/s)	**velocity of hot fluid (m/s)**	**cold fluid inlet (°C)**	**cold fluid outlet (°C)**	**hot fluid inlet (°C)**	**hot fluid outlet (°C)**	**NET. (°C)**
0.1060895	0.0303113	25	33.825984	40	33.860891	32.913582
0.1060895	0.0303113	25	39.709974	50	39.768151	39.189303
0.1060895	0.0303113	25	45.59396	60	45.675412	43.465024
0.1060895	0.0252594	25	33.2765	40	33.310282	32.638578
0.1060895	0.0252594	25	38.794167	50	38.850471	37.730963
0.1060895	0.0252594	25	44.311834	60	44.390659	42.823349
0.1060895	0.0202075	25	32.630053	40	32.662337	32.315006
0.1060895	0.0202075	25	37.716755	50	37.770562	37.770562
0.1060895	0.0202075	25	42.803475	60	42.87878	42.06834

7.3 Square fins

velocity of cold fluid (m/s)	velocity of hot fluid (m/s)	cold fluid inlet (°C)	cold fluid outlet (°C)	hot fluid inlet (°C)	hot fluid outlet (°C)	NET. (°C)
0.1060895	0.0303113	25	28.045295	40	37.242228	31.642071
0.1060895	0.0303113	25	30.075491	50	45.403713	36.403713
0.1060895	0.0303113	25	32.105687	60	53.565198	40.498165
0.1060895	0.0252594	25	27.968769	40	36.801412	31.526734
0.1060895	0.0252594	25	29.947949	50	44.66902	35.87789
0.1060895	0.0252594	25	31.927158	60	52.536628	40.229046
0.1060895	0.0202075	25	27.862313	40	36.185578	31.365735
0.1060895	0.0202075	25	29.770521	50	43.64263	35.609558
0.1060895	0.0202075	25	31.67873	60	51.099682	39.853382

7.4 Y-shape fins

velocity of cold fluid (m/s)	**velocity of hot fluid (m/s)**	**cold fluid inlet (°C)**	**cold fluid outlet (°C)**	**hot fluid inlet (°C)**	**hot fluid outlet (°C)**	**NET. (°C)**
0.1060895	0.0303113	25	33.679626	40	33.697857	32.976632
0.1060895	0.0303113	25	39.466043	50	39.496429	38.294387
0.1060895	0.0303113	25	45.25246	60	45.295	43.612141
0.1060895	0.0252594	25	33.157123	40	33.174679	32.715217
0.1060895	0.0252594	25	38.595205	50	38.624465	37.858695
0.1060895	0.0252594	25	44.033287	60	44.074251	43.002173
0.1060895	0.0202075	25	32.550295	40	32.567	32.411597
0.1060895	0.0202075	25	37.583825	50	37.611666	37.352661
0.1060895	0.0202075	25	42.617354	60	42.656333	42.293726

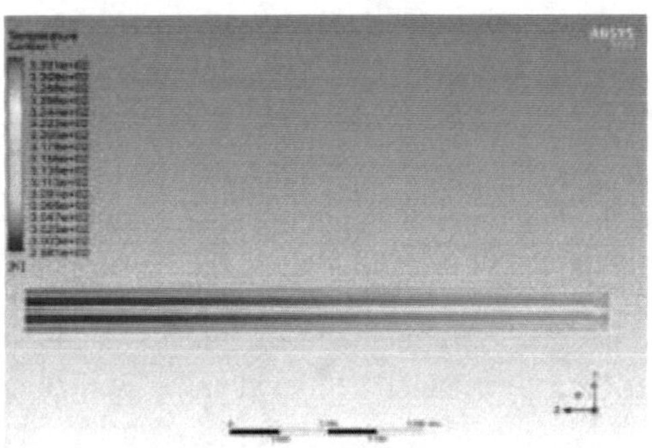

Figure 2. Contour of moodel 1

Figure 3. Contour of moodel 2

Figure 4. Contour of moodel 3

Figure 5. Contour of moodel 4

The performance of heat exchanger with fins and without fins are compared. And the rate of heat transfer vs Reynolds number graphs is plotted for every model. The efficiency of heat exchanger and the amount of heat transfer between cold water and hot water depends upon the design and exact configuration of the heat exchanger. And this heat transfer happens due to difference in temperature between hot and cold fluid. From the below graphs i.e. between heat transfer rate(Q) and Reynolds number (Re) it is understood that heat transfer rate increases as the Reynolds number increases, this applies only for model 2, model 3, and model 4 whereas in case of model 1 slight change is observed i.e. heat transfer rate slightly decreased with increase in Reynolds number. It is clearly understood that the Quad tubed Heat Exchanger of triangular finned hold the maximum heat transfer. And also, it is clearly indicated that heat transfer rate increases as increase in Reynolds number of 40°C of hot water at outer tube and 25 °C of cold water at inner tube. From Figure.2, it can be seen that as the inlet temperature is increased from 40 °C to 60 °C for triangular fins, 80% increase in heat transfer was observed. From Figure.3, for 50 °C as inlet temperature of hot water at outer tube and 25 °C of cold water at inner tube, the maximum heat transfer occurs in both Y shaped and triangular fins with slight changes between them and also it can be observed that in the graph the line of triangular fin intersected the line of Y-shaped fin. The optimum heat transfer can be seen at Re=11000 for both Triangular and Y-shaped fins.

From Figure.4, by considering the 60 °C of hot water as inlet temperature at outer tube and 25 °C of cold water at inner tube it is clearly understood that the maximum heat transfer occurred at Y shaped fin quad tube heat exchanger. This is due the reason that surface area for Y-shape fin is much higher as compared to that of other fins considered in this work. Since the convection heat transfer directly proportional to surface area, higher the surface area higher will be the heat transfer. This results in higher heat transfer by Y-shape fins.

As we discussed, from above figures it is clearly understood that at low temperature triangular finned Quad Tubed Heat Exchanger holds the maximum efficiency and at slightly higher temperature Y-shaped fin Quad Tube Heat Exchanger showed the maximum efficiency. Also, the contours of every model are compared to understand the maximum heat transfer between the fluids.
The above shown contours also indicate that maximum heat transfer efficiency occurred at quad tubed heat exchanger with triangular and Y shaped fins.

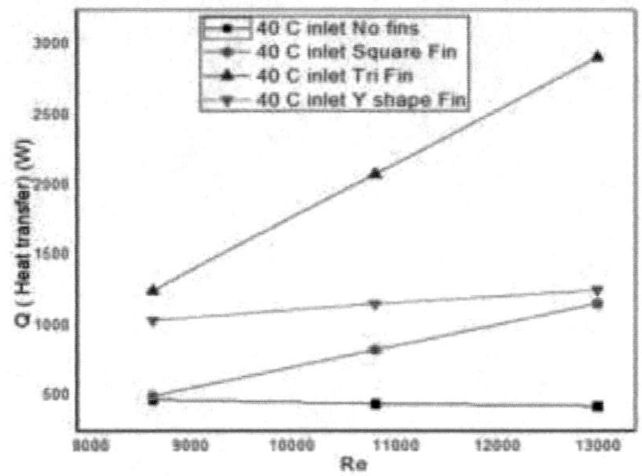

Figure 6. Heat transfer rate vs Reynolds number at 40°C

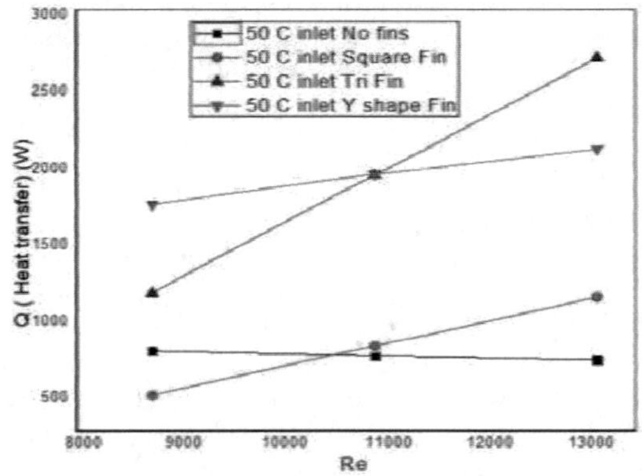

Figure 7. Heat transfer rate vs Reynolds number at 50°C

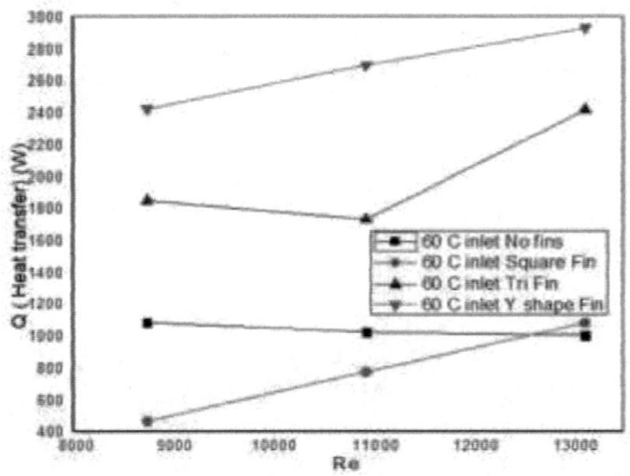

Figure 8. Heat transfer rate vs Reynolds number at 60°C

It is clearly understood that the Quad tubed Heat Exchanger of triangular finned hold the maximum heat transfer. And also, it is clearly indicated that heat transfer rate increases as increase in Reynolds number of 40 °C of hot water at outer tube and 25 °C of cold water at inner tube. From Figure.2, it can be seen that as the inlet temperature is increased from 40 °C to 60 °C for triangular fins, 80% increase in heat transfer was observed. From Figure.3, for 50 °C as inlet temperature of hot water at outer tube and 25 °C of cold water at inner tube, the maximum heat transfer occurs in both Y shaped and triangular fins with slight changes between them and also it can be observed that in the graph the line of triangular fin intersected the line of Y-shaped fin. The optimum heat transfer can be seen at Re=11000 for both Triangular and Y-shaped fins.

From Figure.4, by considering the 60 °C of hot water as inlet temperature at outer tube and 25 °C of cold water at inner tube it is clearly understood that the maximum heat transfer occurred at Y shaped fin quad tube heat exchanger. This is due the reason that surface area for Y-shape fin is much higher as compared to that of other fins considered in this work. Since the convection heat transfer directly proportional to surface area, higher the surface area higher will be the heat transfer. This results in higher heat transfer by Y-shape fins.

As we discussed, from above figures it is clearly understood that at low temperature triangular finned Quad Tubed Heat Exchanger holds the maximum efficiency and at slightly higher temperature Y-shaped fin Quad Tube Heat Exchanger showed the maximum efficiency. Also, the contours of every model are compared to understand the maximum heat transfer between the fluids.

The above shown contours also indicate that maximum heat transfer efficiency occurred at quad tubed heat exchanger with triangular and Y shaped fins.

8. Conclusion

The heat transfer study was done on different types of fins of Quad Tubed Heat Exchanger by considering parallel flow with K-epsilon method where hot water flowing in outer tube and cold water was flowing in inner tubes. It is well known that high heat transfer occurs when the inner tube with cold water had more contact area with hot water at outer tube. From the results it can be concluded that an increase of 152% and 145% in heat transfer found for Y and triangular shaped fins respectively as compared to square fin for same inlet conditions. It is the reason being the surface area for triangular and Y shapes fins is much higher. Therefore, increasing the use of Y-shaped fin in industries and other applications can save time and money with its effective efficiency.

9. References

1. Yuntao Jiang, Yitai Ma, Minxia Li, Lin Fu "An experimental study of transcritical CO2 water–water heat pump using compact tube-in-tube heat exchangers" Energy Convers. Manage. 76 (2013) 92–100.
2. Saud Ghani, Seifelislam Mahmoud Ahmad Gamaledin, Mohammed Rashwan, Muataz Ali Atieh "Experimental investigation of double-pipe heat exchangers in airconditioning applications" Energy Build. 158 (2018) 801–811.
3. B.H. Salman, H.A. Mohammed, K.M. Munisamy "Characteristics of heat transfer and fluid flow in microtube and microchannel using conventional fluids and nanofluids" A review, Renew. Sustain. Energy Rev. 28 (2013) 848–880.
4. N. Srivastava, M.A. Burns, Analysis of non-Newtonian liquids using a microfluidic capillary viscometer, Anal. Chem. 78 (5) (2006) 1690–1696.
5. J. Mozley "Predicting dynamics of concentric pipe heat exchangers" Ind. Eng. Chem. 48 (1956) 1035–1041.
6. T. Ma, W.-X. Chu, X.-Y. Xu, Y.-T. Chen, Q.-W. Wang "An experimental study on heat transfer between supercritical carbon dioxide and water near the pseudocritical temperature in a double pipe heat exchanger" Int. J. Heat Mass Transfer 93 (2016) 379–387.
7. K.S. Syed, Muhammad Ishaq, Zafar Iqbal, Ahmad Hassan "Numerical study of an innovative design of a finned double-pipe heat exchanger with variable fin-tip thickness" Energy Convers. Manage. 98 (2015) 69–80.
8. N. Targui, H. Kahalerras "Analysis of fluid flow and heat transfer in a double pipe heat exchanger with porous structures" Energy Convers. Manage. 49 (2008) 3217–3229.
9. N. Targui, H. Kahalerras "Analysis of a double pipe heat exchanger performance by use of porous baffles and pulsating flow" Energy Convers. Manage. 76 (2013) 43–54.
10. R.L. Mohanty, S. Bashyam, D. Das "Numerical analysis of double pipe heat exchanger using heat transfer augmentation techniques" Int. J. Plast. Technol. 18 (2014) 337–348.
11. H. Arya, M.M. Sarafraz, O. Pourmehran, Maziar Arjomandi "Heat transfer and pressure drop characteristics of MgO nanofluid in a double pipe heat exchanger" Heat Mass Transfer (2019), https://doi.org/10.1007/s00231-018-02554-1.
12. V. Nikkhah, M.M. Sarafraz, Faramarz Hormozi "Application of spherical copper oxide (II) water nano-fluid as a potential coolant in a boiling annular heat exchanger" Chem. Biochem. Eng. Q. 29 (2015) 405–415.

13. M.M. Sarafraz, Faramarz Hormozi, S.M. Peyghambarzadeh, Nabil Vaeli "Upward flow boiling to DI-water and CuO nanofluids inside the concentric annuli" J. Appl. Fluid Mech. 8 (2015) 651–659, https://doi.org/10.18869/acadpub.jafm.67.223.19404.
14. Ebru Kavak Akpinar "Evaluation of heat transfer and exergy loss in a concentric double pipe exchanger equipped with helical wires" Energy Convers. Manage. 47 (2006) 3473–3486.
15. W. Ahmad, K. Saifullah Syed, M. Ishaq, A. Hassan, Z. Iqbal "Numerical study of conjugate heat transfer in a double-pipe with exponential fins using DGFEM" Appl. Therm. Eng. 111 (2017) 1184–1201.
16. M. Sheikholeslami, M. Gorji-Bandp, D.D. Ganji "Effect of discontinuous helical turbulators on heat transfer characteristics of double pipe water to air heat exchanger" Energy Convers. Manage. 118 (2016) 75–87.
17. Saman Pourahmad, S.M. Pesteei "Effectiveness-NTU analyses in a double tube heat exchanger equipped with wavy strip considering various angles" Energy Convers. Manage. 123 (2016) 462–469.
18. S. Eiamsa-ard, S. Pethkool, C. Thianpong, P. Promvonge "Turbulent flow heat transfer and pressure lose in a double pipe heat exchanger with louvered strip inserts" Int. Commun. Heat Mass Transfer 35 (2008) 120–129.
19. M. Sheikholeslami, D.D. Ganji "Heat transfer improvement in a double pipe heat exchanger by means of perforated turbulators" Energy Convers. Manage. 127 (2016) 112–123.
20. R.M. Manglik, P.P. Fang "Effect of eccentricity and thermal boundary conditions on laminar fully developed flow in annular ducts" Int. J. Heat Fluid Flow 17 (1995) 298–306.

IOP Conf. Series: Materials Science and Engineering 1013 (2021) 012030 doi:10.1088/1757-899X/1013/1/012030

A CFD Study on fly ash nanofluid heat transfer behavior in a circular tube

Praveen Kanti[1], K.V.Sharma[2,*], C.G.Ramachandra[3], M Gurumurthy[4]

[1]Department of Mechanical Engineering, Jyothy institute of technology, Bangalore, Karnataka, India

[2]Center for Energy Studies, Department of Mechanical Engineering, JNTUH College of Engineering, Kukatpally, Hyderabad, Telangana, India

[3]Department of Mechanical Engineering, Presidency University, Bangalore, Karnataka, India

[4]Department of Mechanical Engineering, Jyothy institute of technology, Bangalore, Karnataka, India

* Author correspondence E-mail: praveenkanti87@gmail.com

ABSTRACT

STAR CCM+ software is used to investigate the forced convection heat transfer of fly ash/Water nanofluids flowing through a horizontal circular tube in the turbulent flow regime under the constant heat flux of 707 W/m^2. The volume concentrations of nanofluid are 0.3, 0.5, 1.0, and 1.5 vol. %, used in this analysis. The results revealed that both the Reynolds number and concentration intensify the Nusselt number of fly ash nanofluids. The highest heat transfer coefficient augmentation is about ~60% observed at 1.5% concentration in comparison to the water.

1.0 Introduction

The most important technological issues affecting many sectors, such as power plants, manufacturing processes, transportation, and electronics, are heating and cooling. Classical heat transfer rate enhancement approaches, such as usage of fins, heat surface vibration, fluid injection or suction, and the utilization of electrical or magnetic fields, practically fulfill the requirements of heat flux increments [1–2]. Common heat transfer fluids such as water, ethylene glycol, and glycerine appear to also have low conductivity contrast to metals and metal oxides. Hence, fluids containing suspended solid particles would have greater heat transfer capabilities relative to traditional heat transfer fluids [3-4].

Practical implications of fluids with a millimeter or micrometer-sized suspension particles display several issues, such as particle instability, oxidation, and flow path congestion, and produce additional pressure drop [5–8]. Modern developments in the processing of nanosized metal or non-metallic particles have permitted the development of a new category of fluid called nanofluid coined by Choi in 1995 [8].

Fotukian and Nasr [9] studied much diluted (< 0.24% volume) water-based copper oxide nanofluid turbulent convective heat transfer efficiency. In their analysis, it was described that the improvement in nanofluid concentration had a marginal impact on heat transfer amplification.

The convective heat transfer for water-dependent Al_2O_3 and CuO nanofluids was investigated by Heris et al. [10] through a circular coil in the laminar flow and noticed that alumina nanofluids displayed significant amplification in contrast to copper oxide nanofluids.

Variyenli. [11] examined the heat transfer behavior plate heat exchanger using fly ash nanofluid experimentally and numerically by changing fluid mass flow rate and temperature. ANSYS FLUENT is used to model heat transfer characteristics. The findings revealed that the nanofluid increased the overall heat transfer coefficient of about 6-20% compared to water.

In the present work, an efficient single-phase model was adapted to determine the heat transfer coefficient of fly ash nanofluid of volume concentrations 0.3 to 1.5% flows through a copper tube with a constant heat flux of 707 W/m^2 at various mass flow rates.

2.0 Methodology

2.1 Model details

The single-phase approach, which was predominantly used for nanofluids, is being introduced in this research [20]. Each of the governing equations was used for simulation, and various experiments were performed on the same basis. The turbulent viscous realizable k-ε, two-layer, all Y+ treatment was used for this research.

The 1kW capacity nichrome heater is mounted on the copper tube surface to provide constant heat flux. Three K- type thermocouples are positioned on the tube surface at equidistance from the pipe inlet to measure the surface temperature and the other two thermocouples are used to measure fluid inlet and outlet temperatures. The heater is wrapped with Asbestos insulation to minimize heat losses. The copper tube inlet and outlet are extended to 5 times of tube inner diameter to get the fully developed flow

2.2 Theoretical analysis

The fly ash nanofluid simulation results were used to calculate the heat transfer coefficient and Nusselt number using Eqs. (1) and (2) respectively.

$$h = \frac{Q}{A_S(T_S - T_b)} \quad (1)$$

Where, $T_b = \frac{T_i + T_o}{2}$, and, $T_s = \frac{T_1 + T_2 + T_3}{3}$

Nusselt number of fly ash nanofluid was given by,

$$Nu_{nf} = \frac{h_{nf} D}{k_{nf}} \quad (2)$$

The Reynolds number and Prandtl number of the fly ash nanofluid are estimated using Eqs. (3) and (4), respectively. The properties measured at bulk temperature T_b.

$$Re_{nf} = \frac{\rho_{nf} V D}{\mu_{nf}} \quad (3)$$

$$Pr_{nf} = \frac{\mu_{nf} C_{pnf}}{k_{nf}} \quad (4)$$

The Dittus–Boelter [12] equation is applicable for the range of $Re > 10^4$ and $0.6 < Pr < 200$

$$Nu = 0.023\, Re^{0.8}\, Pr^{0.4} \quad (5)$$

Below are the correlations to determine the Nusselt number of nanofluids.

Pak and Cho [16] for TiO_2 nanoparticles

$$Nu = 0.021 Re^{0.8} Pr^{0.5} \quad (6)$$

$10^4 < Re < 10^5$

Duangthongsuk and Wongwises [13] for TiO_2 nanoparticles

$$Nu = 0.07 Re^{0.707} Pr^{0.385}\, \varphi^{0.074} \quad (7)$$

$3000 < Re < 18000,\ 0 < \varphi < 2\%$

2.3 Thermophysical properties

Thermal conductivity and viscosity of fly ash nanofluid for all concentrations are determined using the KD2 Pro analyzer and LVDV-II Pro Brookfield Programmable digital viscometer with the accuracy of ±5% and ±1%, respectively. Table 1 lists the thermophysical properties of fly ash nanoparticles. The test conditions of the present work are described in Table 2. The experimental values of fly ash nanofluid thermal conductivity contrasted to Patel et al. [14] and Maiga et al.[15] corresponding to viscosity.

Pak and Cho [16] relation used to determine the density of nanofluid is given by

$$\rho_{nf} = (1-\varphi)\rho_{bf} + \varphi\rho_{nf} \quad (8)$$

Xuan and Roetzel [17] relation used to determine the fly ash nanofluid specific heat is given by

$$c_{p\,nf} = \frac{(1-\varphi)\,(\rho\, c_p)_{bf} + \varphi\,(\rho c_p)_{np}}{(1-\varphi)(\rho)_{bf} + \varphi\,(\rho)_{np}} \quad (9)$$

Maiga et al.[15] as a function of concentration is

$$\frac{\mu_{nf}}{\mu_{bf}} = (1 + 7.3\varphi + 123\varphi^2) \quad (10)$$

Patel et al. [14]

$$\frac{k_{nf}}{k_w} = \left(1 + 0.135\left(\frac{k_p}{k_f}\right)^{0.273}\left(\frac{T}{20}\right)^{0.547}\left(\frac{100}{d_p}\right)^{0.234}\varphi^{0.467}\right) \quad (11)$$

Table 1 Thermophysical properties of fly ash nanoparticle and 1 vol.% nanofluid

Sl. No	Property	Fly ash nanoparticle	1 vol%. nanofluid
1.	Density, kg/m^3	1920	1005
2.	Specific heat, J/kg K	514	4108.29
3.	Viscosity, kg/m s	-	0.00089
4.	Thermal conductivity, W/m K	1.7	0.715

Table 2 Testing conditions

Sl.No	Constraints	Standards/Range
1.	Tube material	Copper
2.	The inner and outer diameter of the tube	0.01 m and 0.013m
3.	Heating material	Nichrome wire
4.	Length of the tube	2 m
5.	Mass flow rates	2, 3, 4, 5, and 6 LPM
6.	Reynolds number range	4500-16000

2.4 Boundary conditions

The volume concentrations of fly ash nanofluids from 0.3 to 1.5% are used as the input condition. In this analysis, water is also used to compare the results of nanofluids using a CFD approach. It was assumed that Reynolds number was increased as input data from 4500 to 16000 at each iteration run. Table 3 illustrates the boundary conditions of the present study.

Table 3 Boundary conditions

Sl.No	Constraints	Standards/Range
1.	The inlet temperature of nanofluid	30°C
2.	Uniform heating of the copper tube	45 W
3.	Constant heat flux	707 W/m^2

2.5 CFD model

STAR-CCM+ 13.06.012 software creates the geometry, meshes the modeling, and also solves the turbulent heat transfer equations. The number of elements in the meshed model is 1445625. Figs. 1(a) and 1(b) illustrate the CAD model and meshed model.

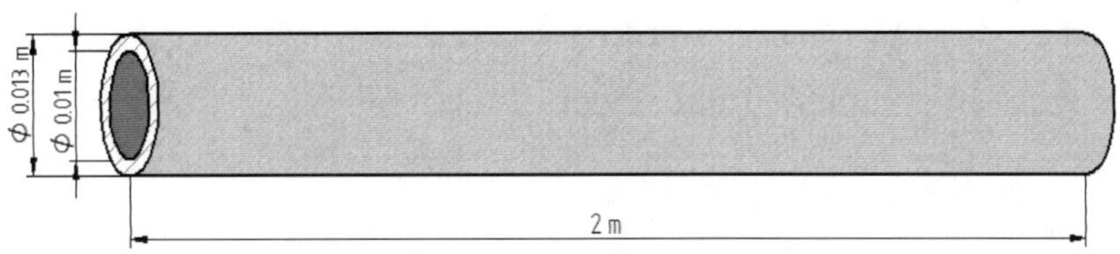

Fig. 1(a) CAD model

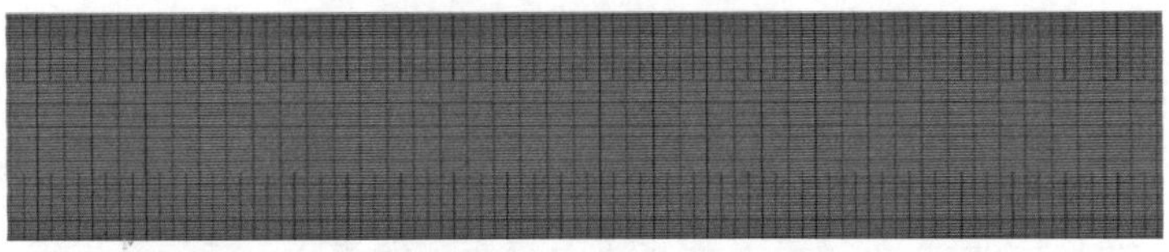

Fig. 1(b) Meshed model of the copper tube.

3. Results and discussions

3.1 Thermophysical properties

The thermal conductivity and viscosity of fly ash nanofluids at volume concentration (0.3–1.5%) and base fluid were measured at 30°C temperature. The thermal

IOP Conf. Series: Materials Science and Engineering 1013 (2021) 012030 doi:10.1088/1757-899X/1013/1/012030

conductivity and viscosity of fly ash nanofluids with particle loading are significantly enhanced. The highest deviation between the thermal conductivity data from the experiment and the values obtained by Patel et al. [14] is -4.0%. Likewise, measured dynamic viscosity data has the maximum variance with the values estimated by Maiga [15] correlation is 2.89%. Such variations are due to the presence of a surfactant and even smaller spherical particle form due to the influence of the ball milling effect [21, 22, 23, 24].

3.2 Heat transfer coefficient (HTC)

Figs. 2 and 3 display the influence of Reynolds number on heat transfer coefficient and Nusselt number. It is found that from both the figures, the intensification of heat transfer and Nusselt number with the concentration of fly ash nanofluids and Reynolds number. This is due to the increment in the nanoparticles' number in the base fluid which enhances the Prandtl number and nanofluid thermal conductivity in contrast to the water. Also, Brownian movement and dispersion stability also responsible [18-19]. The largest augmentation in the HTC and Nusselt number is about 60% and 36% in the context of Reynolds number tested with a concentration of 1.5%. Figs. 4(a) and 4(b) illustrate the velocity plots for 2 and 6 LPM for the concentration of 1.5 %. It is observed that fluid velocity is maximum at the center of the tube.

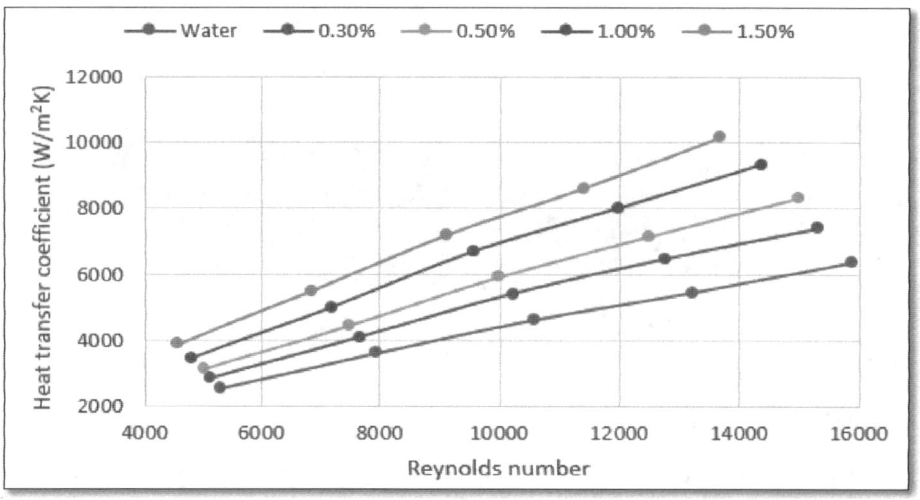

Fig. 2. Impact of Reynolds number on HTC of fly ash nanofluid

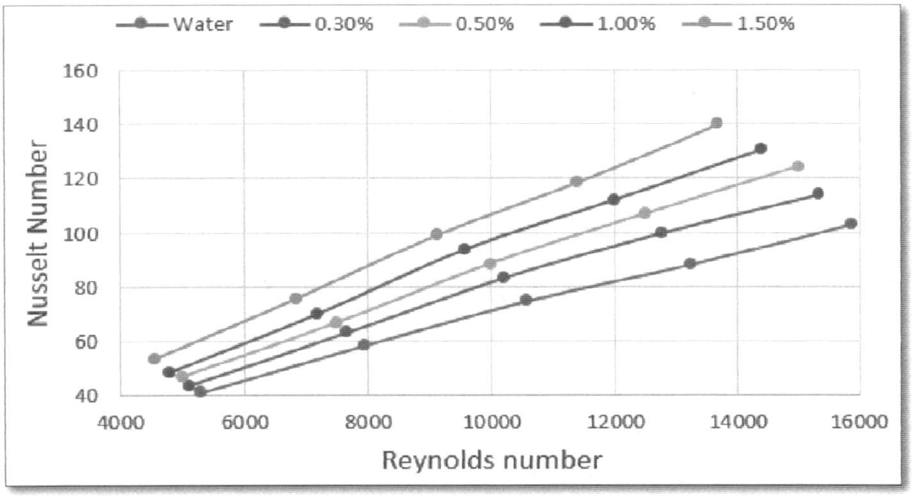

Fig. 3. Reynolds number against fly ash nanofluid Nusselt number

Fig. 4(a). Velocity plot for the concentration of 1.5% at 2 LPM

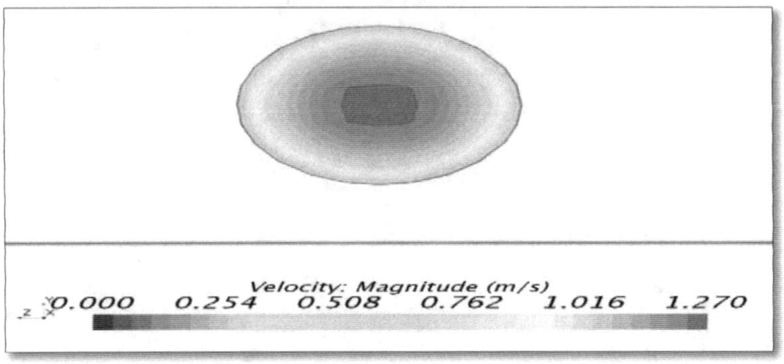

Fig. 4(a). Velocity plot for the concentration of 1.5% at 6 LPM

The Nusselt number for water determined using Eq. (2) compared with the Dittus–Boelter [12] equation in order to assess the reliability of numerical data is illustrated in

Fig. 5. A decent agreement between the computational data of the water Nusselt number and the data determined with Eq. (5) under various Reynolds numbers.

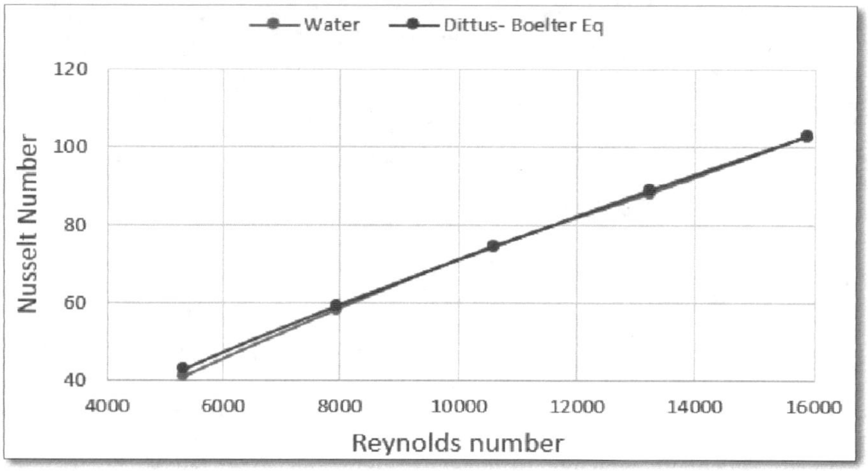

Fig. 5. Comparison of water Nusselt number with Eq. (5)

The Nusselt number of fly ash nanofluid determined using computational values is compared with the values estimated from various correlations. Fig. 6 depicts the comparison between the Nusselt number of fly ash nanofluid and that of estimated from Pak and Cho [16], and Duangthongsuk and Wongwises [13] correlation for 0.5% concentration. The fly ash contains TiO_2 has its chemical composition, hence comparison made with these two correlations. The largest deviation in Nusselt number of fly ash nanofluid and that of determined using Pak and Cho [16], and Duangthongsuk and Wongwises [13] correlation was 17.9% and 55%, respectively.

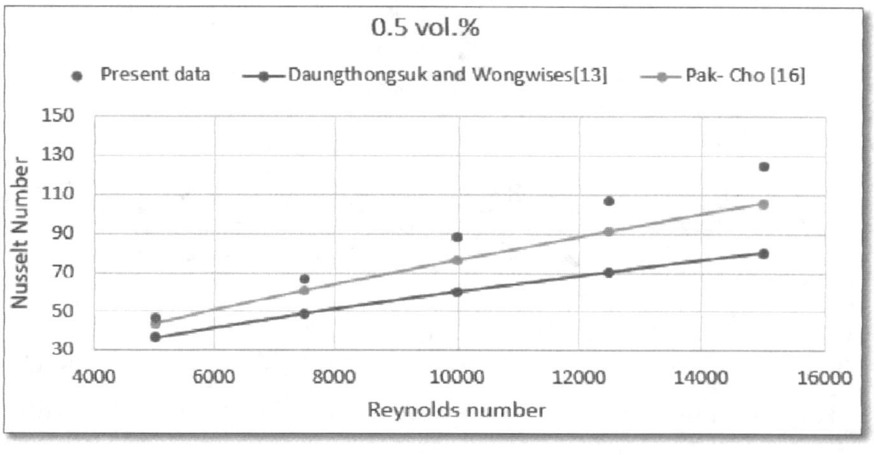

Figure 6. Comparison of Nusselt number of fly ash nanofluids with Eq. (6) and (7).

Hussein et al. [20] performed the convective heat transfer characteristics of SiO_2 nanofluid for various concentrations numerically. The fly ash contains nearly 60% of silica along with other metal oxides in different mass ratios. Hence, the comparison is made with silica nanofluid. The Nusselt number of fly ash nanofluid contrasted to data of Hussein et al. [20] for 1.0% volume concentration is plotted in Fig. 7. It is witnessed that fly ash nanofluid shows a higher Nusselt number contrast to that of silica nanofluid at the same concentration and Reynolds number.

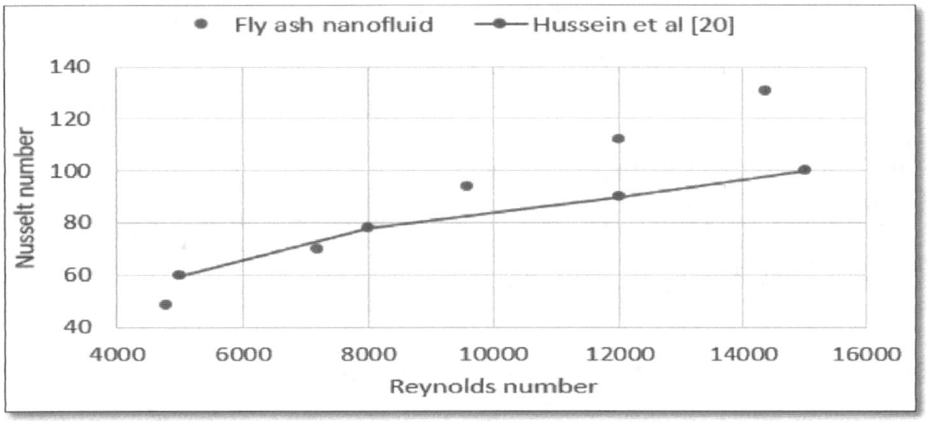

Fig. 7. Comparison of Nusselt number of fly ash nanofluid with Hussein et al. [20]

4. Conclusions

Throughout this work, the numerical simulation approach was adopted to analyze the fly ash nanofluid heat transfer characteristics under constant heat flux for various flow conditions.

1. The maximum deviation between the experimental viscosity and thermal conductivity values and that obtained from Eq. (10) and Eq. (11) is 2.9% and -4.0%, respectively.
2. The largest augmentation in the HTC was about 60% observed at a 1.5% concentration of fly ash nanofluid contrasted to water.
3. The maximum amplification in the Nusselt number was 36% obtained for a concentration of 1.5% of fly ash nanofluid contrasted to water.
4. Finally, the Reynolds number effect on nanofluid heat transfer is more powerful than the effect on volume concentration of fly ash nanofluids.

References

1. A. S. Ahuja, Augmentation of Heat Transport in Laminar Flow of Polystyrene Suspension, I. Experimental and Results, J. Appl. Phys., vol. 46, pp. 3408–3418, 1975.
2. B. M. Da Silva Miranda and N. K. Anand, Convective Heat Transfer in Channel with Porous Baffles, Numer. Heat Transfer A, vol. 46, pp. 425–452, 2004.
3. A. E. Bergles, Recent Development in Convective Heat Transfer Augmentation, Appl. Mech. Rev., vol. 26, pp. 675–684, 1973.
4. H. Masuda, A. Ebata, K. Teramat, and N. Hishinuma, Alteration of Thermal Conductivity and Viscosity of Liquid by Dispersing Ultra-fine Particles (Dispersion of Al_2O_3, SiO_2 and TiO_2 Ultra-fine particles), Netsu Bussei (Japan), vol. 4, pp. 227–233, 1993.
5. P. Kebliski, S. R. Phillpot, S. U. S. Choi, and J. A. Eastman, Mechanism of Heat Flow in Suspension of Nano-Sized Particle (Nanofluids), Int. J. Heat Mass Transfer, vol. 45, pp. 855–863, 2002.
6. H. Xie, J. Wang, T. Xi, and Y. Liu, Thermal Conductivity of Suspensions Containing SiC particles, Int. J. Thermophys., vol. 23, pp. 571–580, 2002.
7. B. X. Wang, L. P. Zhou, and X. F. Peng, A Fractal Model for Predicting the Effective Thermal Conductivity of Fluid with Suspension of Nanoparticles, Int. J. Heat Mass Transfer, vol. 46, pp. 2665–2672, 2003.
8. S. U. S. Choi, Enhancing Thermal Conductivity of Fluid with Nanoparticles, in D. A. Siginer and H. P. Wang (eds.), Developments and Applications of Non-Newtonian flows, FED-vol., vol. 66, p. 99, American Society of Mechanical Engineers, New York, 1995.
9. S. M. Fotukian and M. Nasr Esfahany, Experimental Study of Turbulent Convective Heat Transfer and Pressure Drop of Dilute Cu-Water Nanofluid Inside a Circular Tube, Int. Comm. Heat Mass Transfer, vol. 37, pp. 214–219, 2010.
10. S. Z. Heris, M. N. Esfahany, and S. G. Etemad, Experimental Investigation of Convective Heat Transfer of Al_2O_3-Water Nanofluid in Circular Tube, Int. J. Heat Fluid Flow, vol. 28, pp. 203–210, 2007.

11. H. İ. Variyenli, "Experimental and numerical investigation of heat transfer enhancement in a plate heat exchanger using a fly ash nanofluid," Heat Transfer Research., vol. 50, no. 15, pp. 1477–1494, 2019.
12. F. W. Dittus, L. M. K. Boelter, "Heat Transfer in Automobile Radiators of the Tubular Type," University of California Publications on Engineering., vol. 2, pp. 443–461, 1930.
13. Duangthongsuk W, & Wongwises S. An experimental study on the heat transfer performance and pressure drop of TiO_2–water nanofluids flowing under a turbulent flow regime. International Journal of Heat and Mass Transfer. 2010:53:334–344.
14. H. Patel, T. Sundararajan, S. K. Das, "An experimental investigation into the thermal conductivity enhancement in oxide and metallic nanofluids," J. Nanopart. Res., vol.12, pp. 1015–31, 2010.
15. S. E. B. Maiga, S. J. Palm, C. T. Nguyen, G. Roy, N. Galanis, "Heat transfer enhancement by using nanofluids in forced convection Flows," Int J Heat Fluid Flow., vol. 26, pp. 530–46, 2005.
16. B. C. Pak, Y. L. Cho, "Hydrodynamic and heat transfer study of Dispersed fluids with submicron metallic oxide particles," Exp. Heat Transf., vol. 11, pp.151–170, 1998.
17. Y. Xuan and W. Roetzel, "Conceptions for heat transfer correlation of nanofluids," International Journal of Heat and Mass Transfer., vol. 43, no. 19, pp. 3701–3707, 2000.
18. M. K. Moraveji, M. Hejazian, "CFD Examination of Convective Heat Transfer and Pressure Drop in a Horizontal Helically Coiled Tube with CuO/Oil Base Nanofluid," Numerical Heat Transfer, Part A: Applications, vol. 66, no. 3, pp. 315–329, 2014.
19. M. Hejazian, M. K. Moraveji, "A Comparative Analysis of Single and Two-Phase Models of Turbulent Convective Heat Transfer in a Tube for TiO_2 Nanofluid with CFD," *Numerical Heat Transfer, Part A: Applications.* vol. 63, no. 10, pp. 795–806, 2013.

20. A. M. Hussein, K. V. Sharma, R. A. Bakar, K. Kadirgama, "The Effect of Nanofluid Volume Concentration on Heat Transfer and Friction Factor inside a Horizontal Tube," *Journal of Nanomaterials.*, pp. 1–12, 2013. doi:10.1155/2013/859563
21. Praveen Kanti, Viswanatha Sharma Korada, C.G. Ramachandra, P.H.V. Sesha Talpa Sai. "Experimental study on density and thermal conductivity properties of Indian coal fly ash water-based nanofluid", International Journal of Ambient Energy, 2020
22. Kanti P, Sharma KV, Ramachandra CG, Bhramara P. Stability and thermophysical properties of fly ash nanofluid for heat transfer applications. . Heat Transfer. pp. 1–18, 2020. https://doi.org/10.1002/htj.21849.
23. Kanti P, Sharma KV, Ramachandra CG, Azmi WH. Experimental determination of Thermophysical Properties of Indonesian Fly Ash Nanofluid for Heat Transfer Applications. Particulate science technology. pp. 1–16, 2020. https://doi.org/10.1002/htj.21849.
24. Kanti P, Sharma KV, Ramachandra CG, Minea AA. Effect of ball milling on the thermal conductivity and viscosity of Indian coal fly ash nanofluid. Heat Transfer. 2020;1–16. https://doi.org/10.1002/htj.21836

List of symbols

A_s : surface area of the tube, m^2

C_p : specific heat, J/kg K

D : diameter of a tube, m

d_p : Diameter of nanoparticle, nm

h : heat transfer coefficient, W/m^2 k

K_p : Thermal conductivity of nanoparticle, W/m K

L : Length of the tube, m

LPM : Litre per minute

m : mass, kg

Nu : Nusselt number

IOP Conf. Series: Materials Science and Engineering 1013 (2021) 012030 doi:10.1088/1757-899X/1013/1/012030

Re : Reynolds number

T_b : bulk temperature °C

T_i : fluid temperature at inlet, °C

T_o : fluid temperature at outlet, °C

T_s :Average surface temperature, °C

V : velocity of the fluid, m/s

Greek symbols

ρ : Density , kg/m^3

φ : Volume concentration

μ : Viscosity, kg/m.s

Subscript

bf : base fluid

nf : nanofluid

np : nanoparticle

p : particle

Numerical study on fly ash–Cu hybrid nanofluid heat transfer characteristics

Praveen Kanti[1*], K.V.Sharma[2,] C.G.Ramachandra[3], Gurumurthy M[4], B M Raghundana Raghava[5]

[1]Department of Mechanical Engineering, Jyothy institute of technology, Bangalore, Karnataka, India

[2]Center for Energy Studies, Department of Mechanical Engineering, JNTUH College of Engineering, Kukatpally, Hyderabad, Telangana, India

[3]Department of Mechanical Engineering, Presidency University, Bangalore, Karnataka, India

[4]Department of Mechanical Engineering, Jyothy institute of technology, Bangalore, Karnataka, India

[5]Simulation Engineer, Dheya Technology, Bangalore, Karnataka, India

*Corresponding author email: praveenkanti87@gmail.com

ABSTRACT

The current numerical study is aimed to examine the forced convection heat transfer of fly ash-Copper (80:20% by volume) water-based hybrid nanofluids flowing in a horizontal circular copper tube under a constant heat flux of 7962W/m^2 using STAR CCM+ software. The volume concentrations of 0.5% and 1% are considered for the analysis within the Reynolds number range of 6900-26500. The findings show that the heat transfer coefficient and the Nusselt number of hybrid nanofluid at a concentration of 1 vol.% are increased by about 66.0% and 36.67% compared to that of water.

Introduction

Hybrid nanofluids, in which two separate nanoparticles are dispersed in a base fluid, combine the advantages of increasing the thermal conductivity of both constituents and aim to maintain viscosity at low numbers [1-3]. Numerically and experimentally numerous studies have studied different kinds of nanofluids in laminar and turbulent flow conditions. The first research work on convective heat transfer of nanofluids in a circular tube was reported by Pak and Cho [4]. Hussein et al. [5] examined the influence of the cross-sectional area on heat and flow characteristics. They observed that in contrast with the circular tube and elliptical tube, the flat tube has a substantial enhancement in heat transfer and pressure drop.

Madhesh et al. [6] examined the augmentation in heat transfer and rheological characteristics of Cu-TiO_2 hybrid nanofluids flows in the tubular heat exchanger for the volume concentration of 0.1% to 1.0%. The findings showed that the hybrid nanofluid heat transfer coefficient improved by 59.3% for 0.7% of the hybrid nanocomposite volume concentration.

Sundar et al. [7] investigated the heat transfer characteristics of Fe_3O_4 / MWCNT hybrid nanofluids under turbulent conditions in a flat tube. Results showed a 32% augmentation in heat transfer with a penalty of 1.2 times increment in pumping power for volume concentration of 0.3% below 20,000 Reynolds contrast to water.

Moghadesi et al. [8] examine the influence of Al_2O_3 nanofluid and Al_2O_3-Cu hybrid nanofluid flowing through a horizontal circular tube under laminar forced convective heat transfer using CFD software. In contrast to alumina nanofluid and water, the average Nusselt number improvement for hybrid nanofluids at 0.1 vol. % was 4.73% and 13.46%, respectively.

The current research aims to study the heat transfer coefficient and Nusselt number of water-based fly ash–Copper (80:20% by volume) hybrid nanofluids for turbulent flow using the CFD method. The hybrid nanofluids volume concentrations of 0.5 and 1.0% flow in a circular copper tube having an inner and outer diameter 16mm and 19mm with a constant heat flux of 7962 W/m^2.

2.0 Nanofluid Preparation

The first step in the hybrid nanofluid preparation is the water suspension and stabilization of nanoparticles. Kumar and Arasu [9] used Eq. (1) to calculate the volume concentration of hybrid nanofluid.

$$\varphi(\%) = \frac{\left(\frac{m_{np1}}{\rho_{np1}}\right)+\left(\frac{m_{np2}}{\rho_{np2}}\right)}{\left(\frac{m_{np1}}{\rho_{np1}}\right)+\left(\frac{m_{np2}}{\rho_{np2}}\right)+\left(\frac{m_{bf}}{\rho_{bf}}\right)} \tag{1}$$

Using a two-step method, fly ash and copper nanoparticles in the ratio of 80:20% by volume are weighed with a digital electronic balance to prepare the hybrid nanofluid in 0.5 and 1.0 vol. % using water. This mixture is subjected to magnetic stirring for 30 min. The stability of the mixture is obtained by adding Triton X-100 surfactant 0.25% by weight of nanoparticles. After preparing the suspension, an ultrasonic probe is run for 180 min to break any agglomeration. The average size of hybrid consisting of fly ash and Cu particles is also evaluated with Zeta sizer found to be 15 nm. The maximum and minimum zeta potential values of 36.6 mV and 34.1 mV are observed at 0.5% and 1.0% volume concentration respectively.

2.1 Thermophysical properties

LVDV-II Pro Brookfield digital viscometer and KD2 Pro analyzer were used to determine the dynamic viscosity and thermal conductivity of hybrid nanofluids respectively. Tables 1 and 2 displays the thermophysical properties of nanoparticles and test conditions.

Takabi and Salehi [10] developed a correlation given by Eq. (2). The experimental values of density for the hybrid nanofluid are compared with the values estimated with Eq. (2).

$$\rho_{hnf} = \varphi_{fnp}\,\rho_{fnp} + \varphi_{Cu}\,\rho_{Cu} + (1 - \varphi_{hnf})\rho_{bf} \quad (2)$$

Where $\varphi_{hnf} = \varphi_{fnp} + \varphi_{Cu}$

The fly ash–Cu hybrid nanofluid specific heat have been estimated using the mixture relation proposed by Takabi and Salehi [10] given in Eq. (3)

$$(c_p)_{hnf} = \frac{(\rho\varphi c_p)_{fnp} + (\rho\varphi c_p)_{Cu} + (1-\varphi_{hnf})(\rho\varphi c_p)_{bf}}{\rho_{hnf}} \quad (3)$$

Where $\varphi_{hnf} = \varphi_{fnp} + \varphi_{Cu}$

The measured viscosity values of hybrid nanofluid are then compared with relations proposed by Wang et al. [11] shown in Eq.(4).

The Eq. (4) proposed by Wang et al. [11] is given by

$$\frac{\mu_{nf}}{\mu_{bf}} = (1 + 7.3\varphi_{hnf} + 123\varphi_{hnf}^{\ 2}) \quad (4)$$

The experimental thermal conductivity values of hybrid nanofluid are compared with the Maxwell model [12]. Eq. (5) is the Maxwell model [12] used to estimate the thermal conductivity of mono nanofluids. In the present study to assess the thermal conductivity of hybrid nanofluid, Eq. (6) resulted from the Maxwell equation, used to determine the thermal conductivity of fly ash- Cu (80:20% by volume) hybrid nanofluid.

$$\frac{k_{nf}}{k_w} = \frac{k_p + 2k_w - 2\varphi_p(k_w - k_p)}{k_p + 2k_w + \varphi_p(k_w - k_p)} \quad (5)$$

For hybrid nanofluid,

$$\frac{k_{hnf}}{k_{bf}} = \frac{\frac{(\varphi_{FA}k_{FA}+\varphi_{Cu}k_{Cu})}{\varphi}+2k_{bf}+2(\varphi_{FA}k_{FA}+\varphi_{Cu}k_{Cu})-2\varphi k_{bf}}{\frac{(\varphi_{FA}k_{FA}+\varphi_{Cu}k_{Cu})}{\varphi}+2k_{bf}-(\varphi_{FA}k_{FA}+\varphi_{Cu}k_{Cu})+\varphi k_{bf}} \quad (6)$$

Table 1 Thermophysical properties of fly ash nanoparticle

Sl. No	Property	Fly ash nanoparticle	Copper nanoparticle
1.	Density, kg /m^3	1920	8933
2.	Specific heat, J / kg- K	514	385
3.	Thermal conductivity, W/m-K	1.7	400

Table 2 Test conditions

Sl.No	Constraints	Standards/Range
1	Heating material and Tube material	Nichrome wire and copper
2.	Mass flow rates	5, 8, 11, 14, and 16 LPM
3.	Insulating material	Asbestos rope
4.	Thermocouples on the tube surface	5
5.	Thermocouples used to measure fluid inlet and outlet temperatures	2, (T_i and T_o)

2.2 Governing equations and CFD model

In this work, the single-phase method, incompressible fluid, the turbulent viscous realizable k-ε, two-layer, all Y+ treatment was used. The one kW capacity Nichrome heater is mounted on the tube surface to provide uniform heat flux. Five K- type thermocouples are positioned on the tube surface at equidistance from the pipe inlet to measure the surface temperature and the other two thermocouples are used to measure fluid inlet and outlet temperatures. The heater is wrapped with Asbestos insulation to minimize the heat losses to the surroundings. The tube inlet and outlet are extended to five times of tube inner diameter to get the fully developed flow

2.3 Theoretical analysis

The simulation results of nanofluid were used to calculate the heat transfer coefficient and Nusselt number using Eqs. (7) and (8) respectively

$$h_{hnf} = \frac{Q}{A_s(T_s - T_b)} \quad (7)$$

Where, $T_b = \frac{T_i+T_o}{2}$, $T_s = \frac{T_1+T_2+T_3+T_4+T_5}{5}$

Nusselt number of water and hybrid nanofluid is given by,

$$Nu_{hnf} = \frac{h_{hnf} D}{k_{hnf}} \tag{8}$$

The Reynolds number and Prandtl number of water and hybrid nanofluid are estimated using Eqs. (9) and (10), respectively. The thermal properties have taken at the bulk temperature T_b.

$$Re_{hnf} = \frac{\rho_{hnf} V D}{\mu_{hnf}} \tag{9}$$

$$Pr_{hnf} = \frac{\mu_{hnf}\, C_{phnf}}{k_{hnf}} \tag{10}$$

The Dittus–Boelter [13] equation is applicable for $Re > 10^4$ and $0.6 < Pr < 200$ is given by

$$Nu = 0.023\, Re^{0.8}\, Pr^{0.4} \tag{11}$$

The Nusselt number correlations for nanofluids are listed below.

Pak and Cho [4]

$$Nu = 0.021 Re^{0.8} Pr^{0.5} \tag{12}$$

Duangthongsuk and Wongwises [14]

$$Nu = 0.07 Re^{0.707} Pr^{0.385}\, \varphi^{0.074} \tag{13}$$

Maiga et al. [15]

$$Nu = 0.085 Re^{0.71} Pr^{0.35} \tag{14}$$

2.4 Boundary conditions

The fly ash - Copper hybrid nanofluid 0.5 and 1.0 vol. % and the inlet temperature of the fluid in all the cases is 30°C used as input conditions. Water is also used to validate and compare the hybrid nanofluid results. The boundary conditions of the analysis presented in Table 3. STAR-13.06.012 (r8 Double Precision) software meshes the modeling of the test segment. Meshed model of test segment shown in Fig.1. The STAR CCM+ program is to solve turbulent heat transfer equations using a finite volume method (FVM).

Table 3 Boundary conditions

Sl.No	Constraints	Value
1.	The inlet temperature of nanofluid	30°C
2.	Uniform heating of the copper tube	600 W
3.	Heat flux	7962W/m^2
4.	Tube length and inner diameter	1.5m and 16mm

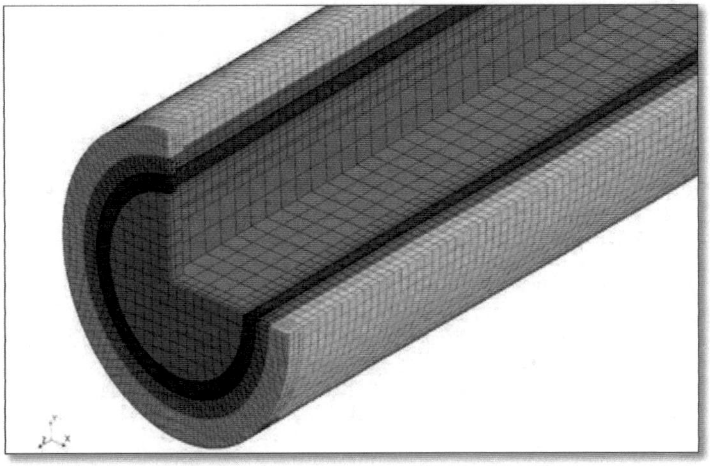

Fig. 1 Meshed model of the test section

3. Results and Discussions

3.1 Thermophysical properties

Thermal conductivity and viscosity of fly ash–Cu hybrid nanofluid measured with KD2 Prometer and LVDV-II Pro Brookfield Programmable digital viscometer respectively. The Maxwell model [12] underestimates the fly ash-copper hybrid nanofluid thermal conductivity. Similarly, the Wang model [11] also underestimates the experimental viscosity values of fly ash– copper hybrid nanofluid. The thermal conductivity and dynamic viscosity values of studied hybrid nanofluid at 30°C for different concentrations are presented in Table 4.

Table 4. Thermal conductivity and dynamic viscosity values of fly ash –Cu hybrid nanofluid at 30°C

Concentration	Thermal conductivity (W/m K)	Dynamic viscosity (mPa-s)
0.5 vol.%	0.7	0.975
1.0 vol.%	0.752	0.99

3.2 Heat transfer characteristics

The influence of Reynolds number on the heat transfer coefficient of fly ash-Cu hybrid nanofluid is plotted in Fig. 2. The heat transfer coefficient of hybrid nanofluid is found to intensify with the concentration of particles and Reynolds number. It is because of fly ash-Cu hybrid nanofluid has good thermal conductivity [7-8]. The presence of copper particles even in small amount enhance the hybrid nanofluid thermal conductivity.

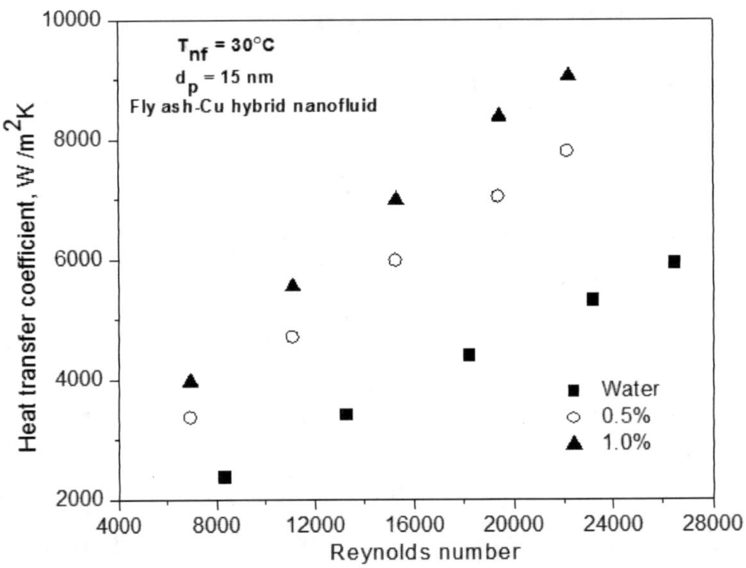

Fig. 2. Reynolds number on the heat transfer coefficient of hybrid nanofluid

To validate the results of a present study, The Nusselt numbers of water determined with Eq. (8) are shown in contrast with that of determined with the Dittus–Boelter [13] equation in Fig. 3. A fair agreement between the Nusselt number of water and that of estimated with Dittus –Boelter [13], i.e., Eq. (11) confirms the reliability of the data.

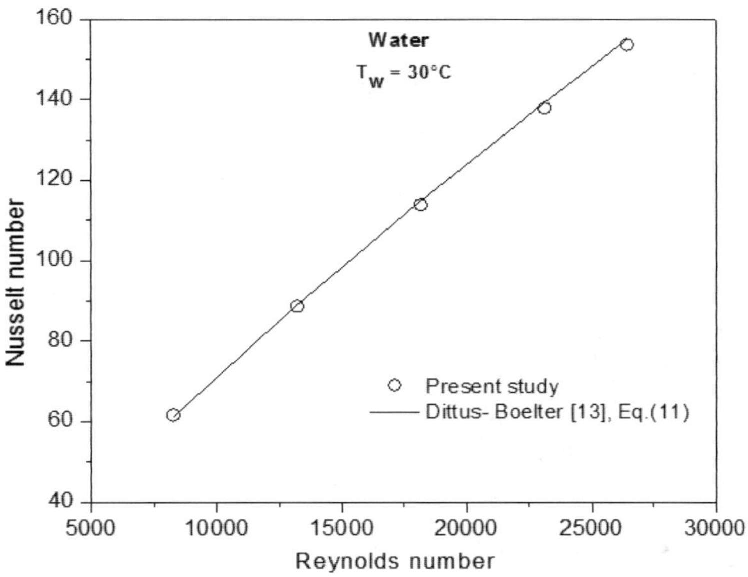

Fig. 3. Nusselt number of water estimated from Eq. (8) in contrast with correlations

The Nusselt number of hybrid nanofluids was determined using Eq. (8), and it is illustrated in Fig. 4. The Nusselt number of nanofluid enhances with particle concentration and Reynolds number. Since the thermal conductivity of fly ash –Cu hybrid nanofluid is more in contrast to water. The micro convection, the surface area of nanoparticles contributes to the thermal conductivity of hybrid nanofluid's improvement [5-8].

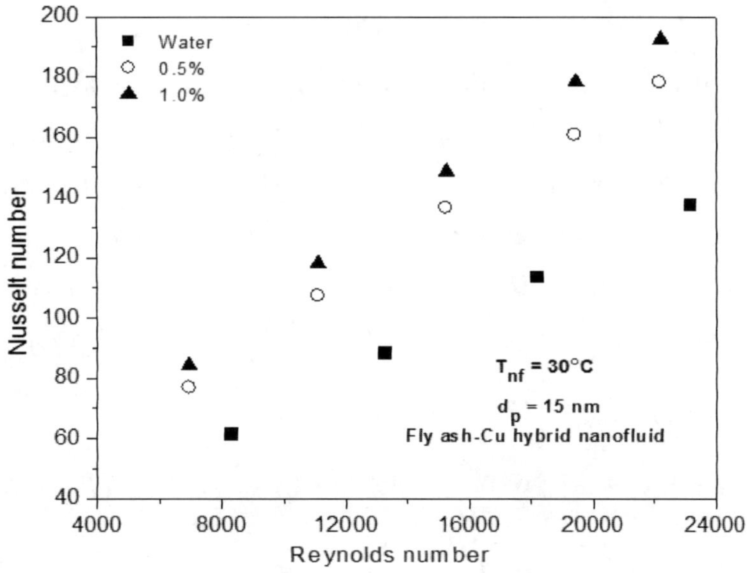

Fig. 4 Nusselt number of hybrid nanofluid variation with Reynolds number

The Nusselt number of fly ash –Cu hybrid nanofluids are compared with Pak and Cho [4], Duangthongsuk and Wongwises [14], and Maiga et al. [15] correlations available for the Nusselt number estimation of nanofluids is depicted in Fig. 5. Eqs. (12), (13), and (14) underestimate the Nusselt number of hybrid nanofluid. These variances due to the improved thermal properties of fly ash-Cu hybrid nanofluid [17].

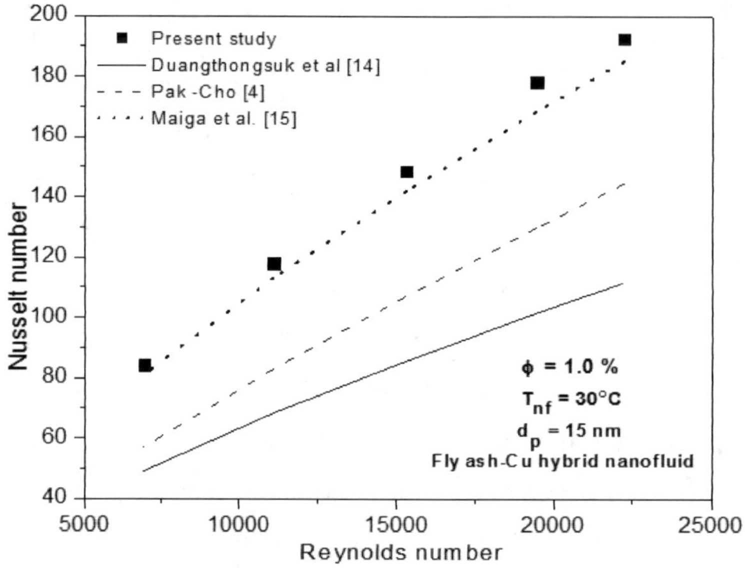

Fig. 5 Nusselt number of hybrid nanofluids comparison with that of correlations.

The fly ash –Copper hybrid nanofluid Nusselt number is compared with that of Cu-TiO_2 hybrid nanofluids used by the Madhesh et al. [6] at a 1.0 vol. % portrayed in Fig. 6. It is found that the Nusselt number of hybrid nanofluid enhanced with both particle loading and the Reynolds number in both the works. The deviation between the Nusselt numbers of both the hybrid nanofluids at lower Reynolds number is insignificant.

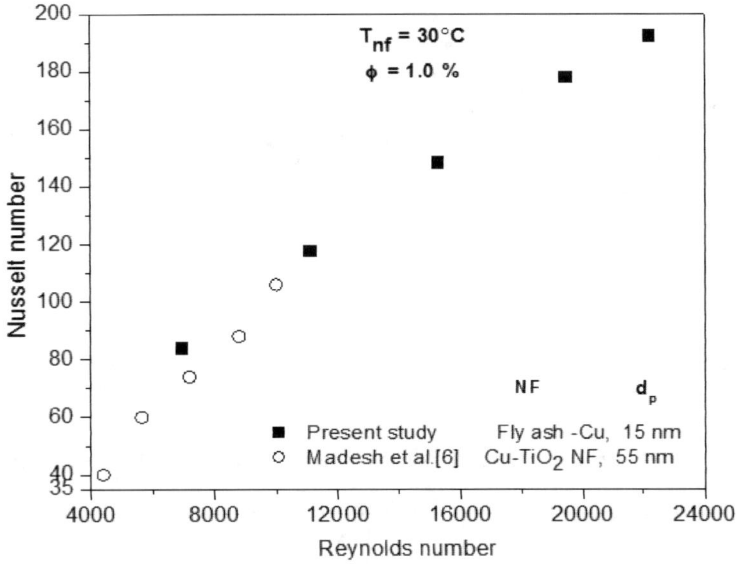

Fig. 6 Comparison of the Nusselt number of hybrid nanofluids with Madhesh et al. [6].

4. Conclusions

The following conclusions are taken from the work under consideration.

1. The correlations available in the literature failed to estimate the thermal conductivity and viscosity values of fly ash –Cu (80:20% by volume) hybrid nanofluid.
2. The addition of nanoparticles to the base fluid enhances the convective heat transfer. The maximum amplification in the heat transfer coefficient of 66.0% was observed at a concentration of 1.0 % of hybrid nanofluid compare to water.
3. The correlations available in the literature for estimation of Nusselt number of nanofluids were not suitable to estimate that of hybrid nanofluid.
4. Finally, fly ash-Cu hybrid nanofluid has exhibits better heat transfer characteristics compared to the base fluid, and also it is cost-effective.

References

1.	Minea AA. Pumping power and heat transfer efficiency evaluation on Al_2O_3, TiO_2 and SiO_2 single and hybrid water-based nanofluids for energy application. Journal of Thermal Analysis and Calorimetry. 2019. doi:10.1007/s10973-019-08510-3.

2.	Moldoveanu GM, Huminic G, Minea AA, Huminic A. Experimental study on thermal conductivity of stabilized Al_2O_3 and SiO_2 nanofluids and their hybrid. Int J Heat Mass Transf. 2018;127:450–7.
3.	Sidik NAC, Adamu IM, Jamil MM, Kefayati GHR, Mamat R, Najafi G. Recent progress on hybrid nanofluids in heat transfer applications: a comprehensive review. Int Commun Heat Mass Transf. 2016;78:68–79.
4.	Pak BC, Cho YL, Hydrodynamic and heat transfer study of Dispersed fluids with submicron metallic oxide particles. Exp. Heat Transf. 1998;11:151–170.
5.	Hussein AM, Sharma KV, Bakar RA & Kadirgama K. The effect of cross sectional area of tube on friction factor and heat transfer nanofluid turbulent flow. International Communications in Heat and Mass Transfer. 2013;47:49–55.
6.	Madhesh D, Parameshwaran R, Kalaiselvam S. Experimental investigation on convective heat transfer and rheological characteristics of Cu–TiO_2 hybrid nanofluids. Exp Therm Fluid Sci. 2014;52:104–15.
7.	Sundar LS, Singh MK, Sousa ACM. Enhanced heat transfer and friction factor of MWCNT–Fe_3O_4/water hybrid nanofluids. Int Commun Heat Mass Transf. 2014;52:73–83.
8.	Moghadassi A, Ghomi E, Parvizian F. A numerical study of water based Al_2O_3 and Al_2O_3-Cu hybrid nanofluid effect on forced convective heat transfer. Int J Therm Sci. 2015;92:50–7
9.	Kumar DD, Arasu AV. A comprehensive review of preparation, characterization, properties and stability of hybrid nanofluids. Renew Sustain Energy Rev. 2018;81:1669–89.
10.	Takabi B, Salehi S. Augmentation of the heat transfer performance of a sinusoidal corrugated enclosure by employing hybrid nanofluid. Adv Mech Eng. 2014;20:10-14
11.	Wang X, Xu X, Choi SUS. Thermal conductivity of nanoparticle-fluid mixture. J. Thermophys. Heat Transf. 1999;13 (4):474–480.
12.	Maxwell JC. A Treatise on Electricity and Magnetism. UK: Clarendon, 1973.

13.	Dittus FW, Boelter LMK. Heat Transfer in Automobile Radiators of the Tubular Type. University of California Publications on Engineering. 1930;2:443–461.
14.	Duangthongsuk W, & Wongwises S. An experimental study on the heat transfer performance and pressure drop of TiO_2–water nanofluids flowing under a turbulent flow regime. International Journal of Heat and Mass Transfer. 2010:53:334–344.
15.	Maiga SEB, Palm SJ, Nguyen CT, Roy Galanis G, Heat transfer enhancement by using nanofluids in forced convection Flows. Int J Heat Fluid Flow. 2005;26:530–46.
16.	Takabi, B., & Shokouhmand, H. Effects of Al_2O_3–Cu/water hybrid nanofluid on heat transfer and flow characteristics in turbulent regime. International Journal of Modern Physics C, 2015;26(04):1-12

List of symbols

A_s : surface area of the tube, m^2

C_p : specific heat, J /kg K

D : diameter of a tube, m

d_p : Diameter of nanoparticle, nm

h : heat transfer coefficient, W/m K

K_p : Thermal conductivity of nanoparticle, W/m K

L : Length of the tube, m

LPM : Litre per minute

m : mass, kg

NF : Nanofluid

Nu : Nusselt number

Pr : Prandtl number

Re : Reynolds number

T_b : bulk temperature ^oC

T_i : fluid temperature at inlet, ^oC

T_o : fluid temperature at outlet, ^oC

V : velocity of the fluid, m/s

Greek symbols

ρ : Density , kg/m^3

φ : Volume concentration

μ : Viscosity, mPa.s

Subscript

bf : base fluid

Cu : Copper

FA : Fly ash

hnf : hybrid nanofluid

np : nanoparticle

w :Water

A REVIEW ON ADSORPTION CHILLING SYSTEMS AND VIABLE FUTURE FOCUS IN REFRIGERATION

Aalderik Jude J[1], Dr. Devendra Dandotiya[2] and Dr. Bhaskar Pal[3]
[1,2,3] Department of Mechanical Engineering, Presidency University, Bengaluru, India.
E-mail: devendradandotiya@presidencyuniversity.in

Abstract: - The adsorption refrigeration system has gained significant insight in recent years as it has paved an important way of utilizing low-grade thermal energy and eco-friendly refrigerant efficiently. However, it is yet to be commercially viable due to its low Coefficient of Performance (COP) and specific cooling power (SCP) in comparison to the traditional chilling technologies. This paper reviews various validation in incorporating different strategies to maximize the performance and cooling power of the system, out of which, prominent stages include-waste heat recovery, mass recovery, multi-bed, multi-stage, and revamp functionality.

This study acknowledges different parameters and allied substitution, for instance, low evaporative temperature, low generative temperature functioning can be controlled by infusing the mass recovery technique. Furthermore, other variables like appropriate cycle time and switching time are discussed to attain optimum performance of the system as it Is regarded that the adsorption chiller greatly relies on the functionality of the system. Besides, to the above-stated variables, an adsorbent material is employed and this assures a considerably desirable performance.

This study further extended in briefing the economic criterion, increasing the power to mass ratio, and also improving the knowledge regarding the potential performance of the system in the long run when the energy- driven is either solar or waste heat energy source.

Keywords: Adsorption, COP, Waste Heat recovery, Refrigeration

1. Introduction

This paper comprises of the discussion on the Adsorption chilling technology. Firstly, a glimpse idea of conventional refrigeration technology is discussed. Secondly, the need for an alternative cooling system is explained. Later, an introduction to the adsorption mechanism and a brief analogy between the two chilling systems is discussed. Other performance criteria and future scope to the technology in combination with sensors is broadly explained.

VAPOR COMPRESSION REFRIGERATION CYCLE (VCR CYCLE)

The most commonly used refrigeration cycle is the VCR cycle and regarded as the conventional refrigeration cycle. There are mainly four processes involved in this cycle. Initially, the evaporator absorbs the heat from the surroundings at a certain pressure and temperature and the refrigerant is heated up to saturated vapor line, therefore increasing temperature and the refrigerant undergoes a reversible adiabatic process under compression with the help of the compressor, thereby the pressure of the refrigerant increases and attains a superheated state. A constant pressure process takes place after de-superheating of refrigerant in vapor state in the condenser chamber, where all the heat absorbed is emitted out and a saturated liquid refrigerant enters the throttle section which neither produces nor consumes work, and hence pressure decreases in this isenthalpic process. After this process, the temperature is attained which is lesser than the initial temperature of the surrounding, and the refrigerant goes back to the evaporator and the cycle continues. This cycle provides the necessary refrigeration effect. [1]

IOP Conf. Series: Materials Science and Engineering 1013 (2021) 012032 doi:10.1088/1757-899X/1013/1/012032

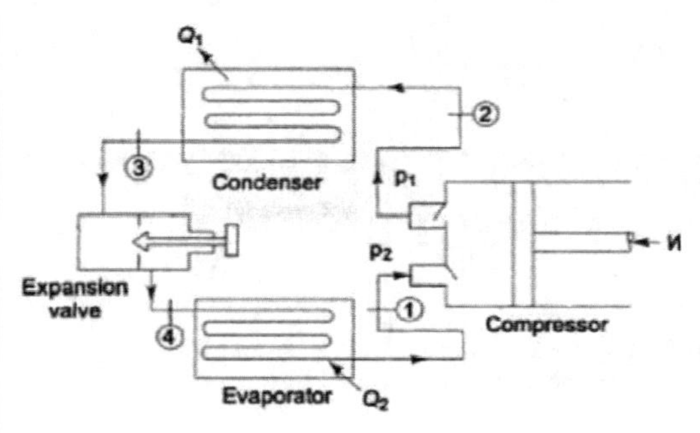

Figure 1. Vapor Compression Refrigeration System (VCRS)

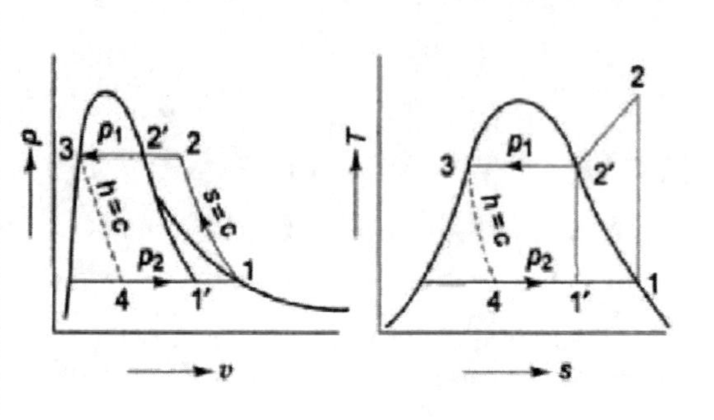

Figure 2. p-v and T-s diagram of VCRS

NEED FOR AN ALTERNATIVE COOLING SYSTEM

The traditional VCR cycle uses high-grade energy and environmentally hazardous refrigerant. According to the U.S Energy Information Administration (EIA), as in 2019, electricity use for cooling the interior of buildings by the U.S. residential and commercial sectors was about 380 billion kilowatt-hours (kWh), which was equal to about 10% of total U.S. electricity consumption in that year. Similarly, for a country like India, which constitutes about 17.7% of the world population, as of 2020, consumption of electricity is increasing exponentially over the years.

The air quality index (AQI) in most of the cities in India, is close to 400 which is a severe condition and the impact of cooling systems catering to air pollution is significant. So, there is a top-priority requirement for an alternative solution. The adsorption cooling technology proves to be an efficient way to overcome all the difficulties caused due to the conventional systems by the use of low-grade energy and waste heat recovery process.

Adsorption mechanism is a surface phenomenon in which the accumulation of molecular species occurs at surface- rather than the bulk of solid. The adsorbent solids are present in their finely divided state having large surface area and thus acts as good adsorbent and this property increases with an increase in surface area per unit mass of adsorbent at a given temperature and pressure. During this process of adsorption, the residual forces on the surface of the adsorbent get balanced by adsorbing the adsorbate, thereby decreasing its surface energy, which reflects as an exothermic reaction. This heat generated is called enthalpy of adsorption, which causes a significant increase in the temperature of the eco-friendly refrigerant (generally water), therefore eliminating the use of a compressor in the system; and thus, abating the use of electricity to a large extent. The adsorption chamber acts as a thermal compressor which operates on heat [1].

Adsorption cycle which works on two separate phases- the adsorption phase and the desorption phase. In the adsorption phase, at low pressure, water in the evaporator evaporates, then water in its vapor form moves up to the adsorber and is adsorbed by the adsorbents, thereby increasing the pressure in the adsorber bed. When this attained pressure is more than the condenser pressure, naturally the refrigerant moves to the condenser. In the desorption phase, waste heat or the low-grade energy desorbs the adsorbed water vapor and moves to the condenser to liquefy and reject heat (considered as the waste heat) which later will be used again in the cycle, and then transfers water in liquid form back to the evaporator via the throttle valve and the cycle continues. Most processes are analogous to the traditional process of chilling, except for using the waste heat and adsorber-desorber instead of a compressor, which proves to be effective in the reuse of heat, hence saving a significant amount of thermal energy and power consumption to a large extent respectively [2,3]. Figure 3 depicts the adsorption mechanism and is evident in heat rejection and heat recovery from the figure.

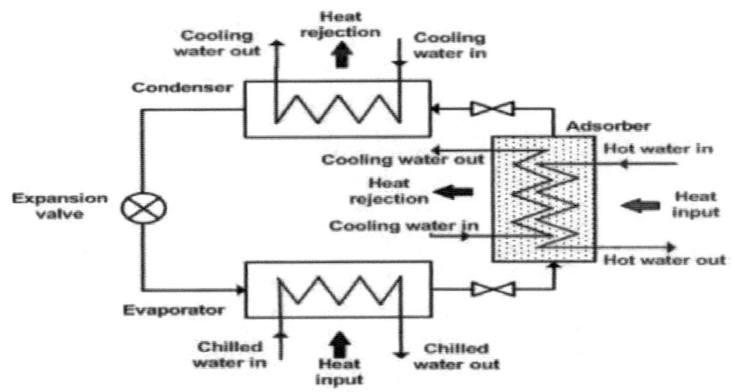

Figure 3. Adsorption refrigeration system mechanism

2. Stages and Performance Parameters

This cycle has its demerits in having a low coefficient of performance and specific cooling power. Researchers are continuously working to combat this by infusing different stages and varying different performance criteria. A generic depiction of the cycles and performance variations are presented below. [3]

A. Heat Recovery

The objective of this cycle is to recover the heat energy by the virtue of temperature difference between the absorber and desorber during the switching between the evaporation and condensation process. The energy in the bed with higher temperature passing to adsorber is transferred directly to the bed with lower temperature while passing to the desorber, also increasing the pressure while passing through absorber to condenser and decreases pressure (depressurization) while desorption. The cooling water flows into the hot adsorber and is heated up, hence the heated water passes to cold desorber, thereby pre- heating it. Then, the cooling water returns and the hot water bypasses the system so that there is no external thermal energy required during this process. This entire heat recovery takes place in the adsorption bed. [1] [3]

B. Mass Recovery

The technique of mass recovery proves to be very effective in increasing the cooling capacity of the chilling system, which is the main objective of any cooling system, thereby increases the Coefficient of Performance (COP). It is seen from the heat recovery process that an increase in temperature pressurizes the refrigerant and attains a higher pressure than the inlet condenser pressure and vice-versa occurs during the desorption process (cooling of the bed), where the adsorbers are not directly connected to the evaporator and condenser. But, in the mass recovery process, fractional pressure change occurs even without any temperature increase or decrease in the bed. Here, adsorber and desorber may be connected and mass of the refrigerant circulates in the bed under pressure drop of either adsorber or desorber. This naturally occurs during the switching process and this pressure drop continuously makes the refrigerant circulate, hence mass is recovered [2,4].

With the equation of specific heat formula:

$$Q = mCp\Delta T$$

The cooling capacity of the system increases, as the mass is continuously present in the bed and heat transfer occurs with a greater percentage than earlier without mass circulation. In the equation, Q is the heat energy in Joules(J), m is mass of the refrigerant in kilograms(kg), Cp is the specific heat capacity in Joules/kilograms

Kelvin(J/kg-K), ΔT is the temperature difference between adsorber and desorber in Kelvin(K). This stage is very effective in places where heat recovery techniques cannot be used due to environmental conditions. [2] [4]

C. Multi-stage Process

This technique is generally incorporated when the basic source of the system comes from solar energy radiation. When there is a variation in climatic conditions, and the source of energy naturally falls short, a multi-stage setup can be used accordingly. Here, the entire setup is made in such a way that the process occurs at two different stages simultaneously. The requirement of two pairs of adsorbent beds with the usual evaporator and condenser is essential and hence this process undergoes the thermodynamic states of the adsorption system – pre-heating, desorption, pre- cooling, and adsorption. It is experimentally proven that this system functions effectively even at a waste heat temperature of 55°C. A combined stage technique increases efficiency by about 20% where the culmination of two identical processes is brought together to function. [1]

D. Multi-bed Process

Out of all the above techniques, the technique of multi-bed increases efficiency and performance rapidly. They have major benefits in maximizing the cooling capacity, improving the quality of instantaneous cooling, etc. From the previously tested experiments and research, it is observed that a multi- bed adsorption chiller reduced the fluctuations in the chilled water outlet, which thereby increased the waste heat recovery efficiency at a greater rate. A combination of a multi-bed adsorber chiller system with infusing the waste-heat recovery technique will increase the system performance exponentially [1,4].

Moreover, other techniques like thermal, cascade cycle, and cycle time can also be parameterized and can be found better performing cycles and techniques by various experiments. However, even though these techniques are efficient in their form, they need to be used at the appropriate conditions of the surrounding. It is also important to always compare the conventional form of chilling systems with adsorption systems whenever a new technique is incorporated to check the performance, so that a better understanding of various factors like Coefficient of Performance (COP), specific cooling power(SCP), economics and cost involved and not leaving behind, the environmental impact it can cause to the society needs to be considered.

3. Working Pairs

Working pairs are the most important factor when we introduce a completely new form of chilling system. Analogous to the refrigerant used in the conventional system, where we parametrize properties like low boiling and melting point, high latent heat of evaporation, less viscous and low specific volume, high thermal conductivity, non-toxic, non-flammable, non-explosive, non-corrosive, easy maintenance, and handling capability, it is advisable to parametrize properties relatively in adsorption chilling systems like high adsorption capacity, a large change of adsorption capacity with small variation in temperature, good compatibility with the refrigerant etc. It is merely impossible to attain all the mentioned properties. Thus, we can use a few common working pairs. The following gives a brief discussion on the working pairs used with their merits and demerits.

A. Activated carbon-methanol systems

Activated-carbon methanol systems prove to be the best user working pairs in adsorption systems due to their large adsorption capacity, which means the affinity of the pair is high with the adsorber bed, high evaporating of latent heat of methanol. The highest COP is achieved with this working pair. On the other side, the functionality of this working pair, though proves to have higher performance is restricted to a working temperature of less than 120°C due to the chemical reaction that can take place between the working pair as the activated carbon helps and speeds up the reaction of methanol to form dimethyl ether at a temperature of about 150°C. Methanol has a high toxic property which makes the pair not suitable to use and also activated carbon does not form a metallic bond and can only covalently bond, thermal conductivity is poor. These properties deficit the use in the chilling systems, as they go against the criteria we have parametrized. [1] [2]

B. Activated carbon-ammonia systems

In the conventional refrigeration systems, we use ammonia to a greater extent, but when activated carbon is paired with ammonia, a desorption pressure of about 1.6MPa is attained, hence a higher-pressure bearing capacity system needs to be designed. Due to this high pressure, it enhances the mass transfer of the working pair and shortens the response time between the switching of the adsorption-desorption process. However, it is still a far thought on using this pair due to the irritant smell of ammonia gas, toxicity, and corrosive properties. Disregarding the chemical and environmental properties, this pair is very well efficient in other thermodynamic and physical properties. So, we must work on eliminating, rather reduce the unsuitable chemical properties for better performance [1].

C. Silica gel-water systems

In the introduction to the adsorption system, silica is one of the firstly regraded adsorbent material. When this adsorbent is paired with environment-friendly water, it is assumed to be the best working pair. But this assumption is disproved, as this pair is a low- temperature working pair that can be driven by about 75°C heat source. Under low pressure, water uptake in silica gel is very little and requires a high evaporative temperature. [1] [4]

D. Calcium chloride- ammonia systems

In chemical adsorption systems, this pair is commonly used due to the large adsorption capacity of calcium chloride. As discussed earlier, ammonia as a commonly used refrigerant has a low boiling point property. However, the other factors of expansion, decomposition, deterioration, and corrosion in the calcium chloride-ammonia adsorption system make it ill-suited for commercial use in more applications, rather it is restricted to very few applications based on specific individual property of the pairs. [1] [3]

Similar pairs and more working pairs are introduced, experimented, and tested for performance. However, specific applications require different working pairs and as engineers and scientists, we must make a decision based on the usage with high efficiency and considering the environmental variables.

4. Future Scope of Adsorption Chilling Technology

Researchers and scientists are continuously working towards the betterment of society. In the field of refrigeration and air-conditioning, it is of prime importance to have a better performing system with greater COP and SCP, not disregarding the economics and environmental hazards, as it has become a necessity to have such cooling systems in most, the developed and developing nations across the globe after rapid urbanization of majority of the parts in the world. The methodology to work on these adsorption systems needs to always be compared with traditional refrigeration technology, to study the merits and demerits of both systems and design and develop a generalized, and almost close to an ideal working cycle with maximum efficiency and performance, also not contributing harmful effluents to the atmosphere.
Specifically, in the field of adsorption technology, different variants are being tried out by researchers, like using different working pairs, which have adsorbent properties and combining them with fin-like structures made of materials with high thermal conductivity, so that heat transfer increases, and therefore cooling capacity also increases proportionally. Recently, researchers have observed that different orientations or arrangements [7] of the adsorbents also cater to differences in performance. Altogether, the culmination of all the stated and other new parameters needs to be considered while designing such systems.

CFD (Computational Fluid Dynamics) analysis needs to be done appropriately with high resolution meshing for accurate results for all variants in materials, orientation, working pairs, etc. Shortly, there are areas on simulating the flow analysis using non-meshing techniques with higher accuracy than the current FEM techniques involved, which also can be incorporated undoubtedly to our adsorption systems.

On the other hand, after industrialization, the world is marching forward towards new technologies in artificial intelligence, machine learning [5], neural networking, etc. These techniques also can be embedded into refrigeration [6] systems, as explained below:

For instance, in our case of adsorption technology, performance depends greatly on various factors, like operating time, operating conditions, switching time between adsorber, and desorber. Initially, we can develop a machine learning algorithm based on supervised learning, where it works on input-output methodology. The following table can give us a better understanding of how it can be performed.

Temperature of surrounding (T)	Technique to be used
< 50°C (less solar radiation)	Multi-stage system
<85°C (low evaporative temperature)	Mass recovery technique
>120°C (high evaporative temperature)	Waste heat recovery technique

Table 1 Input output algorithm of temperature and techniques to be used

In the above table, a temperature sensor measures the surrounding temperature(input) and generates an output of which technique to be used in developing the system. This is the first stage. The values given in the table may vary with the properties of working pairs.

Similarly, another sensor can detect the maximum temperature of adsorber and desorber that is attained between the switching process and can, therefore, compute the switching time and check for maximum efficiency trial with various trial and error datasets collected as shown in the table below.

Maximum temperature of adsorber (°C)	Maximum temperature of desorber (°C)	Time taken In seconds
250	238.9	10.0
253.4	236.1	8.4
255.3	231.9	7.1
252.9	**235.6**	**8.5**

Table 2. Switching time between adsorber and desorber

Two sensors can be embedded into the system, one at adsorber and one at desorber and by measuring the temperature difference between the two and estimating the time taken for switching between adsorber and desorber, we can compute the response time, again with various variables like material, refrigerants, etc. and hence we can plot a graph and check for least switching time between different pairs and beds, and can estimate maximize efficiency. The tables are for illustration purposes only on how future technology can be enabled into adsorption systems. It just gives the idea to work on systems and is not the generated data.

5. Conclusion

This paper reviews the adsorption technology that is currently a point of research in the field of refrigeration and air-conditioning. For any new system to be viable to use, a reference system needs to stand as the fulcrum to compare and estimate results. In that notion, the traditional VCR system was discussed on its working and the thermodynamic cycle. We also got a brief idea of the need for adsorption systems, acting as an alternative and the adsorption mechanism was established broadly. Then, various performance parameters that enable us to provide a better efficient system were discussed on waste heat recovery, mass recovery, multistage and multi-bed systems. Various working pairs also were discussed with their merits and demerits. Conclusively, the paper summarizes on how modern techniques like AI and Machine learning [5] can be incorporated in adsorption systems [6] for better future of the society in making convenient and quick analysis without harming the nature and live in a sustainable environment

References

[1] D.C.WangY. H.LiabD.LiaY. Z.XiabJ.P. Zhanga, "A review on adsorption refrigeration technology and adsorption deterioration in physical adsorption system", Elsevier, https://doi.org/10.1016/j.rser.2009.08.001

[2] Amir Sharafian, Seyyed Mahdi Nemati Mehr, Poovanna Cheppudira Thimmaiah, Wendell Huttema, Majid Bahrami, "Effects of adsorbent mass and number of adsorber beds on the performance of a waste heat-driven adsorption cooling system for vehicle air conditioning applications", Elsevier, https://doi.org/10.1016/j.energy.2016.06.099

[3] Ali Alahmerab, Salman Ajiba, Xiaolin Wangc, " Comprehensive strategies for performance improvement of adsorption air conditioning systems: A review", Elsevier, https://doi.org/10.1016/j.rser.2018.10.004

[4] M.Z.I.Khana1, B.B.Sahab, K.C.A.Alamc, A.Akisawaa, T.Kashiwagia, " Study on solar/waste heat driven multi- bed adsorption chiller with mass recovery", Elsevier, https://doi.org/10.1016/j.renene.2006.02.003

[5] A. Santana, Y. Fukuyama, K. Murakami and T. Matsui, "Machine learning application for refrigeration showcase fault discrimination," 2016 IEEE Region 10 Conference (TENCON), Singapore, 2016, pp. 10-13, doi: 10.1109/TENCON.2016.7847948.

Guo-liang Ding, "Recent developments in simulation techniques for vapour-compression refrigeration systems", International Journal of Refrigeration, Volume 30, Issue 7,2007, ISSN 0140-7007, https://doi.org/10.1016/j.ijrefrig.2007.02.001.

[6] Sourav Mitra, Seung Taek Oh, Bidyut Baran Saha, Pradip Dutt & Kandadai Srinivasan," Simulation Study of The Adsorption Dynamics of Cylindrical Silica Gel Particles" vol 46, Heat Transfer Research.

State of the Art of Phase Change Material (PCM) for the Purpose of Heat Storage and Transfer in Cylindrical and Rectangular/Square Shells/Tubes – A Review

Prashanth S P[1], Dr. Devendra Dandotiya[2] and Dr. Surendra Kumar A M[3]
1,2, 3 Department of Mechanical Engineering, Presidency University. Bengaluru, India.

E-mail: devendradandotiya@presidencyuniversity.in

Abstract: - The state of the art of the PCM's reveals enabled characteristics that opens the door for multiple heat transfer and storage phenomena. The PCM absorbs, stores, and releases the energy during its phase change thus it most widely helps in latent heat transfer techniques. Nowadays Nano-fluid and Nano-phase change materials are replacing the general phase change materials because of its high heat transfer and storage capability. Further, varieties of passive heat transfer techniques such as heat exchangers of shell/tube type of various patterns and geometries are utilized to optimize the thermal energy. The artificial neural network is embedded in the system to recognize the pattern for faster processing. In this article, the PCM heat storage and transfer techniques are critically reviewed in comparison with the Cylindrical and Rectangular/Square shells/tubes based on the effect of parameters such as solidification/melting performance, enthalpy, fluid flow pattern in the liquid molten state, types of PCM used, etc. The numerical/analytical and experimentation techniques are categorized based on the results of various researcher and conclusion is established.

Keywords: Phase change material (PCM), Energy storage, Heat transfer enhancement, Latent heat, natural convection, melting/solidification time.

1. Introduction

In the area of energy sector, storing and transferring the thermal energy without loss is a vital factor. From the earlier days there has been good number of techniques been used such as heat exchangers, chemicals etc., to overcome this issue. As the industries and researchers progress they came across the new composed materials called PCM's, which revealed to be unique for storing and transferring the energy, but since it possesses low thermal conductivity, modifications based on utilization were applied in the form of active and passive techniques to overcome this problem. The PCM is broadly classified under organic and non-organic materials. The prominent PCM materials such as Cu, Al_2O_3, Au, SiC, SiO_2, and TiO_2 [1] are considered for an improved rate of heat transfer. Further, by utilizing various geometries of shells and tubes the heat transfer rate is enhanced. Nowadays Nanofluid and Nanophase change materials are replacing the general phase change materials because of its high heat transfer and storage capability. Further, the artificial neural network is embedded in the system to recognize the pattern for faster processing. The applications of PCM are 1. To dissipate the heat from the photovoltaic panels to maintain its voltage thus generating the electric pulses uninterrupted. 2. To dissipate the heat from electronic equipment's which generates more heat due to 24/7 working conditions such as industrial relays, transmitters, junction elements, etc. 3. To regain solar energy in its absence from the latent heat thermal energy storage (LHTES) system. 4. Portable/central space heaters and greenhouse application etc. This article concentrates on reviewing, how the PCM is utilized to store and transfer the heat in the Cylindrical and Rectangular/Square Tube/shells for different applications.

2. PCM heat transfer in cylindrical shells/tubes.

The containers of the PCM should be optimized based on the type of geometry and material used to store and transfer the heat energy as required. Further, the cylindrical shells/tubes are used to carry the PCM's and sometimes Nanoparticles are dispersed in the PCM's to enhance the heat transfer rate. These are made of different materials and sizes based on the given application. It is subjected to a different rate of heat to identify the PCM solidification/melting temperature. Below are the reviews on different types of techniques/methods used to analyze the heat transfer in cylindrical shells/tubes. Fang et al. [2], developed an index value significantly characterizing the stored energy ratio with respect to LHTES systems. This index value was used to analyze the heat transfer effectiveness of multiple tube tank design as shown in Fig. 1 [3] by using numerical method of conjugate fluid flow and enthalpy technique. This was later compared to stratified water storage (SWS) system on ideal condition. Thus the heat transfer effect of PCM in the container was defined in the form of compact factor (CF). The developed index value was found to be directly proportional at any value of CF, but at any optimal value of CF, quite a percentage of theoretical energy storage was not utilized. Thus it showed that Heat transfer in the working fluid can be further improved.

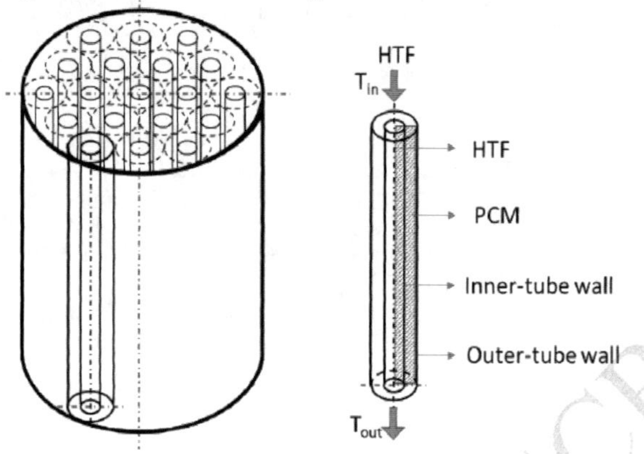

Figure 1. Multiple tube tank design [3].

Jones et al. [4] performed experimental measurements in a cylindrical enclosure melting the n-eicosane material from the side in a well-controlled environment as showing in Fig 2. The authors photographed the melt and ascertained it using an image processing procedure. Comparison between the numerical and experimental methods was done based on the temperature data taken based on the melt location taken. The enthalpy method and multi-block finite volume method of numerical investigation reveals good agreement, with the best at the lower Stefan number 0.1807. The experimental result based on a Stefan number of 0.0836 was recommended for validation against numerical data.

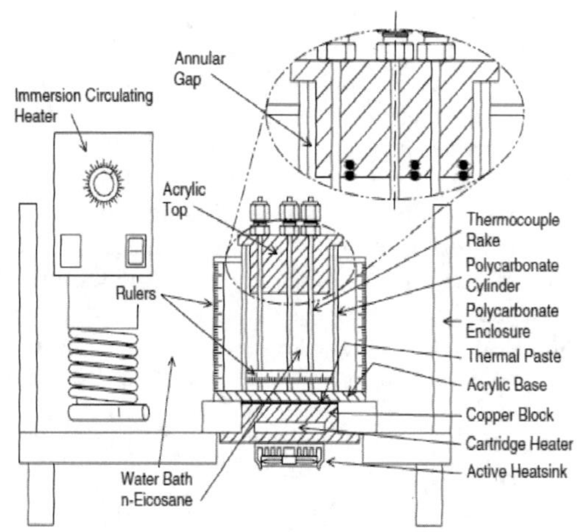

Figure 2. Schematic diagram of cylinder enclosure experimental facility [4]

Regin et al. [5] conducted a series of experiments on the LHTES system considering paraffin wax as PCM filled in horizontal cylindrical capsules. The PCM was melted using hot water around it. The capsule radius effect and working fluid temperature effect on the melting of PCM was investigated. Numerical analysis was carried out by using the enthalpy method to determine the phase change for a range of temperatures instead of constant temperature. The results were analyzed with experimental data, which was processed by the visualization technique considering three stages of melting (natural convection, contact melting, and the phase change temperature range) without disturbing the actual melting process. Both experimental and theoretical results show that heat transfer is significantly enhanced due to variation in the range of temperature. These results were also governed by the immensity of the Stefan number considering the temperature range and capsule radius.

Sparrow and Broadbent [6] conducted experiments on PCM filled inside the vertical tube with top surface being insulated by air space. The PCM was subjected to melt affecting the motions in the liquid melt due to change in the volume and natural convection. Their experimental results showed that the heat transfer rate was increased by fifty percent in comparison to the conduction model. Further, they measured that energy in the melt was twice the pure conduction prediction model. The Dimensionless number group involving Fourier, Stefan, and Grashof numbers were used to correlate the prediction data.

Saitoh and Hirose [7] analyzed the n-octadecane or water in a horizontal circular cylinder capsule as PCM for its melting rate and heat transfer rate for Rayleigh numbers. They obtained numerical solutions to find a solution to high Rayleigh numbers and to improve their resolution against heat transfer parameters. The researchers have shown that based on the position of PCM systems, the storage density and heat transfer rate is affected.

The authors Tay et al. [8] performed experimental investigations based on cold storage applications, by considering its thermal energy with coils of the tube inside a cylinder tank filled with salt hydrate as PCM along with water at -27°C as shown in fig 3. They showed that multiple tube tank design can deliver high storage density with a compact factor of 90% and above. Further by providing the optimized heat transfer techniques, the stored energy can be increased to 70% considering average heat exchanger effectiveness.

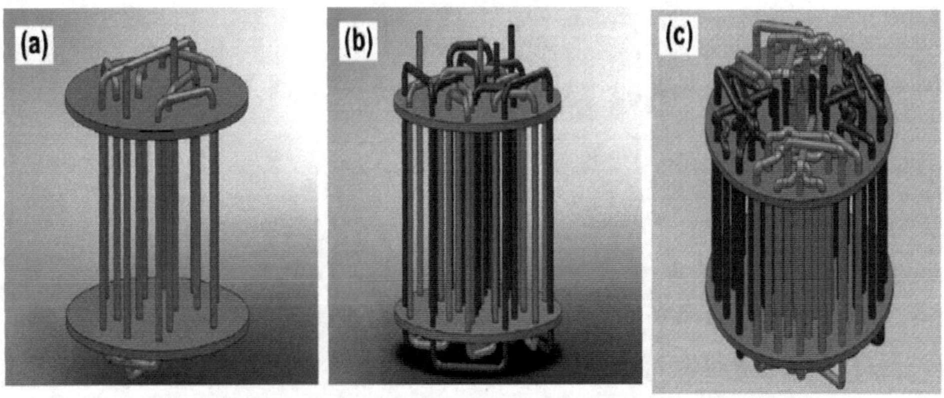

Figure 3. Schematic of multiple tube tank storage system [8]

Rieger and Beer [9] investigated melting ice phenomena inside an isothermal horizontal cylinder experimentally and analytically. The mechanisms governing the heat transfer during the melting process is studied numerically as well as experimentally. In both cases, the heat transfer is said to be increased at the bottom part of the ice body at 8°C. This shows density effect resulting in an upward movement of the solid ice. A close relativeness is said to be found between the numerical and experimental data. Similarly, Rieger et al. [10] followed the same technique to determine the melting heat transfer inside the horizontal tube using n-octadecane as PCM. Thus the phenomena is better understood in phase change process.

Wu and Lacroix [11] performed the computational technique based on body-fitted coordinates to analyze the natural convection melting in an isothermal cylinder mounted vertical. The authors utilized the governing conservation equation considering vorticity, stream function, and temperature. Their analysis shows that the heat transfer rate in the upper surface reduced systematically to zero as convection heat transfer completely raised in the melt. Whereas the bottom surface was observed with intensive heat transfer. The convective flow pattern concerning time was seen to be complicated in comparison to both the top and bottom surface.

The authors Cabeza et al. [12] worked on the storage of solar energy which is the major issue due to drastic changes in climatic conditions. They utilized the different PCM's such as fatty acids, trihydrate and paraffin in the system at the top of a stratified water storage tank as shown in fig 4. This PCM system used multiple cylinders at the top of the water tank. The solar pilot plant was developed to test the behavior of the PCM in a real situation to test the solar system. Plenty tests were carried out by varying the cooling down process, reheating process, and solar penetration with different volume of PCM, and temperature range. It was found that adding the PCM system for the domestic solar water heater is a promising technique. This would allow having a continuous supply of hot water without any external source of energy.

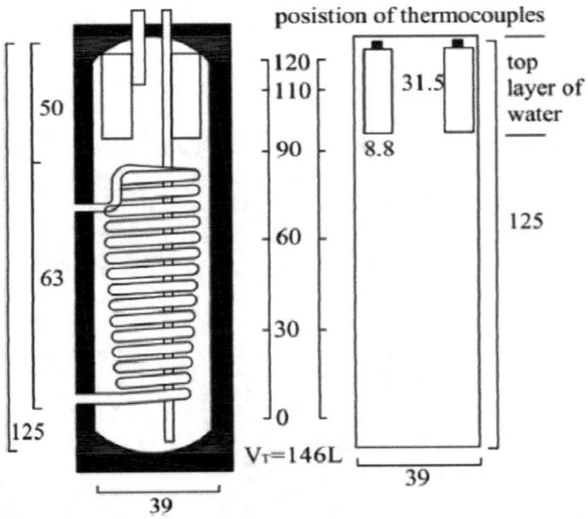

Figure 4. Water tank with PCM modules of different configurations [12]

Sun et al. [13] experimentally and numerically inspected the effect of partial volume fraction of solidification/melting phase change processes in particle-laden fluids. Inspectors observed that similar to the behavior (strong melt convection) of pure wax the slurry with 5% volume fraction behaves. For a volume fraction of 50% and 25%, heat conduction dominated in the melting process, and transition heat transfer was observed respectively. The diffusive flux model was used for numerical analysis which displays reasonable agreement for the partial volume of 5% and 50% for dilute and very dense slurry respectively. Further, they stated that more advanced numerical models are required to forecast to analyze the heat transfer behavior of particle-laden fluids. This study was further testified by Jones et al. [14]

An experimental investigation was performed by Zeng et al. [15] based on Nanoparticle enhanced phase change materials (NePCM's) in a vertical cylindrical cavity – bottom heated. They dispersed multi-walled carbon nanotubes (CNT's) into 1-dodeconal to prepare NePCM samples in liquid phase with various loadings (0 wt.%, 1 wt.%, and 2 wt.%) and found that melting decelerated in the presence of CNT's due to increased viscosity. Natural convection was found to be decreased even though heat conduction was enhanced thus affecting the melting rate for the 1 wt.% sample, but for the 2 wt.%

sample melting rate was completely sealed. The authors declared that there is partial relevancy between experimental and numerical results in terms of melting behavior, this is because of the uncertainty of thermo-physical properties of NePCM's used in the computational models. Hence it is suggested to use measured thermos physical properties of NePCM's instead of using prediction models or correlations.

3. PCM heat transfer in rectangular/square shape shells/tubes.

The different patterns of rectangular/square shells/tubes are used with or without internal/external fins to determine the melting/solidification rate to drive the heat transfer. These arrangements are further utilized with bottom heating or side heating to enhance heat transfer. Apart from the heated sections, other sections are usually well insulated to prevent heat transfer from the respective direction. The horizontal or vertical orientation of the containers are also said to be impacting the heat transfer rate. Below are the studies done by several authors considering the respective geometries.
Ho and Gao [16] conducted melting experiments using n-octadecane as PCM and dispersed Al2O3 as nanoparticles utilized in 25mm square and 60mm long vertical enclosure as shown in figure [5]. The mixture was heated from the vertical sidewalls isothermally insulating other sidewalls. The heat transfer rate was measured across the negligibly subcooled wall which demonstrates that natural convection dominates across the melting region, adding nanoparticles outweighed the dynamic viscosity, the quasi-steady Nusselt number rises to 60% because of increase in 10 wt.% of alumina particles. Thus the particle dispersed heat transfer rate is outweighed by the natural convection heat transfer rate.

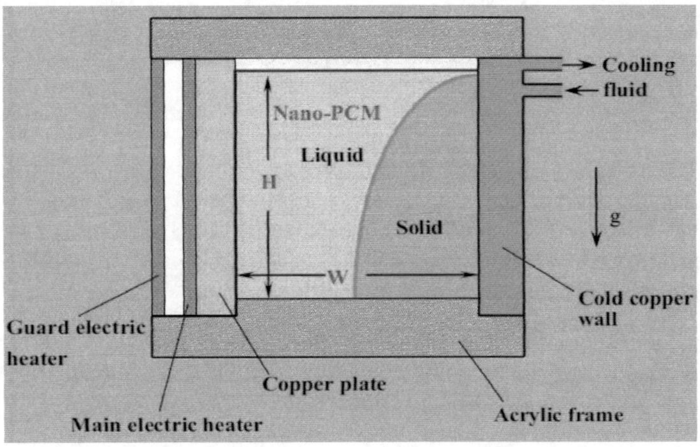

Figure 5. Schematic diagram of experimental test cell.

Motahar et al. [17] experimentally investigated the effect of titanium oxide dispersed as nanoparticles in the n-Octadecane as the PCM in the vertical rectangular enclosure heated on one side to measure its melting rate. Before adding the nanoparticle, PCM is said to have good conductivity, later it is dominated by natural convection. After the dispersion of nanoparticle, the viscosity increased, and thus the heat transfer rate is said to be deteriorated in the form of convection decreasing the melting rate. The melted volume fraction at the Fo number of 0.578 and heat flux of 2500 W/m2 were measured at a different rate of wt.%. The quasi-state Nusselt number reduced with an increase in Bingham number because of an increase in viscous force at odds with buoyant force. The authors developed two universal correlations which were satisfying by considering the uncertainty below ±12%, and ±17%, for the Nusselt number and melted volume fraction respectively.

Bashar and Kamran [18] utilized the rectangular enclosure as shown in figure 6, to experiment on paraffin as PCM with dispersed Nano particles to observe the mixture characteristics. The nanoparticles of aluminum, copper oxide, silver and others were used. After conducting series of experiments they came to the conclusion that enriched-optimum Nano particle-PCM mixture has enhanced thermal behavior in comparison to the pure PCM (CuO-18% and silver-14%). But when the mix ratio is not optimal, it resulted in dense viscosity degrading the characteristics of PCM. The high thermal characteristics was observed at optimal mix ratio for copper oxide-paraffin (0.6:10) and the melt rate/heat flux rate were increased to 25% higher than that of pure PCM.

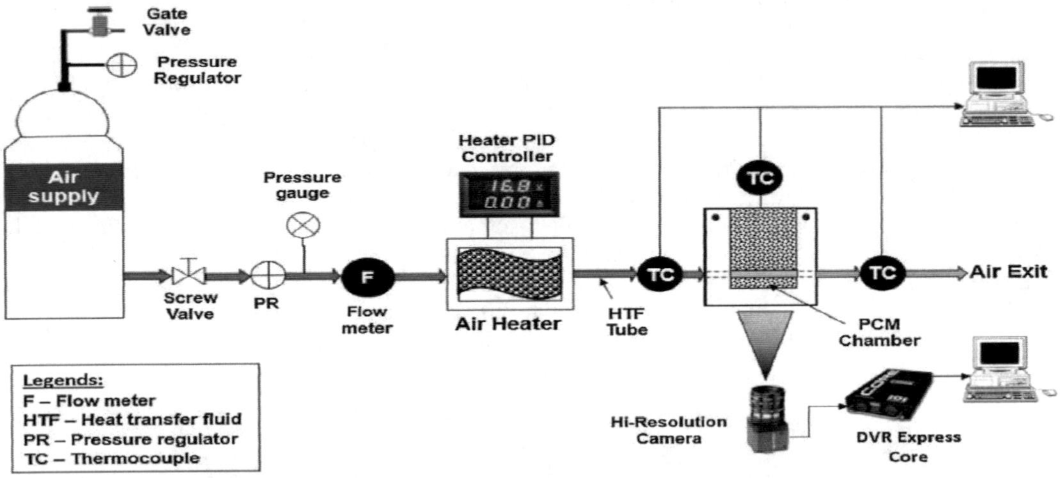

Figure 6. Schematic diagram of the experimental setup. The horizontal heat source (HTF tube) is shown near the bottom of the chamber. [18]

Hoseinzadeh et al. [19] numerically investigated the performance of latent heat thermal storage [LHTS] system consisting of several rectangular PCM slabs by considering the variation in geometries like breadth, length, and width of the wall. In this system, the air is considered as the heat transfer fluid (HTF) to cool/heat the system. The method of enthalpy was used to solve fundamental equation for the melting process in PCM's (CaCl2, 6H2O, and RT25). The general convective heat transfer equation was used to investigate the heat transfer in the air channels. The better results were obtained by using thinner and longer PCM slabs, and by using tiny air channels. The results show that when the outlet air temperature is increased, the flow rate as well as the fraction of molten PCM also increased.

Mosaffa et al. [20] studied the 2-d numerical model using the enthalpy method for a with internal horizontal fins for a rectangular container to predict the solidification temperature of PCM between the melting and solidification interface in the storage. The PCM melting and solidification factor was analyzed for different cell aspect ratio using the derived predication model. The result shows that the temperature of the working fluid or heat transfer fluid (HTF) increased to reduce the solidification time.

Talati and Taghilou [21] utilized the internal finned rectangular container to determine the melting/solidification temperature of the PCM. The authors utilized the lattice Boltzmann method (LBM) to analyze the same. They solved the problem of PCM by the analytical method and compared it against the finite volume method which was later used to predict the accuracy of the system. The freezing time of the PCM was observed half of the time of melting. The analytical solution of the model predicts the required time for solidification of PCM and it is said to be overestimated. Heat loss is reduced at the insulated composite walls of the PCM.

Zivkovic and Fujii [22] performed the numerical analysis on melting time of PCM encapsulated in single rectangular container using simple computational model. They simulated the transient behavior of isothermal phase change and the results were compared with solved test problem as well as with the experimental data showing close agreement. They concluded their work with the following points:
1. The accuracy of the computational data doesn't change even though the containers conduction resistance is omitted.
2. The PCM conduction heat transfer can be ignored the direction of HTF flow as its value was negligible.
3. The prediction of temperature variation within the flat thin container of liquid PCM is found to be having negligible error even though the effect of natural convection is ignored.

Hoseinzadeh and Chamkha [23] worked on the performance of multiple PCMs in different rectangular slabs by maintaining almost constant temperature difference between HTF and PCM. The air was used as the HTF and CaCl2, 6H2O and RT25 as the PCMs. They carried out 2-d numerical investigations of LHTES using enthalpy method and investigated the effect of dimensional parameters such as the PCM slab width and the breadth, apart from this the effect of air flow rate in the exhaust air temperature of system was also measured. Their results seek that using less width PCM slabs, the number of PCM container decreases and good characteristic performance was obtained by using longer and small width PCM slabs. This leads to increased storage volume thus resulting in higher pressure drop in the system. It also shows that when the flow rate of air is increased, the fraction of the molten PCM is also increased.

Krishna et al. [24] analyzed the heat transport phenomena in melting and laminar film condensation in rectangular container of LHTES system. They performed the numerical simulation by employing formulation of modelling for the transient characteristics for PCM with foam with porous structure and is analyzed by finite element method for various dimensions of rectangular container. The results of the simulations show that selection of right aspect ratio of PCM containers is crucial to increase the energy storage and heat transfer. This effect is further deliberated by thermal resistance ratio. The film condensation in the vertical outside walls of containers was considered for the scenario, and the value of best aspect ratio is 1, irrespective of PCM, foam and heat extraction methods. It was also found that utilizing high thermal conductivity foam only doesn't increase the heat storage capacity but appropriate foam type with a large thermal mass or a high thermal conductivity respectively forms the right choice.

Taghilou and Talati [25] presented an approximate analytical/numerical model for the purpose of investigating the solid-liquid interface and temperature distribution of PCM during the solidification of PCM inside a container connected to fins with boundary conditions of time dependent. The numerical and analytical results for the solidification front and temperature distribution are compared with results obtained by the Lattice Boltzmann method. It was found that the use of fin to increase the rate of heat transfer is justifiable in the container with the aspect ratio of less than 1. The highest error rate of 0.5 in aspect ratio was found by using analytical model used for the purpose finding the time required for complete freezing and the lowest was 0.25.

Imen and Mounir [26] investigated the PCM solidification in a finned rectangular heat exchanger including natural convection to exploit solar energy continuously. In the first step they analyzed all the models of solidification and also the temperature/velocity distribution using the numerical model of conservation equations. Their study also showed longitudinal air temperature profiles with the transient evolution. It also shows the effect of number of fins on heat transfer enhancement and the temporal evolution of the solidification front. This study showed the maximum amount of heat extracted during the solidification of the PCM.

Nabeel et al [27] performed experimental and numerical investigation on melting of n-octadecane with CuO nanoparticle suspensions in a square enclosure. The container was thermally insulated and provision was given on one side to heat it at a constant heat flux. They recorded the temperature at different junctions pertained to time-dependent. The finite element method was used to solve the momentum, coupled continuity and energy coupled equations numerically. The loading of nanoparticles into the PCM has a rising effect in thermal conductivity of the PCM/nanoparticle composite, thus augmenting the heat transfer rate results in decreased charging time. But caution should be taken for agglomeration of nanoparticles which increases the viscosity and precipitation. By competing heat transfer mechanisms, the shape of the melting interface is affected highly by the. The flat shape interface is present in the conduction-dominant regime, while curved shape of the interface emphasizes development of natural convection.

Hossein and Babak [28] experimentally investigated the characteristics of lauric acid as PCM in a rectangular thermal storage unit heated from one side. It satisfies thermos-physical properties of PCM as medium temperature PCM. The melt fraction and other heat transfer characteristics such as average Nusselt number, melt front, and soli-liquid interface morphology was studied using image processing technique. The results of the experiments show that the heat conduction was dominant during initial stage of heat transfer, followed by transition from conduction to convection regime, which dominates the heat transfer at later times. Approaching the end of the melting process, bulk temperature of the liquid PCM increases and stratified temperature field appears at upper part of the enclosure which reveals depression of the convection currents.

Soares et al. [29] experimentally investigated the heat transfer with two different PCM (the free-form PCM-Rubitherm the microencapsulated PCM-Micronal) in a vertical stack of rectangular cavities for both melting and solidification processes. The main goal of author was to discuss which type of PCM is better for building applications. They investigated the control temperature value on the hot surface of test sample and the period thermal regulations. The free PCM was found to be better for thermal control of vertical systems due to improved natural convection. But the control temperature effect and thermal regulation period both were reduced. During the discharging process, subcooling palyed an important role during the solidification of the free PCM and its effects can't be neglected.

4. Conclusion.

This paper shows the quench of experimental and numerical studies carried out by researchers considering the effect of PCM's in cylindrical/rectangular containers and their positions. Most of the researchers used the vertical position of cylindrical containers instead of horizontal position as it consumes less ground space, increased storage density ratio, enhanced heat transfer rate from 40% to 60%. In the horizontal cylinders, the heat transfer rate is said to be more at the lower part of the container, this is because of the high storage density effect as the result of the upward movement of the working fluid. The convective heat transfers in both the position of cylinders seemed to be more and as the time progressed the authors found it to be more complicated. In most cases, the diffusive flux model and enthalpy model were used to predict the heat transfer behavior which seemed to be not quite accurate hence there was a difference in results in comparison to experimental and numerical techniques. In the rectangular PCM containers, the dispersion of Nanoparticles increased the convective heat transfer rate in comparison to the natural convection at the beginning, but as time progressed the dispersed particles increased the dynamic viscosity of the working fluid and thus heat transfer rate is degraded for over a while. The heat transfer in the PCM was due to conduction before adding the Nanoparticle in the rectangular enclosure, later it was dominated by natural convection. The cylindrical and rectangular containers are seemed to be more effective in lower and higher storage capacity ratio respectively as well as the heat transfer rate.

References:

[1] Rehman, Tauseef-ur, et al. "A critical review on heat transfer augmentation of phase change materials embedded with porous materials/foams." *Int. J. Heat Mass Transf.* 135 (2019): 649-673.
[2] Fang, Yuhang, Jianlei Niu, and Shiming Deng. "Numerical analysis for maximizing effective energy storage capacity of thermal energy storage systems by enhancing heat transfer in PCM." *Energy and Buildings* 160 (2018): 10-18.
[3] Trp, Anica. "An experimental and numerical investigation of heat transfer during technical grade paraffin melting and solidification in a shell-and-tube latent thermal energy storage unit." *Solar energy* 79.6 (2005): 648-660.

[4] Jones, Benjamin J., et al. "Experimental and numerical study of melting in a cylinder." *International Journal of Heat and Mass Transfer* 49.15-16 (2006): 2724-2738.
[5] Regin, A. Felix, S. C. Solanki, and J. S. Saini. "Latent heat thermal energy storage using cylindrical capsule: Numerical and experimental investigations." *Renewable energy* 31.13 (2006): 2025-2041.
[6] Sparrow, Ephraim M., and J. A. Broadbent. "Inward melting in a vertical tube which allows free expansion of the phase-change medium." (1982): 309-315.
[7] Saitoh, T., and K. Hirose. "High Rayleigh number solutions to problems of latent heat thermal energy storage in a horizontal cylinder capsule." (1982): 545-553.
[8] Tay, N. H. S., M. Belusko, and F. Bruno. "Experimental investigation of tubes in a phase change thermal energy storage system." *Applied Energy* 90.1 (2012): 288-297.
[9] Rieger, H., and H. Beer. "The melting process of ice inside a horizontal cylinder: effects of density anomaly." (1986): 166-173.
[10] Rieger, H., et al. "Heat transfer during melting inside a horizontal tube." (1983): 226-234.
[11] Wu, Yong Ke, and Marcel Lacroix. "Melting of a PCM inside a vertical cylindrical capsule." *International journal for numerical methods in fluids* 20.6 (1995): 559-572.
[12] Cabeza, Luisa F., et al. "Experimentation with a water tank including a PCM module." *Solar Energy Materials and Solar Cells* 90.9 (2006): 1273-1282.
[13] Sun, Dawei, S. Ravi Annapragada, and Suresh V. Garimella. "Experimental and numerical study of melting of particle-laden materials in a cylinder." *International Journal of Heat and Mass Transfer* 52.13-14 (2009): 2966-2978.
[14] Jones, Benjamin J., et al. "Experimental and numerical study of melting in a cylinder." *International Journal of Heat and Mass Transfer* 49.15-16 (2006): 2724-2738.
[15] Zeng, Yi, et al. "An experimental investigation of melting of nanoparticle-enhanced phase change materials (NePCMs) in a bottom-heated vertical cylindrical cavity." *International Journal of Heat and Mass Transfer* 66 (2013): 111-117.
[16] Ho, Ching-Jenq, and J. Y. Gao. "An experimental study on melting heat transfer of paraffin dispersed with Al2O3 nanoparticles in a vertical enclosure." *International Journal of Heat and Mass Transfer* 62 (2013): 2-8.
[17] Motahar, Sadegh, Ali A. Alemrajabi, and Rahmatollah Khodabandeh. "Experimental investigation on heat transfer characteristics during melting of a phase change material with dispersed TiO2 nanoparticles in a rectangular enclosure." *International Journal of Heat and Mass Transfer* 109 (2017): 134-146.
[18] Bashar, Mohammad, and Kamran Siddiqui. "Experimental investigation of transient melting and heat transfer behavior of nanoparticle-enriched PCM in a rectangular enclosure." *Journal of Energy Storage* 18 (2018): 485-497.
[19] Hoseinzadeh, S., et al. "Numerical investigation of rectangular thermal energy storage units with multiple phase change materials." *Journal of Molecular Liquids* 271 (2018): 655-660.
[20] Mosaffa, A. H., et al. "Approximate analytical model for PCM solidification in a rectangular finned container with convective cooling boundaries." *International Communications in Heat and Mass Transfer* 39.2 (2012): 318-324.
[21] Talati, F., and M. Taghilou. "Lattice Boltzmann application on the PCM solidification within a rectangular finned container." *Applied Thermal Engineering* 83 (2015): 108-120.
[22] Zivkovic, B., and I. Fujii. "An analysis of isothermal phase change of phase change material within rectangular and cylindrical containers." *Solar energy* 70.1 (2001): 51-61.
[23] Hoseinzadeh, S., et al. "Numerical investigation of rectangular thermal energy storage units with multiple phase change materials." *Journal of Molecular Liquids* 271 (2018): 655-660.
[24] Kota, Krishna, Louis Chow, and Quinn Leland. "Laminar film condensation driven latent thermal energy storage in rectangular containers." *International journal of heat and mass transfer* 55.4 (2012): 1208-1217.
[25] Taghilou, Mohammad, and Faramarz Talati. "Analytical and numerical analysis of PCM solidification inside a rectangular finned container with time-dependent boundary condition." *International Journal of Thermal Sciences* 133 (2018): 69-81.
[26] Jmal, Imen, and Mounir Baccar. "Numerical investigation of PCM solidification in a finned rectangular heat exchanger including natural convection." *International Journal of Heat and Mass Transfer* 127 (2018): 714-727.
[27] Dhaidan, Nabeel S., et al. "Experimental and numerical investigation of melting of phase change material/nanoparticle suspensions in a square container subjected to a constant heat flux." *International Journal of Heat and Mass Transfer* 66 (2013): 672-683.
[28] Shokouhmand, Hossein, and Babak Kamkari. "Experimental investigation on melting heat transfer characteristics of lauric acid in a rectangular thermal storage unit." *Experimental Thermal and Fluid Science* 50 (2013): 201-212.
[29] Soares, N., et al. "Experimental study of the heat transfer through a vertical stack of rectangular cavities filled with phase change materials." *Applied energy* 142 (2015): 192-205.

Date Palm fibre reinforced composite by multiple resin to enhance the properties of a composite - A Review

Krishnan Rajan Rajthilak[1], Dr. Ramesh. S[2], Dr.G.N. Lokesh[2], Dr. Satish Babu Boppana[2]
1.Research Scholar, School of Engineering, Presidency University, Itgalpura, Rajanukunte, Bengaluru. India
2.School of Engineering, Presidency University, Itgalpura, Rajanukunte, Bengaluru. India.

E-mail: thilak.bemech@gmail.com.

Abstract: - Numerous applications of the composites have been increased in this Era. The Research on polymer matrix composites using Natural fibre is wide. Natural fibres are one of the most abundant resources that exist in nature. In this paper, Date palm fibre is discussed for making of composites as it is one of the most available natural fibres in the world. It is the basic agricultural crop, especially in Middle East countries. Specific strength is the Parameter which makes date palm fibre superior compared to other fibres. To improve proper adhesion, the natural fibres are to be chemically treated to attain the better properties when compared to the untreated fibres. However, resins play a vital role in preparing a Composite. There are multiple resins used as matrices to make a composite. Hence, this paper is to analyse the date palm fibre-based polymer matrix composites for different resins for their Mechanical properties and Thermal properties. The Properties of various Natural fibres are also discussed in this paper.

Key words: Natural fibres, Date palm fibre, Polymers, Multiple Resins.

1. Introduction

Composite materials play a vital role in day to day life for its applications in many fields like (Aerospace, Automobile, Locomotives, structural and building, Industrial applications, sports goods, Prosthetics etc). Synthetic and Traditional are the two main classification of Composites. Synthetic composites are used for its wide range of Applications, whereas the Traditional composites already exists for many years. Natural fibres exist in both animals and plants. Natural fibres dominate the synthetic fibres for its availability, low cost, low density, recyclability & biodegradability. The classification is shown in Fig.1.

The wide varieties of natural fibre include Abaca, Cotton, cane, grass, silk, jute, Kenaf, flax, Hemp, Date palm, banana, sisal, wheat, barley, rape etc. However chemical treatment is needed to improve proper adhesion between the polymer and fibre surface. Chemically treated fibres dominate in many aspects and properties than untreated fibres. The waxy layer removal from the natural fibre makes the rougher surface of natural fibres.

2. Date palm fiber.

Date or date palm is the major agricultural crop in Middle East countries. Date fruits are good in nutrition especially in hot climatic conditions. Date palm trees consists of different residues consists of tree trunks, bunch stalks and leaf stalks. The fibres are extracted from the different parts from Mesh, Rachis, and Petiole and bunch. Date palm wastes rather will pollute the Environment. The waste form Date palm can be utilised for its Eco friendly in nature. Date palm fibre provides superior chemical composition and holds good mechanical strength in contrast with other natural fibres.Fig.2 and Fig 3 shows Date palm classifications.

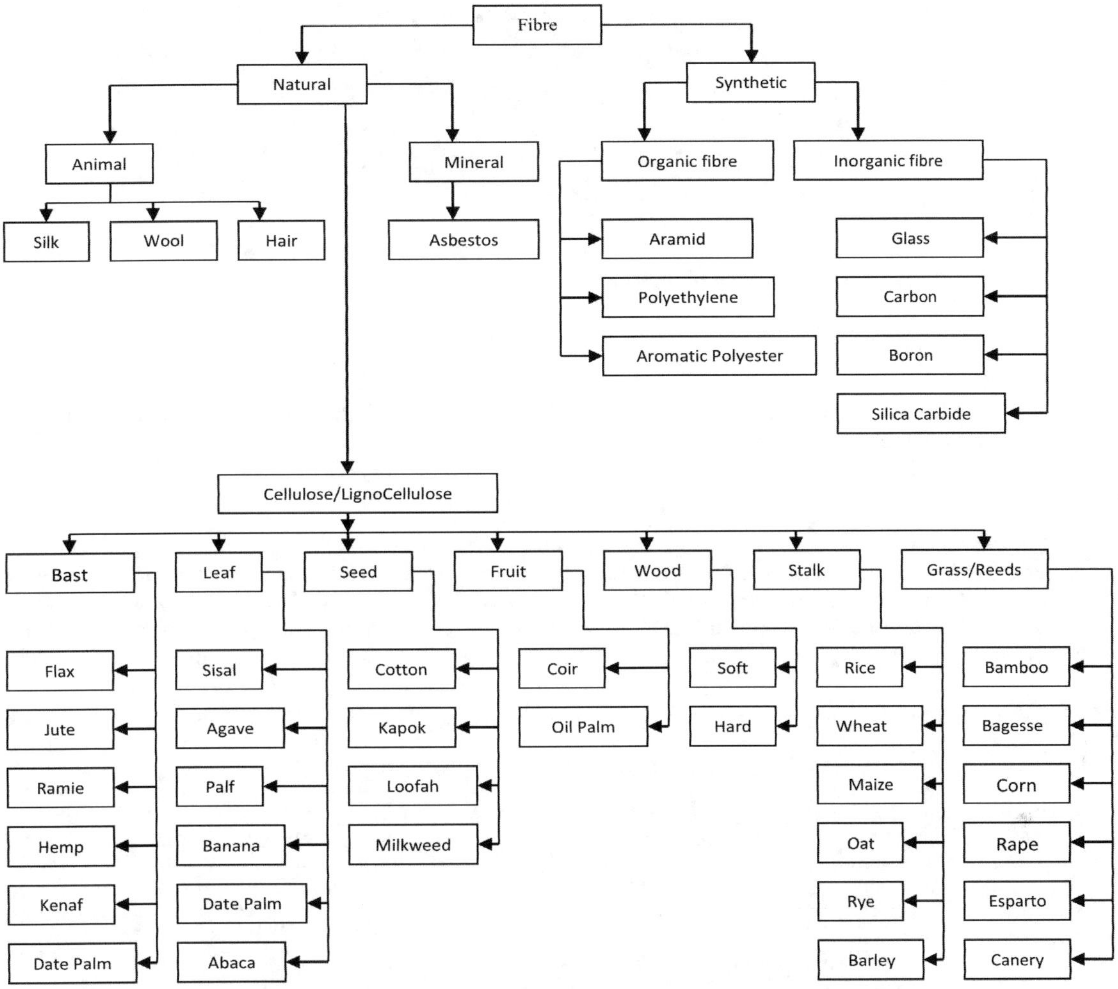

Figure1. Classification of fibres [1]

Figure 2. Date Palm Tree (a) Date Palm fruit (b) [1]

IOP Conf. Series: Materials Science and Engineering 1013 (2021) 012034 doi:10.1088/1757-899X/1013/1/012034

(a) (b) (c)

Figure 3. Date palm tree trunk with leaf rachis (a) and (b-c) Date Palm fibres [1]

2.1 Date palm fibre Reinforced polymer matrix.
Many researchers have reported that, Date palm fibre provides better results when it is chemically treated compared to chemically untreated fibres. It is also observed that Date palm fibre provides better strength when treated with chemical adhesives. The chemical treatment carried out by many researchers for palm date fibres are alkaline treatment, Silane treatment and Acytelation treatment.

2.1.1 Polymer matrix classification.
Polymer matrix is further classified into thermosetting matrix and Thermoset Matrix. Thermoplastic polymers are formed by addition polymerisation at elevated Temperatures. The properties remain same when it is melted and reformed into different shapes. Most used thermoplastics resins are polyethylene, polypropylene, and polystyrene. Thermoset polymers are formed by condensation polymerisation. They have better dimensional stability, stronger and harder than thermosetting polymers. Most used thermosets are epoxies, polyesters, alkyds and amino. Table.1 shows the classification of polymers.

Table 1. Thermoplastics and Thermosets

Polymers	
Thermoplastics	**Thermosets**
Polycarbonate	Phenolic Resins
Acrylic	Polyester Resins
Acetal copolymer Polyoxymethylene	Amino Resins
Nylon	Epoxy Resins
Polyethylene	Silicon Resins
Polyvinyl chloride	Polyurethanes
Polystyrene	
Polypropylene	
Teflon	

3. Discussion on properties with Different Types of fibres.
Many Investigations have been made to identify the most used resins for making the composite. However, the Thermal properties and Mechanical properties of a composite are most essential based on its applications.

3.1 Discussion on Thermal Properties.

Mohamed Hamdy Gheith, et al [2], studied Flexural, thermal, and dynamic mechanical properties of date palm fibres reinforced epoxy composites and reported that date palm fibre with 50% of composition have improved the Thermal stability of composite. The resin used was Epoxy. P.Noorunnisa Khanam,et al[3],studied Improvement of ternary recycled polymer blend reinforced with date palm fibre and reported that polymer ternary blends includes, Recycled low density polyethylene, Recycled high density polypropylene and Recycled polypropylene were used for better thermal stability. N. Saba, et al [4], studied Date palm reinforced epoxy composites: tensile, impact and morphological properties and reported that by using epoxy resin D.E.R 331 based on diglycidyl ether of bisphenol A (DGEBA), and the epoxy hardener (curing agent) Joint-mine 905-3S,concluding that of date palm fibre reinforcement with 50 % composition have the higher Mechanical stability.However Thermal Property was not discucced by the researchers.Ampol Wongsa et al., [5] studied the sisal and coconut fibre incorporated with geo polymer mortar and compared the results with synthetic (Glass fibre) and controlled geopolymer mortars. Researchers concluded that the modest increase in Thermal Conductivity using natural fibre (coconut and sisal) in geopolymer mortars compared to Synthetic glass fibre and controlled geo polymer mortars. Rather the results indicated the thermal conductivity tends to increase with increase in density of geo polymer mortars. However, resins were not used for the making of composite. Fabio da Costa Garcia Filho et al., [6] studied the piassava fibre and their epoxy composites. Researchers used a coating of graphene-oxide to piassava fibre and further Epoxy resin was used for making a composite. The results concluded that the slight increase in Thermal properties by incorporating graphene-oxide coated piassava fibre with epoxy resin compared to a neat Piassava fibre with Epoxy Resin. Sangeetha Dharmalingam et al., [7] studied Amine functionalized Luffa fibre and Epoxy Resin an Interfacial adhesion and Researchers concluded, the increase in thermal stability by Amine functionalized Luffa fibre and epoxy composite compared to unfunctionalized fibre. A. Parre et al., [8] studied the Banana fibre treated with NAOH and untreated Banana fibre. Thermo gravimetric analysis has been carried out with 10 mg of untreated banana fibre and with 1%, 3%, 5%, 7% & 9% NAOH treated Banana fibre. The result concluded that NAOH treated by 5% banana fibre showed the elevated Thermal strength compared to untreated fibre. P. Madhu et al., [9] studied Agave Americana Fibre for Composite Reinforcement on Various Chemical treatments effects. The Agave Americana fibre was treated with four different chemicals NAOH, Potassium permanganate, Benzoyl-peroxide and Stearic acid. The result concluded; surface modified Agave Americana fibre shows better thermal stability. Rather the Researchers included that the chemical treatment of fibre produces better quality fibre compared to untreated fibre for making a composite. Siew Choo Chin et al., [10] studied on Bamboo fibre reinforced composites. The Bamboo fibre was chemically treated with 10% NAOH with soaking duration varied for 0, 24, 48 and 72 hours. Researchers concluded that 10% NAOH chemically treated bamboo fibre soaked to 48 hours exhibits good thermal stability.

The Figure.4 indicates, the Thermal stability comparison of 50% Date Palm fibre with Epoxy [2] and Recycled Ternary polymer blend with Maleic Anhydride [3]. The Results indicates that Recycled Ternary blend with Maleic Anhydride, enhances better Thermal stability.

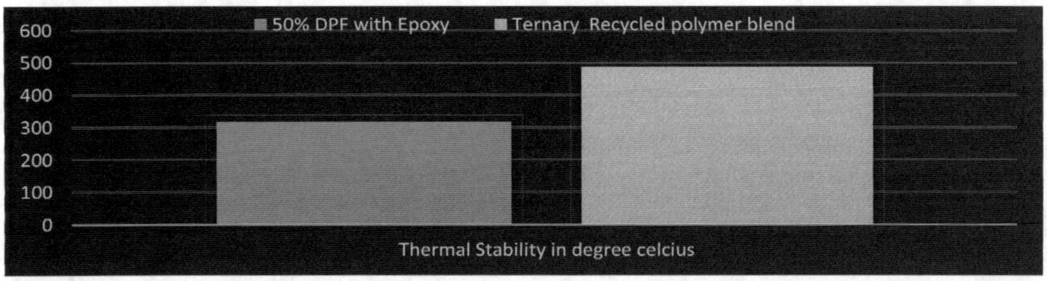

Figure.4 Comparison of Thermal stability of Plain Epoxy and Ternary Recycled Blend composite [2][3]

3.2 Discussion on Mechanical properties.

Mohamed Hamdy Gheith, et al [2], studied Flexural; thermal and dynamic mechanical properties of date palm fibres reinforced epoxy composites and reported that date palm fibre with 50% composition have good flexural strength and modulus.P.Noorunnisa Khanam,et al[3],studied Improvement of ternary recycled polymer blend reinforced with date palm fibre and reported that polymer ternary blends includes, Recycled low density polyethylene, Recycled high density polypropylene and Recycled polypropylene were used for better tensile strength and improved hardness of recycled polymer blend matrix.The resin used was Epoxy. N. Saba, et al [4], studied Date palm reinforced epoxy composites: tensile, impact and morphological properties and reported that by using epoxy resin D.E.R 331 based on diglycidyl ether of bisphenol A (DGEBA), and the epoxy hardener (curing agent) Joint-mine 905-3S,concluding that of date palm fibre reinforcement with 50 % composition have the higher Mechanical stability.Ampol Wongsa et al., [5] studied the sisal and coconut fibre incorporated with geo polymer mortar and compared the results with synthetic (Glass fibre) and controlled geopolymer mortars. The researchers concluded that the flexural strength, tensile strength and compressive strength of sisal and coconut fibre incorporated with geo polymer mortar is increased when compared to Synthetic and Controlled geoploymer mortars. Sangeetha Dharmalingam et al., [7] studied Amine functionalized Luffa fibre and Epoxy Resin an Interfacial adhesion. The conclusion by researchers is that the Tensile strength, Flexural strength, and impact strength is increased in Amine functionalized luffa fibre when compared to unfunctionalized luffa fibre. P. Madhu et al., [9] studied on Agave Americana Fibre for Composite Reinforcement with the Effect of Various Chemical treatments. The researchers concluded that the tensile strength is improved in Agave Americana fibre composite when it is chemically treated. Mohammad Asim et al., [11] studied Effects of date palm fibres loading on mechanical and thermal properties of Date Palm reinforced phenol composite and reported that 50% of date palm fibre improved the impact properties also with better tensile modulus but reduced strength in flexural and Tension. The resin used as matrix for making the composite is Bio-Phenol Resin. H. Khakpour, et al [12], studied Mechanical properties of structural adhesives enhanced natural date palm fibre and reported that Different structural adhesives including Araldite AY 103-1 / HY 991, UHU plus end fest 300, ADEKIT H 9951 T and Eponate were considered to be used as matrix. These are all two-component epoxy adhesives. Two types of UHU adhesive, with dissimilar resin to hardener weight ratio, UHU1 ratio 100R:80H by weight and UHU2 ratio 100R:120H by weight were considered as two different matrices. The results concluded that the date palm fibre could significantly increase the tensile strength of structural adhesives. F. Delzendehrooy, et al [13], studied adhesively bonded lap joints with palm date fibres for strength improvement and reported that UHU plus end fest 300, a solvent-free two-component epoxy based adhesive, an adhesive to mix aluminium substrates and concluded that the overall strength increased by using more qualified fibres.. A.Shalwan,et al [14], studied Influence of date palm fibre and graphite filler on mechanical and wear characteristics of epoxy composites and reported that the thermoset resin epoxy (R246TX) and kinetix (H160 medium) hardener was used as matrix for engineering applications and concluded with improved Mechanical properties, enhanced wear and frictional characteristics. T. Krishnan et al., [15] studied alkaline-treated Sun hemp fibre reinforced polyester composites and concluded higher Impact, Flexural and tensile strength of alkali treated sun hemp fibres compared to untreated fibres. ES Zaini et al., [16] studied rice husk composite. The Polypropylene copolymer is used as a Resin. The composite was made on 15%, 30% and 51% of rice husk, respectively. The result concluded that 15% of rice husk produced good Mechanical Properties. Raghavendra et al, [17] studied different samples of date palm fibres from five different regions and specimens are prepared. Mechanical properties like Tensile, impact and flexural strength are evaluated and found better results. Also, Fourier transforms infrared spectroscopes (FTIR) were used to analyse functional groups.

Figure 5 and 6 indicates the comparison on Tensile strength, Tensile Modulus and % elongation of break, of Recycled Ternary blend with Maleic Anhydride [3] and Plain Epoxy Date composite [4].Table 2 indicates the composition of Ternary blends A1,A2,A3 and A4.The results from the comparison,[3][4], concluded that 50% of Date palm fibre with epoxy and the A3 composition of Recycled Ternary blend composites provides better results.

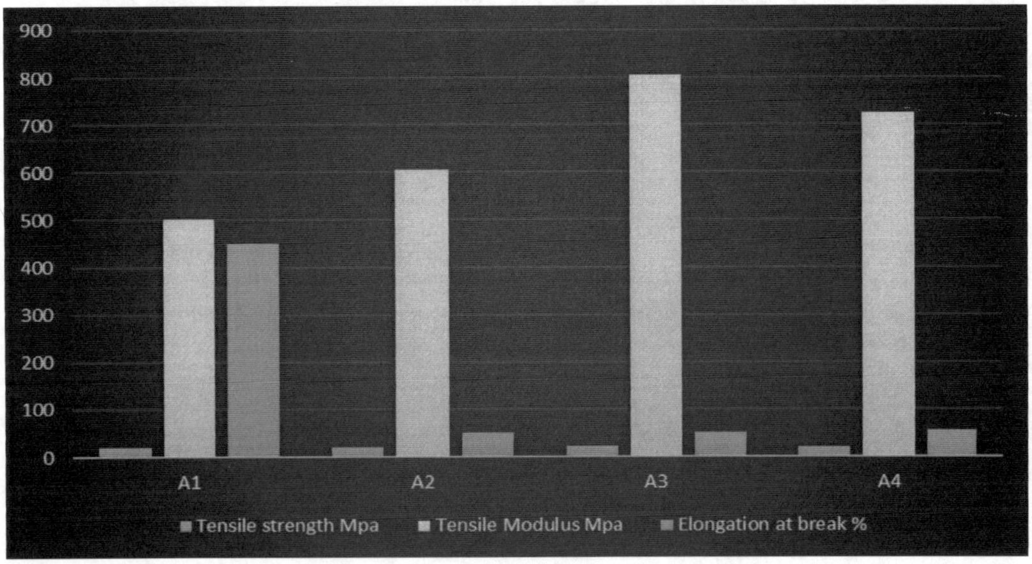

Figure.5. Tensile strength, Tensile Modulus and Elongation at break % of Recycled Ternary Blend composite [3]

Table 2. Composition of Recycled Ternary blend composites [3]

A1	0% of Date palm leaf	20% of Recycled low-density polyethylene	40% of recycled high-density Polyethylene	40% of Recycled Polypropylene	0% of Maleic Anhydride
A2	20% of Date palm leaf	10% of Recycled low-density polyethylene	35% of recycled high-density Polyethylene	35% of Recycled Polypropylene	0% of Maleic Anhydride
A3	20% of Date palm leaf	10% of Recycled low-density polyethylene	35% of recycled high-density Polyethylene	35% of Recycled Polypropylene	1% of Maleic Anhydride
A4	20% of Date palm leaf	10% of Recycled low-density polyethylene	35% of recycled high-density Polyethylene	35% of Recycled Polypropylene	2% of Maleic Anhydride

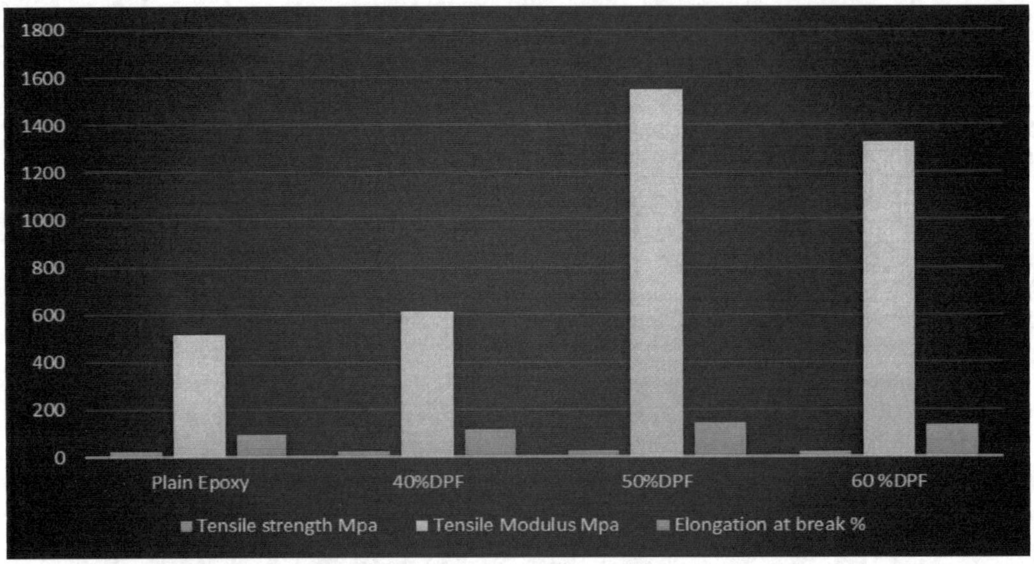

Figure.6. Tensile strength, Tensile Modulus and Elongation at break % of Plain Epoxy Date palm fibre [4].

4. Discussion on properties of hybrid natural fibre composites and its Applications.

K. P. Prashanth, et al [18], studied Synthesis and Characterization Study of Chitosan Based Natural Fibre: Biodegradable Polymer Composite and the Researchers concluded that the pre treated fibres increases the surface smoothness of fibres for the applications in food packaging and tissue engineering. L Prabhu, et al [19], studied Mechanical, Chemical and Acoustical behaviour of Sisal – Tea Waste – Glass Fibre Reinforced Epoxy Based Hybrid Polymer Composites and concluded that the Flexural strength, Impact strength, Tensile strength and Tensile modulus were increased with hybridisation of sisal in glass fibre for its various applications in automobile. R Bhoopathi, et al [20], studied on mechanical strengths of hemp-glass fibre reinforced epoxy composites and results concluded that good tensile strength and flexural strength of composites for its applications in automobiles. J Inbakumar Parivendhan, et al [21], studied Mechanical, wear and thermal behaviour of hemp fibre/egg shell particle reinforced epoxy resin bio composite and concluded with improved Mechanical property on load bearing capability of hybrid composite and increase in egg shell particles improved the initial mass stability for the use of its composites in industrial and structural applications. MR Sanjay, et al [22], studied The hybrid effect of Jute/Kenaf/E-glass woven fabric epoxy composites for medium load applications: impact, inter-laminar strength, and failure surface characterization and concluded with better impact strength of laminate due to hybrid effect and for its applications in automobile industries. GR Arpitha, et al [23], studied an overview on mechanical property evaluation of natural fibre reinforced polymers and concluded that good mechanical properties can be possessed by Natural fibres for its various applications. M Ramesh, et al [24], studied Plant fibre based bio-composites: Sustainable and renewable green materials and concluded that treating of fibres with different alkaline solutions will enhance the various properties which includes Flexural strength, Impact strength, Tensile strength and Fatigue strength for its various applications. M Ramesh, et al [25], studied Evaluation of mechanical and interfacial properties of sisal/jute/glass hybrid fibre reinforced polymer composites and concluded that good Flexural strength, Impact strength and Tensile strength possessed by sisal/jute and glass fibre reinforced hybrid composites. M Ramesh, et al [26], studied Impact behaviour analysis of sisal/jute and glass fibre reinforced hybrid composites and concluded that jute glass fibre reinforced ;polymer composites attains the maximum impact strength compared to sisal jute glass fibre reinforced polymer composites. M Sakthivel, et al [27], studied Production and characterization of luffa/coir reinforced polypropylene composite and concluded with improved mechanical properties when compared to polymer pure for its automotive applications. M Ramesh, et al [28], studied Mechanical property evaluation of sisal–jute–glass fibre reinforced polyester composites and concluded that jute glass fibre reinforced composite possesses maximum tensile strength, Jute and sisal glass fibre reinforced composite possesses maximum flexural strength and sisal glass fibre reinforced composite possesses maximum impact strength. MR Sanjay, et al [29], studied on mechanical properties of banana/e-glass fabrics reinforced polyester hybrid composites and concluded that nine layers of pure glass laminates provides Maximum tensile strength, Flexural strength, and impact strength. M Ramesh, et al [30], studied Biocomposites Reinforced with Animal and Regenerated Fibres and concluded that the animal and regenerated fibres possess low density and good mechanical strength for its few industrial applications. Ramesh, et al [31], studied Life-cycle and environmental impact assessments on processing of plant fibres and its bio-composites: A critical review and concluded the resin utilisation as the major factor for reinforcing with fibres and also concluded that the plant fibre reinforced composite have better applications compared with synthetic fibre composites

5.Conclusions.

- The recent papers by the researchers of the natural fibres concluded that the quality of the composite mainly depends on the proper resins used and the chemical treatment carried out on fibres to make a composite.
- Researchers have chosen multiple resins, in both the thermoplastic polymers and Thermoset polymers.
- However, majority of the researchers reported that Epoxy, the thermoset resin is commonly used, since it has better dimensional stability and good stiffness, better chemical Resistance, Cheaper and widely used upon other resins for Engineering Applications.
- Though researchers reported that the composite from Palm date fibre yield good Flexural properties, Thermal properties and dynamic mechanical properties and tensile strength, Still the Date palm fibre-based composites face several challenges in many of its applications.
- As per the figure 4, the Thermal stability is increased in Recycled Ternary blends compared to plain Epoxy date palm fibre and in contrast from figure 5 and figure 6, the Mechanical properties have minimum deviations in Recycled Ternary blends.
- As per the discussion of various authors, Hybrid natural fibre composites plays a vital role for its vast applications in automobile and industrial applications for its good Mechanical properties.
- The proposed research would be carried out on incorporating 50% of date palm fibre reinforced polymer matrix by blending different types of Resins and to test the Results for the Enhancement of the Mechanical and Thermal Properties.

References.

1. W Ghori, N Saba, M Jawaid and M Asim, "A review on date palm (phoenix dactylifera) fibres and its polymer composites", Material Science and Engineering ,2018.

2. Mohamed Hamdy Gheith, Mohamed Abdel Aziz, Waheedullah Ghori, Naheed Saba, Mohammad Asim, Mohammad Jawaid, and Othman Y. Alothman, "Flexural, thermal and dynamic mechanical properties of date palm fibres reinforced epoxy composites", Journal of Material science and Technology, June 2018.
3. P. Noorunnisa Khanam and M.A. AlMaadeed, "Improvement of ternary recycled polymer blend reinforced with date palm fibre" Materials and Design, April 2014.
4. N. Saba, Othman Y. Alothman, Zeyad Almutairi, M. Jawaid, and Waheedullah Ghori, "Date palm reinforced epoxy composites: tensile, impact and morphological properties", Journal of Material science and Technology, July 2019.
5. Ampol Wongsa, Ronnakrit Kunthawatwong, Sakchai Naenudon, Vanchai Sata, Prinya Chindaprasirt, "Natural fibre reinforced high calcium fly ash geopolymer mortar", Construction and Building Materials, January 2020.
6. Fabio da Costa Garcia Filho, Fernando Santos da Luz, Michelle Souza Oliveira, Artur Camposo Pereira, Ulisses Oliveira Costa, Sergio Neves Monteiro, "Thermal behaviour of graphene oxide-coated piassava fibre and their epoxy composites", Journal of Material Research and Technology, March 2020.
7. Sangeetha Dharmalingam, Omkumar Meenakshisundaram, Vijayakumar Elumalai & Rajendra S. Boopathy, "An Investigation on the Interfacial Adhesion between Amine Functionalized Luffa Fibre and Epoxy Resin and Its Effect on Thermal and Mechanical Properties of their Composites", Journal of Natural fibres, February 2020.
8. A. Parre, B. Karthikeyan, A. Balaji, R. Udhayasankar, "Investigation of chemical, thermal and morphological properties of untreated and NAOH treated banana fibre", Journals of Materials Today, June 2019.
9. P. Madhu, M.R. Sanjay, Mohammad Jawaid, Suchart Siengchin, Anish Khan, Catalin Iulian Pruncu, "A new study on effect of various chemical treatments on Agave Americana fibre for composite reinforcement: Physico-chemical, thermal, mechanical and morphological Properties", Journal Pre-Proof Polymer Testing, February 2020.
10. Chin S C, Tee KF, Tong FS, Ong HR, Gimbun J, "Thermal and Mechanical Properties of Bamboo Fibre Reinforced Composites", Journal of Materials Today, December 2019.
11. Mohammad Asim, Mohammad Jawaid, Anish Khan, and Abdullah M. Asiri, "Effects of date palm fibres loading on mechanical, and thermal properties of Date Palm reinforced phenolic composites", Journal of Material science and Technology, January 2020.
12. H.Khakpour, M.R. Ayatollahi,%0 A.Akhavan-Safar, and L.F.M.da Silva, "Mechanical properties of structural adhesives enhanced natural date palm fibre", Journal pre proof, Composite structures, January 2020.
13. F. Delzendehrooy, M.R. Ayatollahi, A. Akhavan-Safar, and L.F.M. da Silva, "Strength improvement of adhesively bonded lap joints with palm date fibres", Journal Pre proof, Composites part B, February 2020.
14. A. Shalwan, and B.F. Yousif, "Influence of date palm fibre and graphite filler on mechanical and wear characteristics of epoxy composites", Materials and Design, March 2014.
15. T. Krishnan, S. Jayabal, V. Naveen Krishna, "Tensile, flexural, impact, and hardness properties of alkaline-treated Sun hemp fibre reinforced polyester composites", Journal of Natural Fibres, July 2018.
16. ES Zaini, MD Azaman, MS Jamali, KA Ismail, "Synthesis and characterization of natural fiber reinforced polymer composites as core for honeycomb core structure: A review", Journal of Sandwich Structures & Materials, 2018.
17. S. Raghavendra, G.N. Lokesh, "Evaluation of mechanical properties in date palm fronds polymer composites", AIP Conference Proceedings 2057, 020021 (2019).
18. K. P. Prashanth, H. G. Hanumantharaju, G. N. Lokesh, "Synthesis and Characterization Study of Chitosan Based Natural Fibre: Biodegradable Polymer Composite", Procedia Materials Science 5, (2019).
19. L Prabhu, V Krishnaraj, S Gokulkumar, S Sathish, M Ramesh, "Mechanical, Chemical and Acoustical Behaviour of Sisal – Tea Waste – Glass Fibre Reinforced Epoxy Based Hybrid Polymer Composites", Materialstoday: Proceedings 16 (2), 653-660,(2019).
20. R Bhoopathi, M Ramesh, MN Kumar, PS Balaji, G Sasikala, "Studies on mechanical strengths of hemp-glass fibre reinforced epoxy composites", IOP Conf. Series: Materials Science and Engineering 402, 1-8, (2018).
21. J Inbakumar Parivendhan, S Ramesh, "Mechanical, wear and thermal behaviour of hemp fibre/eggshell particle reinforced epoxy resin bio composite", Transactions of the Canadian Society for Mechanical Engineering 42 (3), 280-285, (2019).
22. MR Sanjay, GR Arpitha, P Senthamaraikannan, M Kathiresan, "The hybrid effect of Jute/Kenaf/E-glass woven fabric epoxy composites for medium load applications: impact, inter-laminar strength, and failure surface characterization", Journal of Natural fibres,(2018).
23. GR Arpitha, B Yogesha, "An overview on mechanical property evaluation of natural fibre reinforced polymers", Materials Today: Proceedings 4 (2), 2755-2760, (2017).
24. M Ramesh, K Palanikumar, KH Reddy, "Plant fibre-based bio-composites: Sustainable and renewable green materials", Renewable and Sustainable Energy Reviews 79, 558-584, (2017).
25. M Ramesh, K Palanikumar, K Hemachandra Reddy, "Evaluation of mechanical and interfacial properties of sisal/jute/glass hybrid fibre reinforced polymer composites", Transactions of the Indian Institute of Metals 69 (10), 1851-1859, (2016)
26. M Ramesh, K Palanikumar, K Hemachandra Reddy, "Impact behaviour analysis of sisal/jute and glass fibre reinforced hybrid composites", Advanced Materials Research 984, 266-272, (2014).
27. M Sakthivel, S Vijayakumar, S Ramesh, "Production and characterization of luffa/coir reinforced polypropylene composite", Procedia Materials Science 5, 739-745, (2014).
28. M Ramesh, K Palanikumar, KH Reddy, "Mechanical property evaluation of sisal–jute–glass fibre reinforced polyester composites", Composites Part B: Engineering 48, 1-9, (2013)

IOP Conf. Series: Materials Science and Engineering 1013 (2021) 012034 doi:10.1088/1757-899X/1013/1/012034

29. MR Sanjay, GR Arpitha, L Laxmana Naik, K Gopalakrishna, B Yogesh, "Studies on mechanical properties of banana/e-glass fabrics reinforced polyester hybrid composites", Journal of Material and Environmental Science 7 (9), 3179-3192, (2016).
30. M Ramesh, C Deepa, MR Sanjay, S Siengchin, "Biocomposites Reinforced with Animal and Regenerated fibres", Hybrid Fibre Composites, 10.1002/9783527824571.ch11.
31. M Ramesh, C Deepa, LR Kumar, MR Sanjay, S Siengchin, "Life-cycle and environmental impact assessments on processing of plant fibres and its bio-composites: A critical review", Journal of Industrial Textiles, 1-25.

Power Generation to the Coastal Areas with an Approach to Wave Energy Device-A Prototype

Rupesh P L [1], Hitesh Kumar Rana[2], Soma Shashidar[3], Satya Sai Darshan M[4] and Varun Teja G[5]
1,3,4,5Department of Mechanical Engineering, Veltech Rangarajan Dr Sagunthala R&D Institute of Science and Technology, Chennai, India.
2 Indosta Motors, Greater Noida, India.

E-mail: rupeshkumar221@gmail.com

Abstract: - Ocean cover more than 70% of earth surface and ocean waves carries enormous wave power. The substantial part of world electricity can be supplied by utilizing the largest source of untapped clean energy (wave energy). Wave energy is more predictable compare to wind energy and can be forecast 24 hours ahead. In this work, the concept of utilization of wave energy in the power generation has been delivered. The approach to wave energy has been done by using the combination of a buoy and a piston. The work described in this paper aims at the design and fabrication of a wave energy device. A numerical study of the power production by the fabricated device has been done for the implementation of this device in the Indian coastal areas for a clean and sustainable power production in future.

Keywords: Wave energy; Buoy-Piston-Cylinder; Trough; Crest; Energy Density;

1. Introduction

In a developing (or) a developed nation, the energy consumed to perform an action, fabricate things or resident a building is referred as energy consumption. The graph shown in figure 1 resembles the level of energy production in India for the past 10 years from 2005-2019 [1-5]. In India, the utilities generate electricity of 13,720 kWh at gross but in the country around 15,470 kWh of electricity has been generated by utilities and non-utilities. In 2018-19, the consumption of electricity per capita was 11,810 kWh at gross.

1.1 Electrical energy -Generation & Consumption Scenario

In India, the power utilization used for home applications has significantly increased from the year 2000. The level of family units with access to power has expanded from 55 to greater than 80% in 16 years duration from 2001-2017. In 2014, it has been observed that the electrical usage for one month by an Indian family devoured about 90 kW of electricity on a normal; which is sufficient to operate four-cylinder lights, four roof fans, a TV, a little fridge, and little kitchen machines with run of the mill use hours and effectiveness levels in India. This is three-fourths of the normal month to month family unit utilization in China, a tenth of that in the USA, and 33% of the world average [6-8]. In this post, we investigate information on India's private power and the aberrations in access and utilization across states. We additionally uncover a few irregularities between various sources, highlighting the requirement for better information.

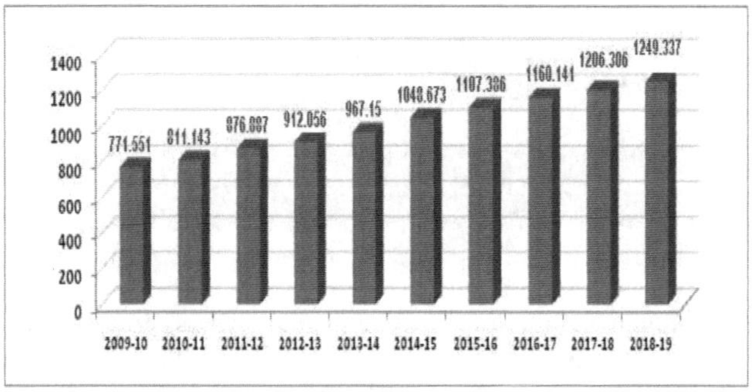

Figure 1. Electric power Generation (Billion Units) for 10 years

All states show extensive increment altogether private power utilization as of late as indicated by information gathered by the Central Electricity Authority (CEA) from dispersion organizations. Somewhere in the range of 2004 and 2015, states like Assam, Bihar, Chhattisgarh, and Jharkhand with low introductory family jolt indicated a high development pace of their private power use (about 11%-16%). States with higher family unit jolt like Delhi, Punjab, Haryana, and Tamil Nadu developed at lower, yet considerable, rates (6%-8%), with high supreme numbers [9-10].

1.2 Utilization of Renewable energy in India

Renewable power frameworks use renewable energy sources to create power with low ozone depleting substance emanations. Clean inexhaustible electrical vitality age frameworks are getting progressively available to the normal mortgage holder either introduced at home and associated with the lattice or bought as Green Power. These sources, for example, the sun, wind and water, are renewed normally however are not accessible persistently. In remain solitary frameworks, back-up power can be provided from capacity batteries as well as generators. The pie chart shown in Fig.2 represents the utilization of various sources of renewable energy in the current scenario.

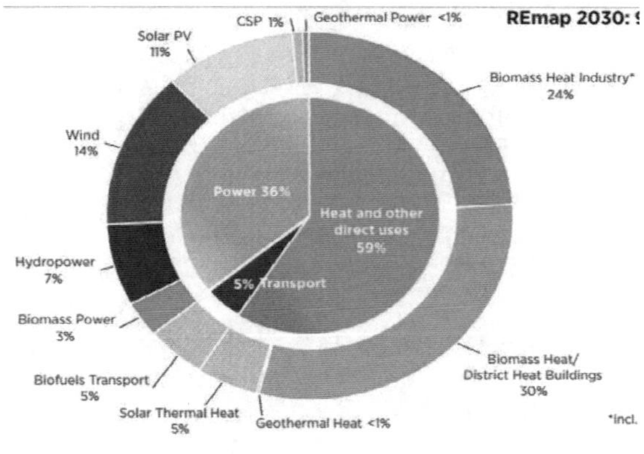

Figure 2. Renewable Usage- Current Scenario

2. Alternate Energy Source

2.1 Wave Energy Consumption Device

Now a days people more depend on the convectional fuel due to which the convectional fuel rate level become lower and also the rate of pollution is becoming very high. About 60-70% of the world covered with the ocean and also India covered about 50-60% with water and by implementing the ocean wave technology we can generate electricity nearby coastal area. Due to which we will less depend on the convectional fuel and rate of pollution will also decreases. In ocean wave energy concept, we are utilizing the kinetic energy of wave and convert into electricity.

The ocean energy technology has attained popularity in recent era due to its like resemblance with hydro power plants. The areas where tidal force could be grown monetarily are moderately not many in light of the fact that a solid tidal vitality concentration (leading to a tidal scope of five meters or more) has been viewed as required for the expense of power to be serious.

The devices used for wave energy utilization has been broadly classified into different types as shown in Fig. 3. Method used for the utilization, location and the system used for power production were the factors by which the wave energy concepts were classified [12-13]. On & Off shore and even shore line can be the locations of wave energy devices. Turbines, Hydraulic & Pneumatic rams and Pumps can be used for production of power from wave. n Over topping Device; Point absorber Buoys; Oscillating water Column; Surface Attenuators are few of the concepts used most for wave energy utilization. Fig.3. depicts six approaches used for wave energy utilization.

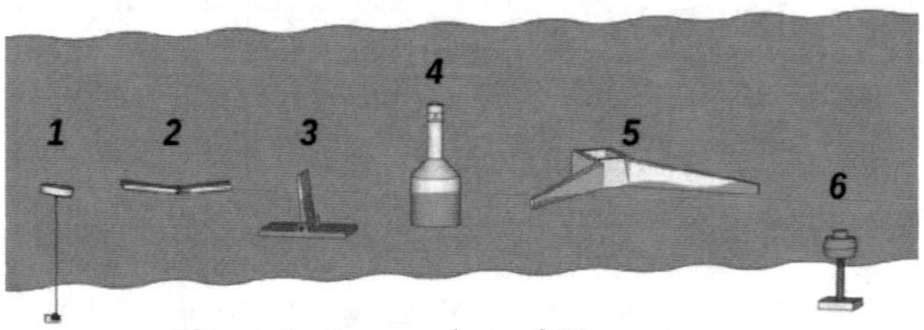

Figure 3. Approaches of Wave energy

When compared to the other sources of renewable energy such as solar, hydro and hydel power which have been well established, generation of power using energy from the waves is not a commercial technology which has been used at wide [11]. However, there have been attempts to use this source of energy since at least 1890 mainly due to its high-power density. Now a days, it has been evident that the utilization of electricity with the help of waves has been spread all over the world.

2.2 Technology applied for the current prototype

The technology applied for the prototype developed in the present work utilizes a buoy, float arm and a piston-cylinder as shown in below Fig. 4.

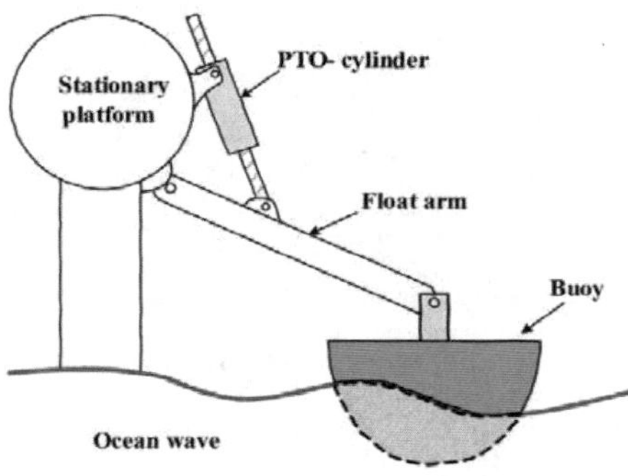

Figure 4. Technology proposed

The technology applied for the prototype described in this work deals with the conversion of wave energy from the ocean wave using a hemispherical cavity called as buoy into the mechanical energy achieved with the reciprocating motion of piston inside the cylinder. The mechanical energy achieved

IOP Conf. Series: Materials Science and Engineering 1013 (2021) 012035 doi:10.1088/1757-899X/1013/1/012035

can be used to pressurize a fluid which in turn drives a turbine shaft connected to a generator thereby producing electric power.

3. Design & Fabrication

The device employed in the present work to produce power from the energy of the waves has been designed using CATIA and the model of the device also been fabricated using the conventional manufacturing techniques.

3.1 Components Required

The components required for the construction of oscillating water column has been listed below:

- Float
- Piston
- Cylinder
- Turbine

The above-mentioned components are responsible to convert the wave energy into mechanical energy (or) desirable power output. The float is used to transmit the wave motion into the reciprocating movement of piston in cylinder. The piston thereby increases the pressure of the working fluid trapped inside the cylinder. The high-pressure fluid flows out a nozzle with high velocity and hits the blades of the turbine. Due to the impulse force given by the working fluid on the turbine blades, the runner rotates and in turn the shaft to which runner is attached also rotates.

3.2 3d model of components

The components required for the conversion of wave energy into pressure energy has been modelled and designed using CATIA-a design software package. The 3D view of the components has been shown in below Fig.5.

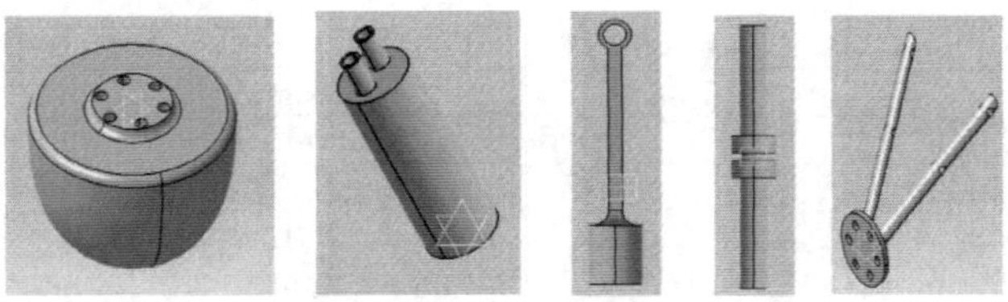

Figure 5. 3D Model of the components (a) Buoy (b) Cylinder with valves (c) Piston rod with piston (d) Oscillating bar (e) Twin rod

The components modelled above was assembled together to form an arrangement for energy conversion as shown in fig.6. The arrangement shown in fig.6. Includes a buoy (or) float coupled with the piston rod using twin rod and oscillating bar. As the wave hits the float, buoy starts oscillating. The oscillating motion of the buoy makes the piston rod to actuate through twin rod. The actuated motion of piston rod makes the piston to move from the bottom dead center to the top dead center of the cylinder. As the wave passes away from the float, the process is reversed. The piston retards back to the bottom dead center from the top dead center as the piston rod is retarded.

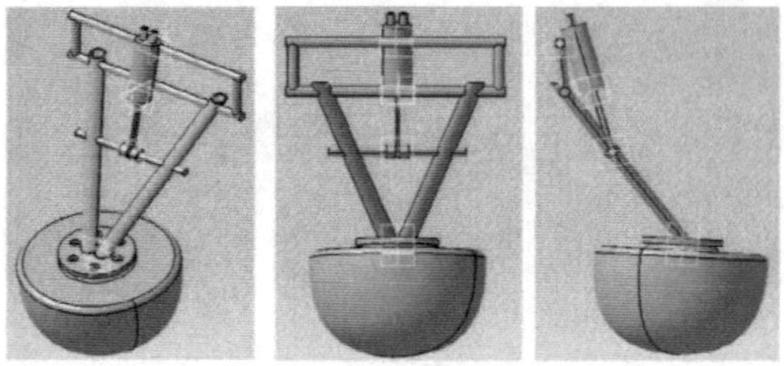

Figure 6. Buoy-Piston- Cylinder arrangement

3.3 Complete Setup

This work concentrates on development of a prototype where four buoy-piston cylinder arrangement has been fixed to a frame. The prototype developed has been modelled by using CATIA package and it has been shown in below Fig. 7.

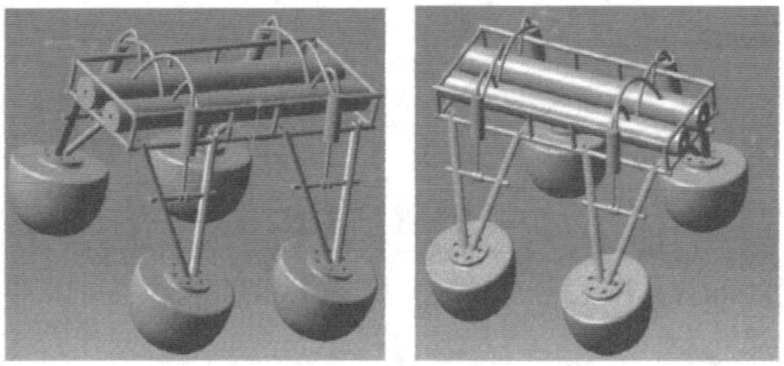

Figure 7. 3D model Complete Set up

3.4 Actual setup

The components which have been modelled were fabricated using the materials as listed in Table I. The components have been assembled to fabricate the prototype through several mechanical processes such as welding, adhesive usage, drilling etc. Fig. 8. depicts the fabrication model of the wave energy plant as modelled in Fig.7. The various parts of the model have been identified in the below Fig. 8.

Table 1. Parts and material used

Parts	Material
Float	Low density material filled with air
Piston	Steel rod of diameter 25 mm
Cylinder	PVC of diameter 10 mm
Inlet and outlet pipe	PVC of diameter 5 mm
Nozzle	Convergent Hose of diameter 0.5 mm
Turbine	3D printed

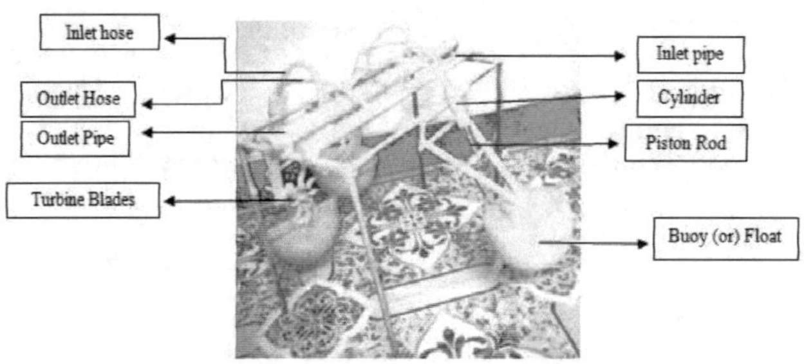

Figure 8. Model of Wave energy plant

3.5 Working of The Prototype

The schematic diagram of the working has been depicted in below fig.9.

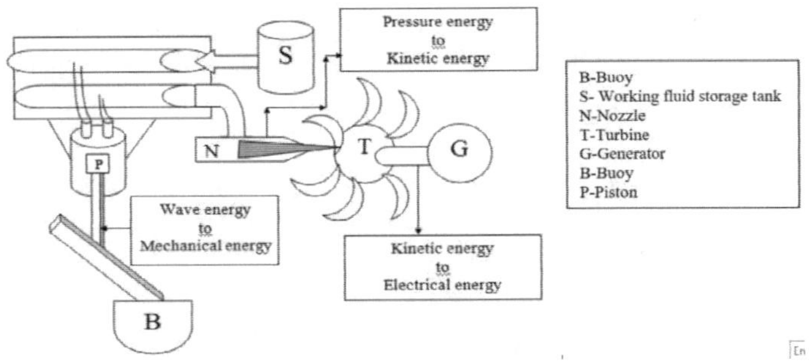

Figure 9. Working principle-Schematic Diagram

The wave motion from trough to crest is utilized in this wave energy device. The rise of the wave is called as crest and the fall of the wave is called as trough. The device which is proposed in the current research use the energy derived from the crest of the wave. The trough and the crest of the wave make the float (or) buoy to rise and fall. A stationery platform has been made in such a way that it stands on the sea floor rigidly and the platform is provided with links which connects the buoy-piston-cylinder arrangement firmly with it. The motion of the float is transferred into mechanical energy which in turn converts into electrical energy through the idea of hydraulics.

4. Numerical Analysis

4.1 Energy Utilization from Wave

The energy utilized from the wave can be calculated by the prediction of Wave power (P) [14] from equation (1) as given below:

$$P = \frac{\rho g^2 T H^2 l}{32\pi} \qquad (1)$$

Where,
ρ = Sea Water density = 1025 (kg/m^3)
g = Acceleration due to gravity = 9.81 (m/s^2)
T = Wave period (s)
H = Wave height (m)
l = Wave front (m)

In the expression of the formulae given above in (1), it is clearly evident that the wave power depends on the following parameters: Wave height (H) and Wave period (T). The Wave height (H) is an alternating parameter which changes based on the state. The varying wave height across different state wise cost has been tabulated in Table II. It was observed from Table II that the minimum height of the wave observed was 0.2 m and the maximum height of the wave was 8 m.

Table 2. Wave energy level across the India

State Wise Coast	Wave Height (m)
Tamil Nadu	0.5-1
Pondicherry	0.2-3.5
Visakhapatnam	0.2-5.74
Orissa	0.5-3

IOP Conf. Series: Materials Science and Engineering 1013 (2021) 012035 doi:10.1088/1757-899X/1013/1/012035

State Wise Coast	Wave Height (m)
Gujarat	0.2-3
Kerala	0.5-1
Mumbai	0.2-1
West Bengal	0.5-4
A&N Island	0.2-8

Fig.10. depicts the variation of wave height at different states in the form of a bar chart. It has been observed that the maximum wave height of 8 m has been observed in Andaman & Nicobar Island and next to that 5 m of wave height has been observed in Vishakhapatnam.

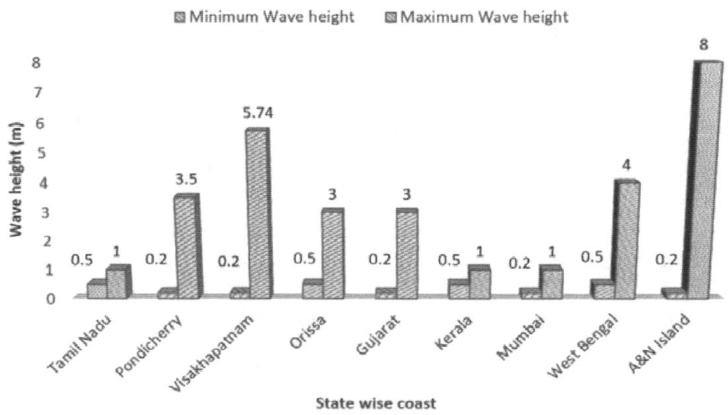

Figure 10. Variation of Wave Height

It has been examined that the wave power derived will be high in the coastal areas of Andaman and Vishakhapatnam as the height of the waves are maximum and moreover wave height is one of the influencing factors for obtaining maximum wave power.

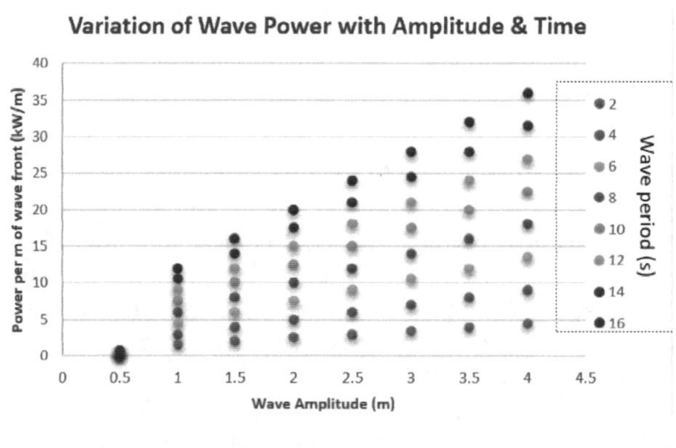

Figure 11. Variation of Wave Power

The variation of wave power along with amplitude and time period of the wave has been projected in Fig. 11. It has been observed that as the amplitude and time period of the wave increases, the power grabbed per meter of wave front increases. At an amplitude of 4.5 m and for 16 seconds, the power grabbed was 40 kW per meter of wave front.

5. Conclusions

The scope of the work mainly focuses on the development of a clean energy source for power generation. The concept of utilization of rare source of renewable energy (Wave energy) has been delivered in the above study. A prototype for the conversion and usage of wave energy was also fabricated in the above work. The relation between the wave height and the wave energy flux was also discussed in the study and it has been proved that height of the wave contributes a major role in the efficient conversion of wave energy into electrical energy. The future scope of the above project deals with the establishment of a real model in the coastal area of Pondicherry for the production of clean and low-cost energy for the household situated in the coast..

References

[1] Leijon, M., Danielsson, O., Eriksson, M., Thorburn, K., Bernhoff, H., Isberg, J., Sundberg, J., Ivanova, I., Sjostedt, E., Agren, O., Karlsson, K.E., and Wolfbrandt, A., "An electrical approach to wave energy conversion," *Renew. Energy*, vol. 31, pp. 1309–1319, 2006.

[2] Baker, N. J. and Mueller,M, "Direct drive wave energy converters (in French)," *Revue des Energies Renouvelables*, vol. 4 ED-02, pp. 1-7, 2007.

[3] Salter, S. H, "Wave power," *Nature*, vol. 249, no. 5459, pp. 720–724, 1974.

[4] Falnes J, Bjarte-Larsson T., "Theoretical and experimental investigation of wave energy conversion by a phase-controlled heaving body," in *Proceedings of the Institution of Mechanical Engineers, Part M: Journal of Engineering for the Maritime Environment*, vol. 220, no. 4, 2006, pp. 175-183.

[5] Mueller, M. A. and Baker, N. J., "Direct drive electrical power take-off for offshore marine energy converters," in *Proc. IMechE, Part A: J. Power and Energy*, vol. 219, 2005, pp. 223-234.

[6] Leijon, M., Bernhoff, H., Agren,O., Isberg, J., Sundberg, J., Berg, M., Karlsson, K. E., and Wolfbrandt A., "Multiphysics simulation of wave energy to electric energy conversion by permanent magnet linear generator," in *IEEE Trans. Energy Convers.*, vol. 20, no. 1, 2005, pp. 219-224.

[7] Polinder, H., Damen,M. E. C.,and Gardner,F., "Linear PM generator system for wave energy conversion," in *AIEEE Trans. Energy Convers.*, vol. 19, no. 3, 2004, pp. 583-589.

[8] Nielsen, K and Meyer, N. I, "The Danish Wave Energy Programme," in *Proc. 3rd European Wave Energy Conference*, Patras, Greece, 30 September – 2 October, 1998.

[9] Kofoed, J. P., Vicinanza, D. and Osaland, E, "Estimation of Design Wave Loads on the SSG WEC Pilot Plant based on 3-D Model Tests," in *Proc. ISOPE 2006.*

[10] Steenstrup, P. R, "Wave Star Energy – new wave energy converter, which is now undergoing sea trial in Denmark," in *Proc. International Conference Ocean Energy*, October, 23-24th, 2006, Bremerhaven, Germany.

[11] Brooking, P. R. M., Mueller, M. A., Baker, N.J., Haydock, L., and Brown, N., "Power conversion in a low speed. reciprocating electrical generator," in *Proc. International Conference on Electrical machines*, 2002, Bruges, Belgium.

[12] Hals J, Bjarte-Larsson T, Falnes J., "Optimum Reactive Control and Control by Latching of a Wave-Absorbing Semisubmerged Heaving Sphere," in *Proc. 21st International Conference on Offshore Mechanics and Arctic Engineering*, 2002, pp. 415-423.

[13] Ran, L., Tavner,P. J.,Mueller,M. A., and Baker,N. J., "Power conversion and control for a low speed, permanent magnet, direct-drive, wave energy converter," in *Proc. Third IET International Conference on Power electronics, machines and drives*, 2006, pp. 17-21.

[14] Kofoed J. P, "Wave Overtopping of Marine Structures – Utilization of Wave Energy," Ph.D. dissertation, Dept. Elect. Eng., Aalborg University, Cambridge, MA, January 17, 2003.

EXPERIMENTAL ANALYSIS OF INFLUENCE OF INJECTION PRESSURE ON COMPRESSION IGNITION ENGINE WITH BIODIESEL AND NANOPARTICLES BLEND

Sharun Mendonca[1*], Thirumaleshwara Bhat[2], Ravikantha Prabhu[1], Rudolf Dsouza[1]
[1]Assistant Professor, Department of Mechanical Engineering, St. Joseph Engineering College, Vamanjoor, Karnataka, India-575028
[2]Professor, Shri Madwa Vadiraja Institute of Technology and Management, Bantakal, Karnataka, India-574115

*corresponding author:Sharunmendonca@gmail.com

Abstract
The scarcity *of* conventional fuel and stringent emission norms made researchers to look after alternative fuels to run *an internal combustion engine*. One of the possible alternatives for the Compression Ignition engine is biodiesel. Although there are some challenges such as high viscosity, low calorific value, carbon deposit on *the injector nozzle*, etc. To overcome these challenge nanoparticles are added which will bring most properties near to virgin diesel. Simarouba a non-edible biodiesel feedstock used and aluminum oxide nanoparticles are added with 50 nm size. Using a probe-type Ultrasonication process nano-biodiesel blend is prepared. To improve stability SDS surfactants are added. The characterization of nanoparticles studied using SEM image. The performance of the engine tested for different dosage levels of nanoparticles mainly 25 ppm, 50 ppm, and 75 ppm. The injection timing, injector pressure, and speed kept constant, whereas injection pressure is varied 200 bar, 225 bar, and 250 bar. There is found a 5.2% increase in brake thermal efficiency for nanoparticles added blend biodiesel at 250 bar compare to 200 bar pressure. For the same thing brake specific fuel consumption decreased by 9%. Most of emissions such as Carbon monoxide, unburnt hydrocarbons decreased, whereas there is a significant amount of increase in oxides of nitrogen emission found.
Keywords : Biodiesel, nanoparticles, transesterification, stability

Notations

CI- Compression Ignition	BTE-Brake Thermal Efficiency
SEM-Scanning Electronic Image	BSFC-Brake Specific Fuel Consumption
SDS- Sodium Dodecyl Sulfate	B20-Diesel(80%)+Simarouba(20%)

1. Introduction
To travel people or to take goods from one place to transportation is required. For transportation most of vehicles uses chemical energy in the form fuel. There are different types of fuels are used to run the engines. High octane number fuels are used in Spark Ignition engines, whereas high Cetane number fuels are used in Compression Ignition engines. Due to many reasons in India, most of the heavy vehicles use Compression Ignition Engines. Due to rapid depletion of fuel it is the need of the hour to shift from conventional fuels to alternative fuels. One of the possible alternatives for diesel is biodiesel because it has most of the properties identical to diesel. The of the major disadvantage in biodiesel is its low calorific value and high viscosity to overcome this challenge biodiesel is used as blend with diesel[1]. Injector deposit is another major challenge to overcome. Injector deposit is another major challenge to overcome. Endurance test on engine after using biodiesel shows in SEM image there is a greasy type of deposition. The higher level of carbon deposits observed at the injector tip after usage of biodiesel without much modification[2]. The biodiesel has prospective to replace conventional diesel fuel[3].
With recent development of nanotechnology, a lot of researchers is using nanoparticles in different applications. The usage of nanoparticles in diesel engines in biodiesel tends to reduce toxicity, improves stability. Application nanoparticles tend to improve performance characteristics and decrease the major pollutions[4]. The major challenge in nanoparticles characterization is in relevance with physical conditions.[5].
Simarouba one of the non-edible oil feedstock used to produce biodiesel from transesterification process. The brake thermal efficiency is found to be increased by 9.14% for biodiesel nanoparticles blend SME2040 (20% biodiesel+40 PPM nanoparticles), this is because graphene nanoparticles having higher thermal conductivity increases combustion efficiency which in turn increases heat transfer coefficient. As a heat transfer coefficient increases, it has directly related with convection heat transfer. As the load increased, there will be an increase in smoke because fuel to air ratio becomes lean to richer. But with the addition of surfactant, smoke is decreasing because there will be an increase in atomization. Also with the addition of surfactant carbon

monoxide will decrease because there decrease in viscosity. The unburnt hydrocarbon is more for simarouba biodiesel compare to diesel because of poor atomization. But unburnt hydrocarbon decreased with the addition of graphene nanoparticles to biodiesel. The higher peak pressure and heat release rate were obtained for SME2040[6]. The three chief components of biodiesel are oleic acid, linolic acid and palmitic acid. To increase specific gravity and Cetane number, the better molar ratio between methanol and acid is used. The transetification process plays important role in determine the fuel properties of biodiesel[7].

2. Biodiesel and nanoparticles blend preparation

The biodiesel used for current experimental work is non-edible oil Simarouba. The main reason using Simarouba seeds for production biodiesel are higher free unsaturated fatty acid which leads to long term stability[8]. The properties of Aqueous Aluminium oxide nanoparticles are shown in Table 1.The properties of diesel, simarouba biodiesel and B20 (Diesel 80%+ simarouba 20%) is shown in Table 2. Properties are found using apparatus in lab where Cetane value is taken from literature. One of issue will arise is aggregation of nanoparticles in fuel after some time. This is related to stability of nanoparticles. It may be overcome by different processes like Ultrasonication process, adding surfactants to the mixture. The anionic surfactants having better thermal conductivity compare to cationic surfactants[6]. Fig 1 shows the characterization of Nanoparticles.

Fig 1 SEM Analysis of Aqueous Aluminium Oxide Nanoparticles

Table 1 Properties of Aqueous Aluminium Oxide Nanoparticles

Parameter	Property
Manufacturer	PLATONIC NANOTECH
Average particle size	30-50 nm
Surface area	120-140 m^2/g
Purity	99.90%
Density	3.97 g/cc
Molecular weight	101.96 g/mol
Melting point	2055^0C
Morphology	Spherical

Table 2 Fuel Properties of Diesel, Simarouba biodiesel and B20

Properties	Unit	Simarouba	Diesel	B20	ASTM	Apparatus used
Kinematic viscosity at 40^0C	cSt	4.68	2-5	2.83	D 445	Redwood Viscometer
Calorific Value	KJ/kg	37950	43000	41900	D 240	Bomb calorimeter
Density	Kg/m^3	880	840	848	D 1298	Hydrometer

Flash point	^0C	160	65	67	D 93	Pensky Martin Apparatus
Cetane number	-	53	44-55	45	D 613	-

The probe type Ultrasonication is the best method of preparation of biodiesel-nanoparticles blend. The simarouba biodiesel blend is taken in the ratio of 80% pure diesel and 20% simarouba biodiesel, which makes B20. The dosing level of Aqueous Aluminum Oxide is added to blend B20 by 25 ppm (by weight) and goes an increment of 25 ppm. Sodium Dodecyl Sulfate (SDS) is used as surfactant[9].

3. Engine specification and test procedure

The engine test was carried out in single cylinder four stroke diesel engine with water cooled. The engine specification is given in Table 3. Initially diesel fuel used, performance, emission and smoke is measured. Subsequently B20, B20+25 ppm Al_2O_3, B20+50 ppm Al_2O_3 and B20+75 ppm Al_2O_3 used and taken readings. During the experiment, Injection timing (23^0bTDC) and Compression Ratio (CR) kept same where Injection Pressure is varied has 200 bar,225 bar and 250 bar. The experiment was conducted at constant speed 1500 RPM. To change injection pressure, different injector nozzles are used.

Table 3 Engine Specifications

Type	Single cylinder four stroke diesel engine
Rated power	5.2 KW
Cooling	Water cooled
Compression ratio	17.5:1
Stroke	110 mm
Bore	87.5 mm
Capacity	661 cc
Dynamometer	Eddy current supply 230 V AC

4. Uncertainty analysis of the experimental data

There will be always measurements errors. By taking 5 trails reading uncertainty of any measured reading was found. The Table 4 shows uncertainties in measured values. The values are obtained by average of 5 readings.

Table 4 Uncertainties in measured values

Measured variable	Accuracy (±)
Load	0.1
Fuel consumption in g	0.1
HC	±1.3
CO	±2.8
NO_x	±2.2
Smoke	±2.2

5. Results and discussion

5.1 Brake thermal efficiency & Brake Specific Fuel consumption

The variation of BTE and BSFC for the test fuel for different loads is shown in fig 2, 3, 4,5,6,7 at Injection pressure 200 bar, 225 bar and 250 bar. Due to high viscosity and lower calorific value there is the decrease in brake thermal efficiency whereas adding nanoparticles increases with dosage 75 ppm with injection pressure increases the Brake thermal efficiency 5.2% whereas brake specific fuel consumption decreases by 9%. There is clearly seen significant increase in brake thermal efficiency and decrease the brake specific fuel consumption when increase of pressure and dosage of nanoparticles.

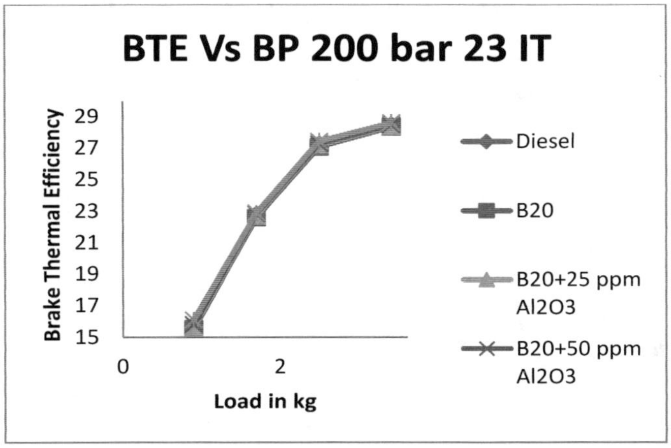

Fig 2 Brake Thermal Efficiency Vs Load at 200 bar Injection Pressure and 23^0bTDC Injection Timing

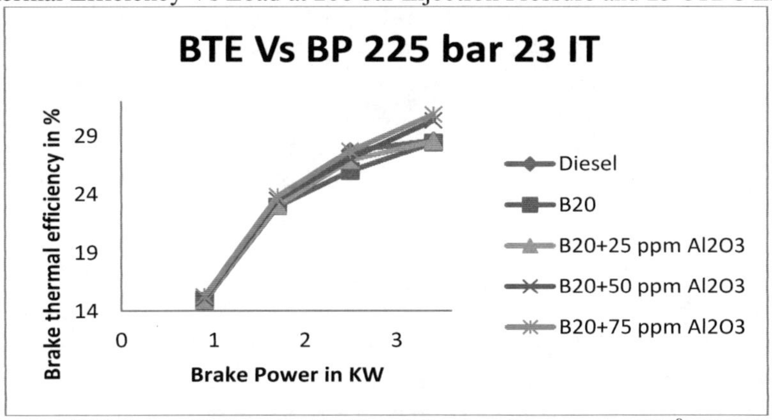

Fig 3 Brake Thermal Efficiency Vs Load at 225 bar Injection Pressure and 23^0bTDC Injection Timing

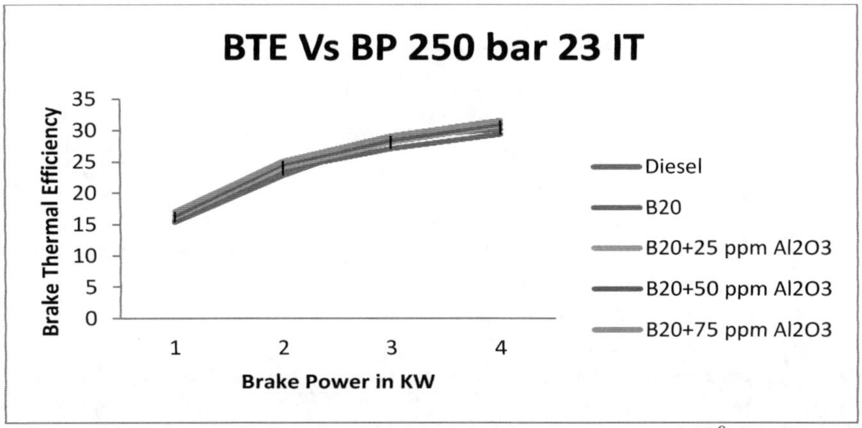

Fig 4 Brake Thermal Efficiency Vs Load at 250 bar Injection Pressure and 23^0bTDC Injection Timing

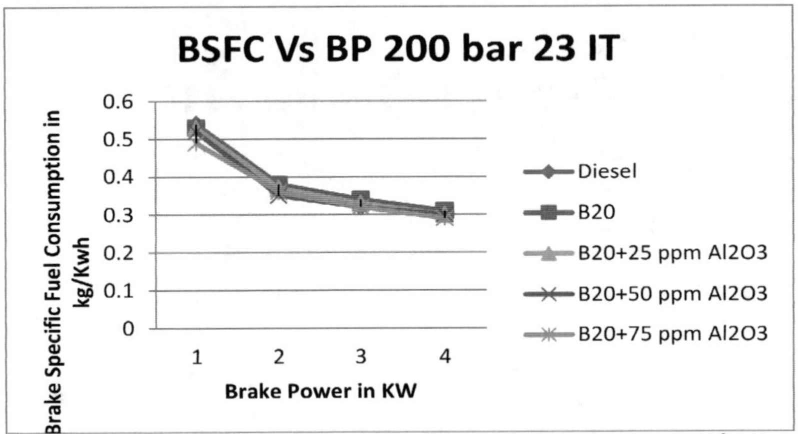

Fig 5 Brake specific fuel consumption Vs Load at 200 bar Injection Pressure and 23^0bTDC Injection Timing

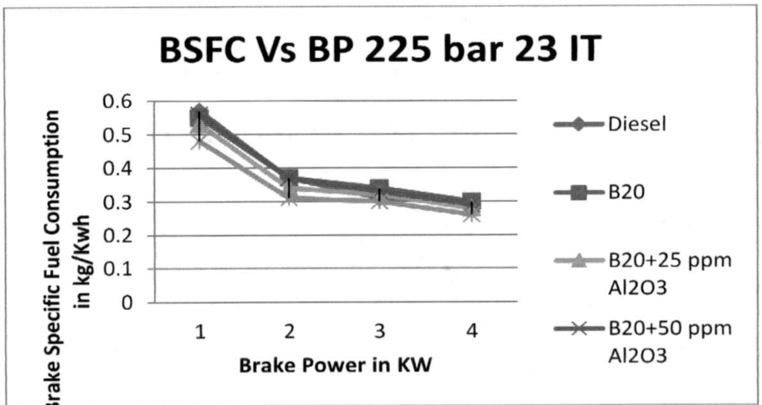

Fig 6 Brake specific fuel consumption Vs Load at 225 bar Injection Pressure and 23^0bTDC Injection Timing

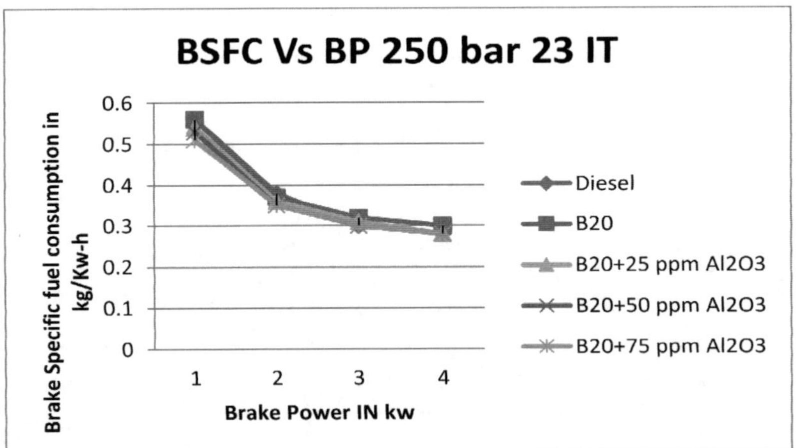

Fig 7 Brake specific fuel consumption Vs Load at 250 bar Injection Pressure and 23^0bTDC Injection Timing

5.2 Emissions

The variation of emissions Vs load is shown in fig 8, 9,10,11,12,13,14,15 and 16. It is clear from graph that addition of nanoparticles with simarouba biodiesel decreases the formation of unburnt hydrocarbon emission because metallic nanoparticles increase the rate heat release rate. Also it can be seen increase in injection pressure decreases the formation hydrocarbon emission because more complete combustion. High injection pressure uses nanoparticles in better way.

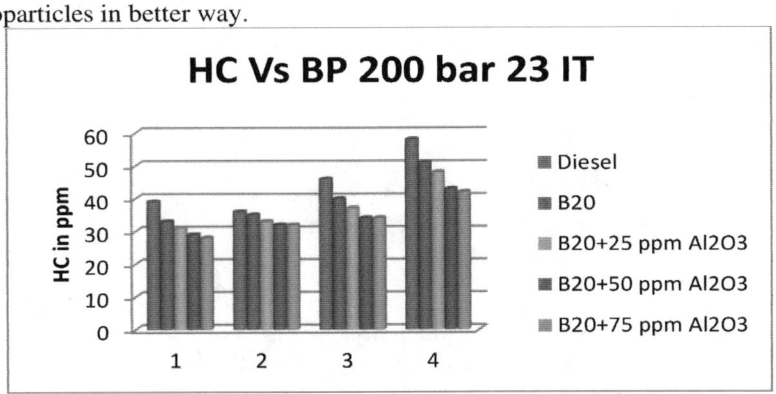

Fig 8 Hydrocarbons Vs Load at 200 bar Injection Pressure and 23^0bTDC Injection Timing

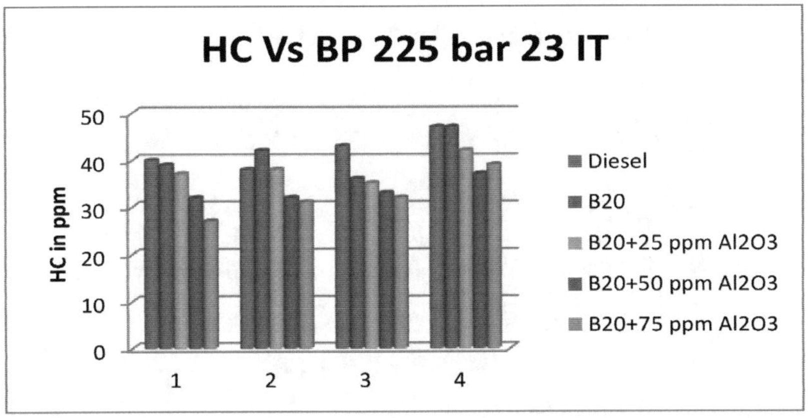

Fig 9 Hydrocarbons Vs Load at 225 bar Injection Pressure and 23^0bTDC Injection Timing

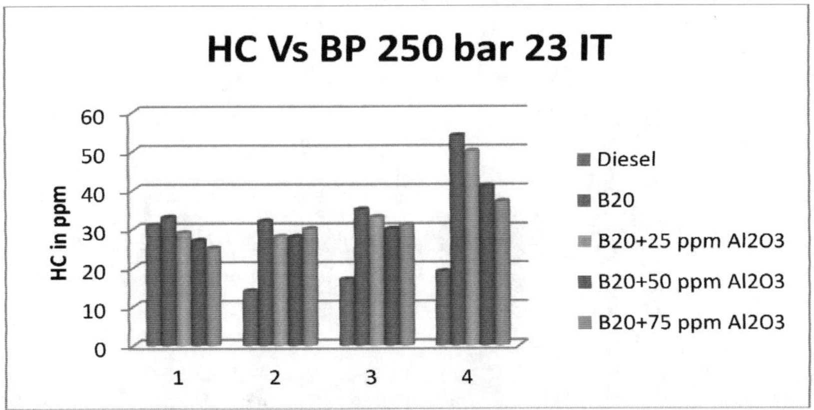

Fig 10 Hydrocarbons Vs Load at 250 bar Injection Pressure and 23^0bTDC Injection Timing

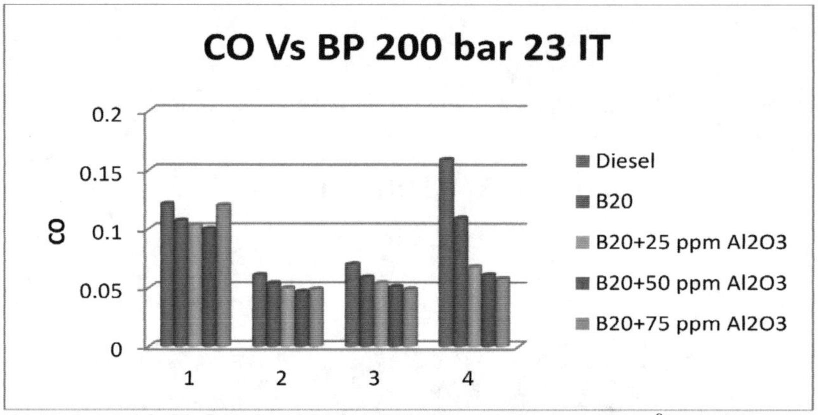

Fig 11 Carbon monoxide Vs Load at 200 bar Injection Pressure and 23^{0}bTDC Injection Timing

The same trend of decrease in Carbon Monoxide emission can be observed with increase in dosage level of aqueous aluminium oxide nanoparticles. But at injection pressure 250 bar there is rapid increase in carbon monoxide is observed for B20+75 ppm Al_2O_3, it may be due to availability less oxygen produces more carbon monoxide in place of carbon dioxide.

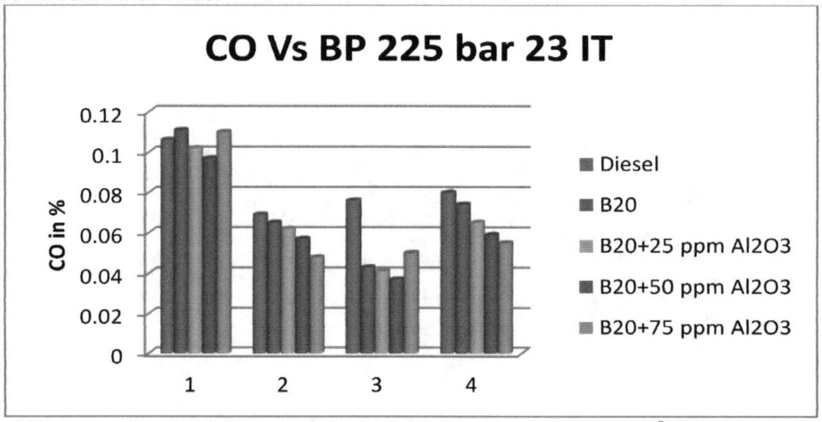

Fig 12 Carbon monoxide Vs Load at 225 bar Injection Pressure and 23^{0}bTDC Injection Timing

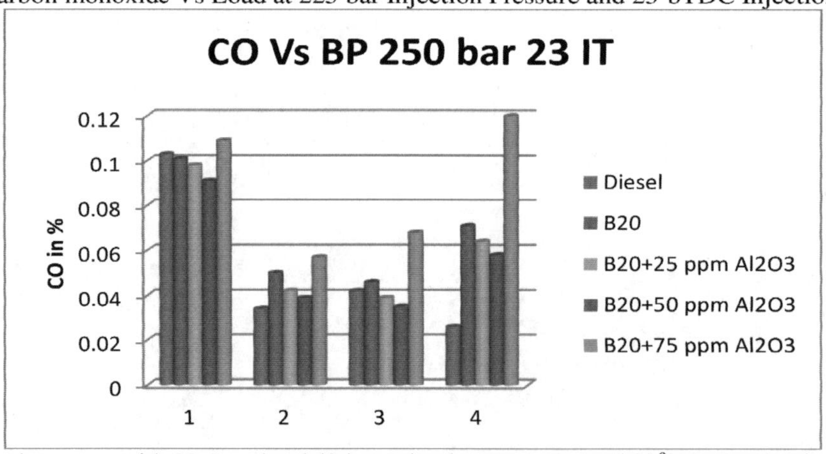

Fig 13 Carbon monoxide Vs Load at 250 bar Injection Pressure and 23^{0}bTDC Injection Timing

The variation between NO_X emissions Vs Load is seen in graph fig 14,15 and 16. It is observed there is almost linear increase in oxide of nitrogen in most of cases. Highest oxides of nitrogen emission are observed at 250 bar pressure for B20+75 ppm Al_2O_3, it is may be due to high temperature inside the combustion chamber during combustion.

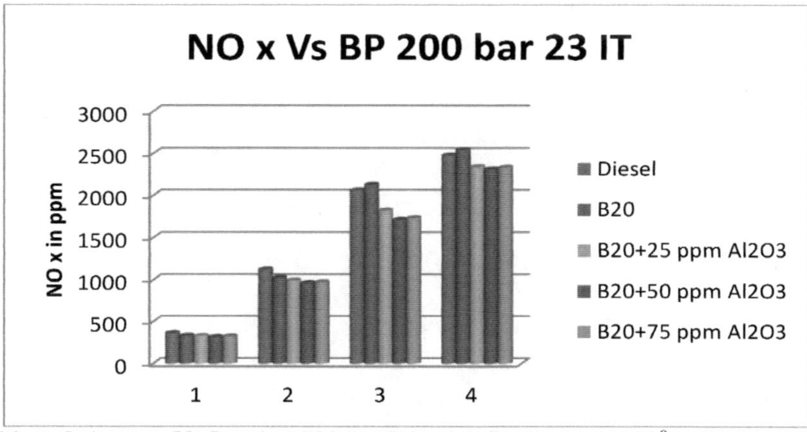

Fig 14 oxides of nitrogen Vs Load at 200 bar Injection Pressure and 23^0bTDC Injection Timing

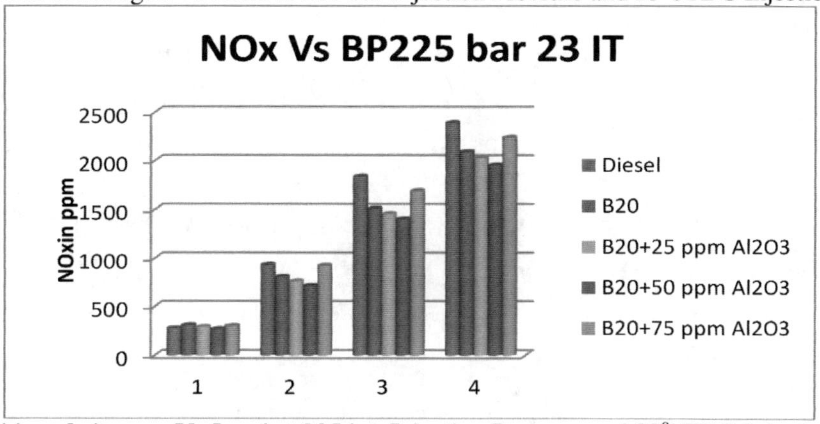

Fig 15 oxides of nitrogen Vs Load at 225 bar Injection Pressure and 23^0bTDC Injection Timing

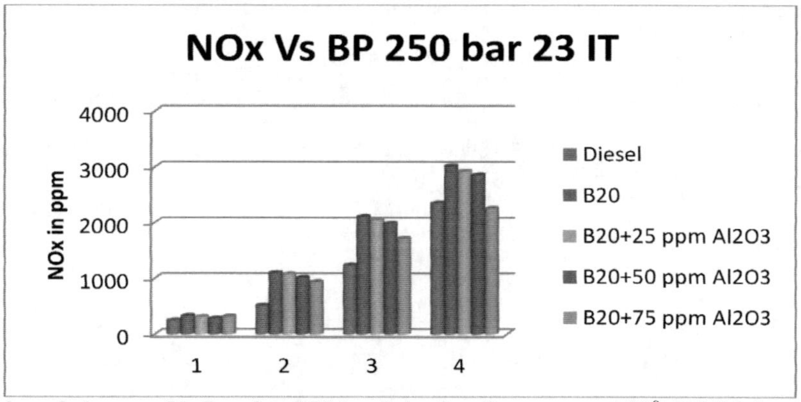

Fig 16 oxides of nitrogen Vs Load at 250 bar Injection Pressure and 23^0bTDC Injection Timing

6. Conclusions

The current study experimentally investigates the influence of injection pressure on CI engine with Aqueous Aluminium Oxide nanoparticles with Simarouba biodiesel. Based on the results obtained following conclusions are made

- The viscosity of simarouba biodiesel found decreasing with addition of aqueous aluminium oxide nanoparticles.
- Engine performance is improved with increase in dosage level of Nanoparticles. But increase in the dosage of nanoparticles above 75 ppm causes problem with stability

- Carbon monoxide and hydrocarbon emissions are decreased with addition nanoparticles and also with increase in injection pressure.
- The oxides of nitrogen emission are increased for higher injection pressure.

The simarouba biodiesel can be used with 20% blend with aqueous aluminium oxide nanoparticles with 75 ppm with injection pressure 250 bar. But need to study about nanoparticles in emission as well as stability of nanoparticles for long term.

References

[1] K. Ved, K. Padam, and C. H. Ocor, "Study of Physical and Chemical Properties of Biodiesel from Sorghum Oil," *Res. J. Chem. Sci.*, vol. 3, no. 9, pp. 64–68, 2013.

[2] A. M. Liaquat, H. H. Masjuki, M. A. Kalam, and I. M. Rizwanul Fattah, "Impact of biodiesel blend on injector deposit formation," *Energy*, vol. 72, pp. 813–823, 2014.

[3] D. Tutunea, M. Bica, C. Cernaianu, and A. Dima, "Correlation Between Physico-Chemical Properties of Biodiesel Blends," *Termotechnica*, vol. 2, pp. 67–70, 2013.

[4] E. Ettefaghi *et al.*, "A novel bio-nano emulsion fuel based on biodegradable nanoparticles to improve diesel engines performance and reduce exhaust emissions," *Renew. Energy*, vol. 125, pp. 64–72, 2018.

[5] S. Bhattacharjee, "DLS and zeta potential - What they are and what they are not?," *J. Control. Release*, vol. 235, pp. 337–351, 2016.

[6] B. M. Paramashivaiah, N. R. Banapurmath, C. R. Rajashekhar, and S. V. Khandal, "Studies on Effect of Graphene Nanoparticles Addition in Different Levels with Simarouba Biodiesel and Diesel Blends on Performance, Combustion and Emission Characteristics of CI Engine," *Arab. J. Sci. Eng.*, 2018.

[7] C. Y. Lin and Y. W. Lin, "Fuel characteristics of biodiesel produced from a high-acid oil from soybean soapstock by supercritical-methanol transesterification," *Energies*, vol. 5, no. 7, pp. 2370–2380, 2012.

[8] M. Sharun, "Influence of Aluminium Oxide Nanoparticles in a Compression Ignition Engine with Simarouba Biodiesel," *Mater. Today Proc.*, 2018.

[9] B. M. Paramashivaiah and C. R. Rajashekhar, "Studies on effect of various surfactants on stable dispersion of graphene nano particles in simarouba biodiesel," *IOP Conf. Ser. Mater. Sci. Eng.*, vol. 149, no. 1, 2016.

The effect of Zinc Oxide on Operation of Compression Ignition Engine with EGR fueled with Waste Cooking Oil Biodiesel

Ashley Lobo[1], Dr. D K Ramesh [2], Dwaipayan Roy Chowdhury[3]

1, 3 Department of Mechanical Engineering, Dayananda Sagar College of Engineering, Bengaluru, Karnataka.
2 Department of Mechanical Engineering, University Visvesvaraya College of Engineering, Bengaluru, Karnataka

E-mail: ashleylobo@live.com

Abstract: - Nanoparticles are being used increasingly along with second and third generation biofuels. One major advantage of nanoparticles is its surface to volume ratio and also the number of active surfaces it possess which make it ideal for application as a catalyst in a chemical reaction or in biodiesel's case an additive. Zinc Oxide (ZnO) has been employed in the current experimentation as an additive because it improves the properties such as the flash and fire point of the fuel in which it is blended in. Also, ZnO disperses well in fluid and provides advanced theoretical density along with not getting clogged in the fuel pump. In the present work Biodiesel derived from Waste cooking oil-Waste cooking oil methyl ester (WCOME) has been used along with ZnO nanoparticle and the amount of ZnO used is 80 ppm. The engine is also fitted with hot Exhaust Gas Recirculation system in order to decrease the Nitrogen Oxides (NOx) emission. Combustion properties such as Heat Release Rate and P-θ diagram have been thoroughly studied. The performance parameters like Brake Specific Fuel Consumption and Break Thermal Efficiency are also calculated and plotted on graphs for evaluation. Exhaust gases from the engine are measured and the results have been plotted on graphs for comparison. The properties of exhaust gases considered are % of Carbon Monoxide, Nitrogen oxides, Unburnt Hydrocarbons & Smoke Opacity. From the results it was revealed that all 3 parameters: Performance, Combustion & Emissions of the engine was improved by using ZnO as additive.

Keywords: Waste Cooking Oil Methyl Ester; Diesel Engine; Zinc Oxide; nanoparticle; performance; emissions; combustion.

1. Introduction

Biodiesel is becoming a preferred substitute for diesel because not only does it reduce Nitrogen Oxide [1, 2] emissions thanks to the higher oxygen amount existing in the fuel but also reduces particulate matter due to it being a short-chain paraffin hydrocarbon, and biodiesel does all this without requiring any external modification of the engine and without negatively affecting the characteristics of the engine [3]. Waste cooking oil (WCO) is a promising feedstock as it is 90% renewable and numerous biodiesel production facilities are presently utilizing it and WCO can be employed with no penalty on efficiency [4–6] and substantial diminution of Smoke opacity [1,7,8], CO [1,2,7,8], HC [1,2] and NOx [1,2] emissions with respect to diesel. Biodiesel has problems like greater density, inferior calorific value and higher BSFC compared to conventional fuels, and to overcome these shortcomings, additives play an important role. Metallic additives is an attractive option which was a popular in the early 2000s but its usage decreased until recently, but it has started increasing due to the accessibility and availability of nanoparticles. Interest in metallic additives is reinvigorated due to the attractive proposition nanoparticles provides, with mixing it with diesel and biodiesel and the nanoparticles acting as catalyst [9, 10]. Metal based additives are effective in reducing because of reaction with H2O vapor by metals in the emissions at the exhaust to yield extremely reactive Hydroxyl ions (OH-), and also the metals behave like oxidation catalyst lowering the temperature of oxidation for soot leading to improved particle burn out. Nanometal oxide particles plays the role of an oxidation catalyst leading to complete combustion and ignition due to higher carbon combustion activation [11]. Zinc Oxide (ZnO) is one such nanoparticle additive which is very promising and not much research has been carried out on it. Zinc oxide has high reactivity because of superior surface area, ZnO stores energy in its crevices, provides higher theoretical densities and stabilization for the mixtures. Zinc oxide is a nano additive which reduces knocking, enhances burning rate, rises the fuel viscosity and makes the fuels

IOP Conf. Series: Materials Science and Engineering 1013 (2021) 012037 doi:10.1088/1757-899X/1013/1/012037

conductive. Since nanoparticles have small diameters their concentration in the fluid is minimal the distribution stability of nanoparticles is good. Nano scale particles settle down rapidly after distribution of nano scale particles and can be suspended for weeks under favorable conditions. Settling rates may be explained using the Stokes–Einstein theory [12]. Zinc Oxide improves the dispersion rate of the blend and has several active surfaces, at the same time circumventing clogging in the fuel injectors [13, 14].The ZnO employed in the current experimentation is passivated with oxygen to achieve a stable oxide coating. Nanoparticles can overcome the density difference because of its better interaction between particles caused by its high surface to volume ratio [14-16]. ZnO boosts the heat/mass transfer and radiative properties of the WCO blends which results in the droplets being ignited at a much lesser temperature in contrast to diesel [10]. Viscosity is increased with ZnO concentration in the fuel and when highly viscous fuel is passed through the injectors, the injection process occurs with the formation of uneven fuel droplets which unfavorably impact the dispersion and fuel penetration in the cylinder [17].

2. Preparation of Fuel

2.1 Transesterification

The authors have run the engine with various concentrations of Zinc Oxide nanoparticle in the fuel blends to test out the most ideal concentration of the additive. From the results, 80 parts per million (ppm) proved to be the most favorable concentration for ideal results.
There were totally 3 blends used: Diesel, 20% Biodiesel+80% Diesel (B20) and B20+80 ppm ZnO. The biodiesel was obtained from the Transesterification process. Transesterification has been used to convert WCO which is predominantly triglycerides to glycerol, using methanol as the alcohol and Potassium Hydroxide as a catalyst. The steps carried out in transesterification is similar to Lobo [18].The fuel properties are given in Table 1.

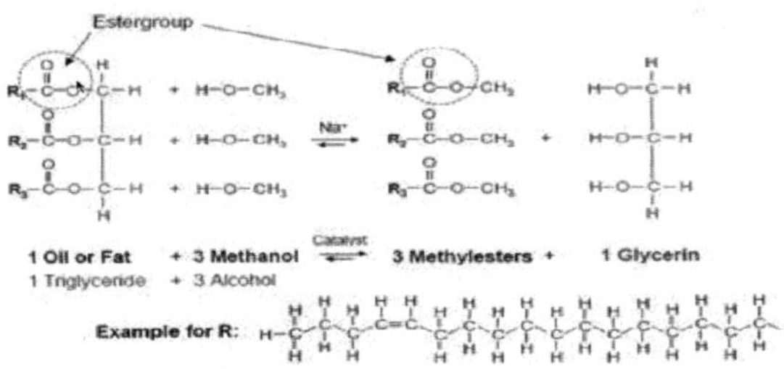

Figure 1. Transesterification of Oil

2.2 Experimental Setup

For experimentation, 80 ppm ZnO Nanoparticle is blended with WCOME and this biofuel is used in a Kirloskar TV2 (shown in Fig. 2) engine the technical specification of the engine is mentioned in Table II. The TV2 engine has been loaded from 0-100% for investigation and all the values were plotted on a graph for easy comparison. In total, three blends are used for experimentation which are Diesel, B20 WCOME (20% WCOME + 80% Diesel) and B20 WCOME+80 ppm ZnO.

IOP Conf. Series: Materials Science and Engineering 1013 (2021) 012037 doi:10.1088/1757-899X/1013/1/012037

Table 1. Properties of Fuels

Properties	Diesel	WCO (Waste Cooking Oil)	WCOME (WCO Methyl Ester)	B20WCOME (20%WCOME+ 80%Diesel)
Kinematic Viscosity (40°C)	3.05	43	5.83	4.01
Heat Value (MK/Kg)	44.5	29	38	42.2
Density (Kg/m^3)	830	914	887	842
Flash Point °C	60	307	150	98
Cetane Number	40	53	51.48	49.18
Iodine Number	6	125	59	17.42

Figure 2. Kirloskar TV-2 Engine used for this experimentation

Table 2. Technical Specifications of the Engine

Manufacturer	Kirloskar oil Engines Ltd
Model	TV-2
No of Cylinder	Two
Type of Engine	Vertical, 4-Stroke cycle, Single acting
Cooling	Water
Fuel	Diesel
HP	16 HP
Starting	Hand Cranking
Bore	87.5mm
Stroke	110mm
Cubic Capacity	1322cc
Compression Ratio	17.5:1
Valve Clearance Inlet	0.18mm
Valve Clearance Exhaust	0.20mm

2.3 Error analysis and uncertainty

Error Analysis and Uncertainty is done for this experimentation similar to Lobo [18] the experiment uncertainty was 1.765%

3. Results and Discussion

3.1 Brake Thermal Efficiency (BTE)

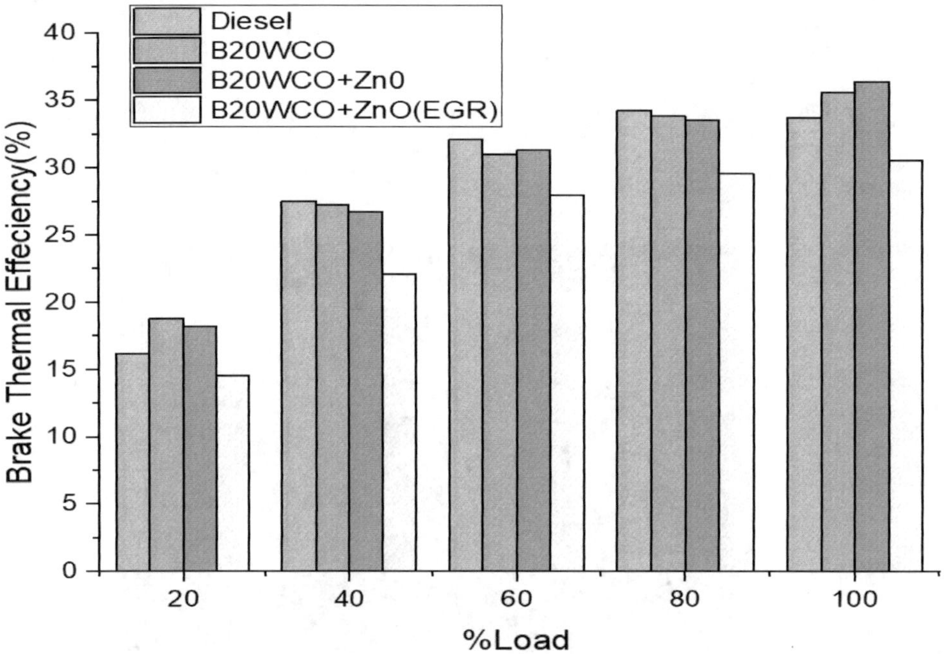

Figure 3. Variation of Brake Thermal Efficiency with Load

Fig. 3 illustrates that Brake Thermal Efficiency of the Waste cooking oil B20 blends decreases at all loads (except at full load) when compared to Diesel. A possible explanation for the reduction of BTE can be that the requirement of oxygen for combustion is reduced because of the oxygen buffer of ZnO [19]. Nanoparticles help to split the H2 atoms from H2O which could have participated in the combustion process along with improving the rate of heat conduction which lowers BTE [20, 21]. When EGR is applied to the nanoparticle blend the BTE decreases by 3% in comparison with diesel and 5% when paralleled to the nanoparticle blend without EGR. There is a fall in BTE, because for EGR to work there has to be a pressure difference between the inlet and exhaust manifolds to make sure the exhaust gases reach the inlet manifold and partially replace the air used for combustion. This is achieved by throttling the air of the inlet flow which adds extra load on the engine because it rises the pumping work. Thus for the same output more work has to be done thereby decreasing the thermal efficiency. This is offset by pumping more fuel but increases the BSFC when EGR is applied [22]. Similar results were obtained for Brake Thermal Efficiency by Prabakaran & B.Vijayabalan [34].

3.2 Brake Specific Fuel Consumption (BSFC)

Fig. 4 shows the Fluctuation of BSFC with Load. There is reduction of 10% for the BSFC of B20 and 12% for B20+ZnO (compared with the baseline Diesel reading). One of the causes for reduction of BSFC can be better combustion caused by shorter delay of ignition as a result of addition of a nanoparticle like ZnO [23]. The combustion is also improved by addition of an oxidizing agent, and

ZnO plays that role of an oxidizing agent when added to the blend, and because ZnO is an oxidizing agent it also improves the Air to Fuel ratio reducing the BSFC [23]. Employing an additive like Zinc Oxide increases the flash point properties and reduced the pour point of the blend, thereby reducing BSFC [24]. Also the lower oxidation temperature caused by the catalyst effect of the ZnO nanoparticle directly affects the decline of BSFC [25]. The same trend in BSFC is backed by studies 50-53.

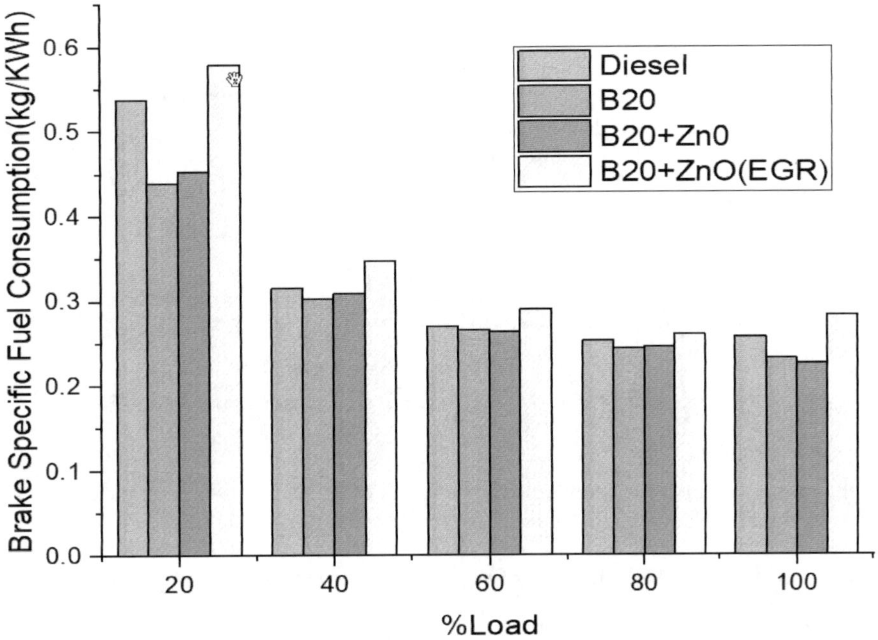

Figure 4. Variation of Brake Specific Fuel Consumption with Load

3.3 Cylinder pressure v/s Crank Angle (P-θ)

Fig. 5 shows the Fluctuation of Cylinder pressure with crank angle. The peak in-cylinder is 11% lesser for B20 when matched with diesel at 100% load and 10% for B20+ZnO as a consequence of superior viscosity of the Zinc Oxide blend [26]. With increase in pressure the delay of ignition reduces and the fuel is better atomized [27]. There is lesser buildup of the fuel in the first phase which causes decrease of the peak pressure as amount of prepared fuel in the first phase is associated with peak pressure in the cylinder. Enhancement of the catalytic property takes place as a result of ZnO because the addition of ZnO enhances the ignition and as an outcome of that is there is accelerated combustion [28-30]. For B20+ ZnO and B20+ZnO (EGR) there is a slight shift in peak pressure away from TDC which is caused by ignition delay as a result of the higher density and viscosity of the B20 & B20+ZnO blend. This higher density and viscosity causes a small delay of atomization of the fuel and thereby a minor delay in combustion of the fuel with oxygen thus causing Ignition Delay. This can be easily overcome by changing the injection parameters like injection timing or injection pressure. When we come to EGR there is decrease of nearly 4% when compared to diesel at 100% load and 1% in comparison to the blend without EGR, the decrease in peak pressure is due to the different composition which EGR introduces mainly Carbon Dioxide. And this subsequently increases as load increases [31]. Comparable results were obtained by Kegl & Hribernik [49].

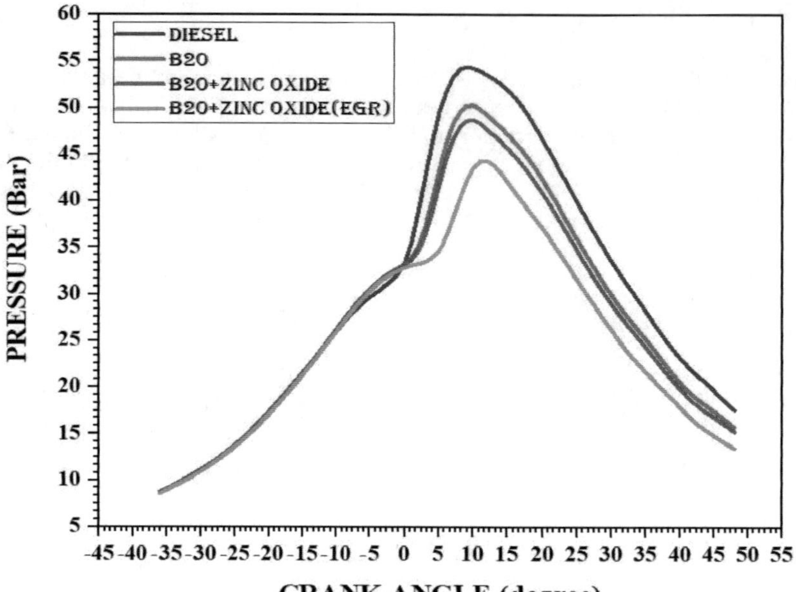

Figure 5. Variation of Cylinder Pressure with Crank Angle

3.4 Heat Release Rate (HRR)

In the Fig. 6 the maximum HRR is represented, the value of HRR is lower (compared to diesel) by 6% for B20 and 11% for B20+ZnO at 100% load. The cause for decline is the lower end temperature of combustion due to lesser calorific value of B20 [32]. The greater cetane number and improved evaporation rate and A-F (Air fuel) mixture of the blends there is shorter ignition delay [32] leading to lesser HRR. Also the better volume to surface area ratio and upgraded ignition properties (due to addition of ZnO) causes better combustion of the blends compared diesel [33] leading to lower HRR. When EGR is applied to the blend there marginal variation in HRR value. Similar results were obtained by Sadhik Basha and Anand [28, 48].

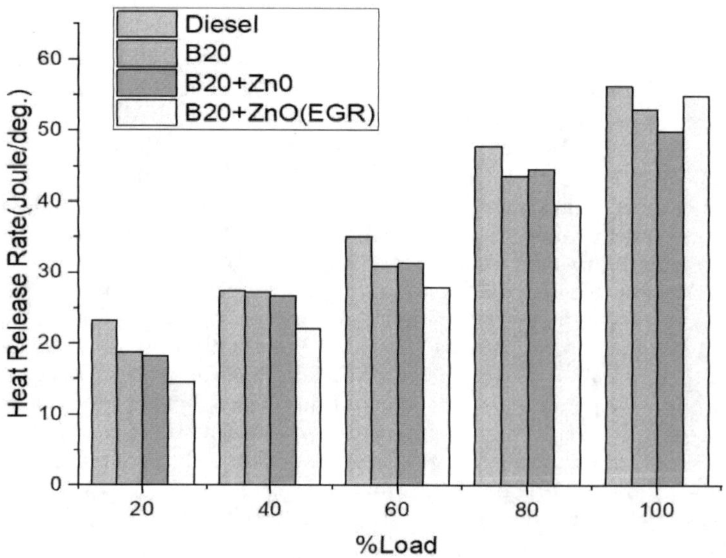

Figure 6. Variation of Heat Release Rate with Load

3.5 Nitrogen Oxides (NOx)

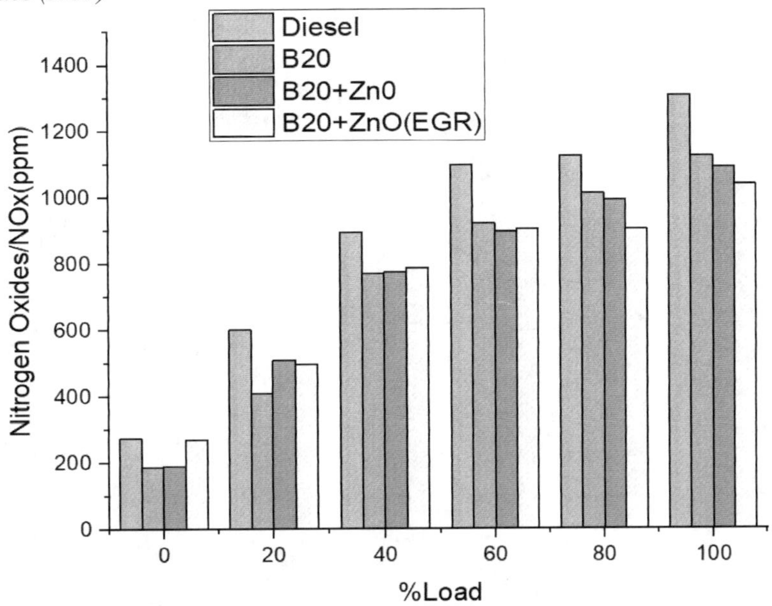

Figure 7. Variation of Nitrogen Oxides (NOx) with Load

Fig. 7 shows the variation of Nitrogen Oxides. The NOx emissons is reduced by 14% for B20 blend and 16% for B20+ZnO. One of the consequences of the blends having higher Calorific value is that the average temperature in the combustion chamber is increased leading to greater amount of oxygen to react and lowering the NOx emissions [28]. With employing thermally stable ZnO there is a decrease of NOx, another reason why NOx is reduced by addition of ZnO is because, ZnO introduces a shorter delay of ignition which results in better A/F ratio and creating an oxygen deficit in the chamber. Due to the poorer iodine number of WCO which is around 59 which warranted the presence of the extra saturated fatty acids in B20 [35]. The cetane number of biodiesel is superior in comparison to diesel due to the lengthier chains of fatty acids and greater degrees of saturation (lower iodine number) which leads to lower NOx emission [36-38].This trend was also observed in studies 54-59. EGR is applied to B20+ZnO blend and the NOx emissons reduces by 20% (when compared to diesel) and 5% when compared to the same blend without EGR. EGR includes replacing the air used for combustion by CO2 and H2O vapor which has higher specific heat capacity than the main components of air which is oxygen and nitrogen leading to lower gas temperatures. Decrease in oxygen content due to it being replaced in EGR also leads to inferior flame temperature and thereby reducing NOx emissons as formation of NOx is a highly temperature dependent phenomenon [39].This decreasing trend of NOx was also observed by Prabakaran, Udhoji & Anurag [60].

3.6 Carbon Monoxide Emissions (CO)

From the Fig. 8 it is clear that for full load the CO (% volume) emissions decrease by nearly 30% for both the blends B20 and B20+ZnO. CO emissions are decreased because greater amount of fuel is being burnt in the chamber and this is elucidated by the presence of ignition delay as depicted in the P-θ diagram. Another plausible explanation can be that more CO is converted to CO_2 as ZnO acts as an oxygen donating catalyst. Prabakaran et al. [34] mentioned that CO emissions can be shrunk by increasing the combustion average temperature. And because of the advanced calorific value of the blends, this could also be a possible explanation for reduction of CO emissions. The decrease in CO emissons is also because by increasing the concentration of additives it leads to enhancement of oxygen owing to ZnO and by addition of biodiesel which increases the air-fuel ratio and gives better and more complete combustion [40].Enrichment of oxygen is another major factor, when the biodiesel is added as it is rich in oxygen, this increase in oxygen content leads to carbon molecules being burned and combusted and thus releases CO2 rather than the poisonous CO. When EGR is applied to B20+ZnO blend there slight increase in CO emissions. Carbon Monoxide tends to increase because of the lower oxygen which is available for combustion which results in rich A/F mixtures at various regions in the combustion chamber. This CO trends are backed by studies 23 & 30.

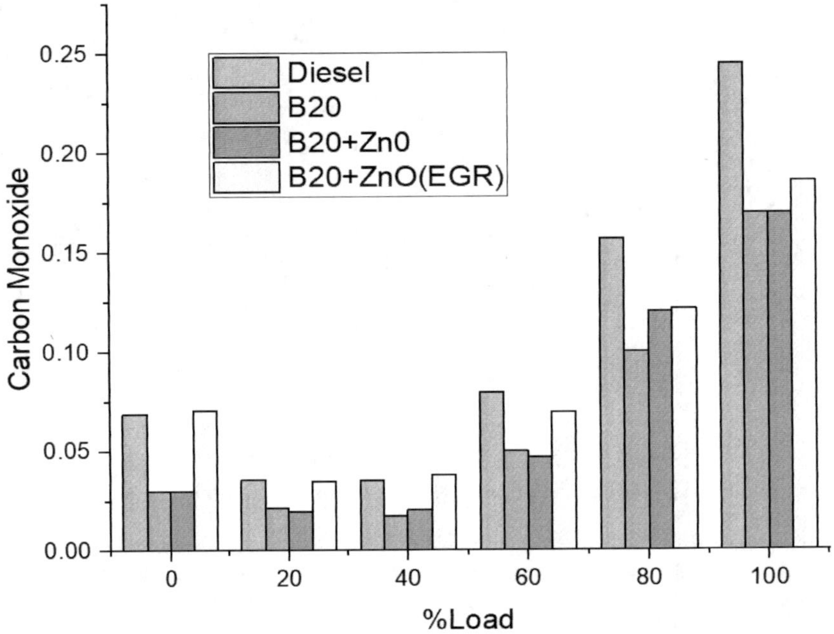

Figure 8. Variation of Carbon Monoxide Emissions with Load

3.7 Hydrocarbon Emissions (HC)

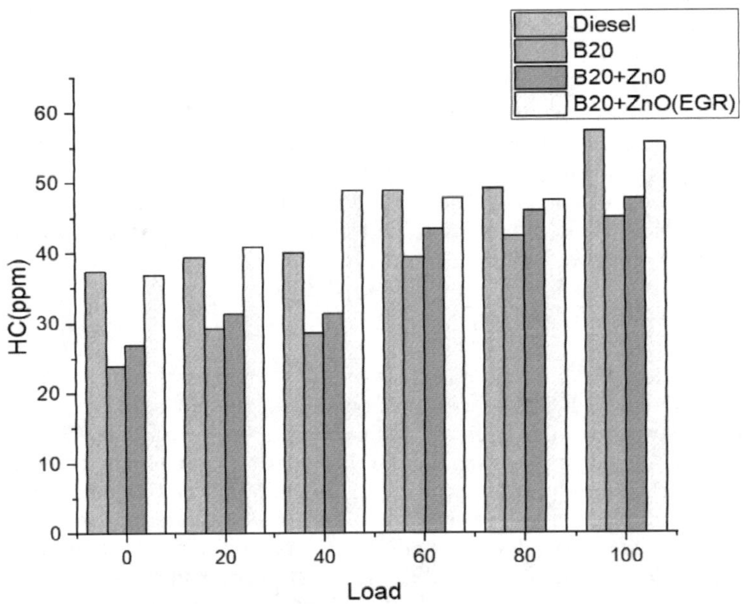

Figure 9. Variation of Unburnt Hydrocarbon Emissions with Load

IOP Conf. Series: Materials Science and Engineering 1013 (2021) 012037 doi:10.1088/1757-899X/1013/1/012037

Fig. 9 shows the Fluctuation of Hydrocarbon emissions with Load. The drop in HC emission can be caused due to the avoidance of undesirable fuel build-up which is brought about by the ZnO nanoparticle and results are as seen in the Fig. 9 [41] .From Fig. 9 it is seen that ZnO is highly operative in tackling HC emissions at 80-100% loads possibly because of the supply of extra oxygen and also by ZnO acting as an oxygen buffer. HC emission drops because of higher O2 quantity of the blends which causes improved overall combustion and higher cetane number which is provided by ZnO which enhances the combustion leading to lower hydrocarbon emissons. Lower excess oxygen content results in improper combustion. But this deficit in oxygen is countermanded when B20 is added because B20 has molecular oxygen which decreases the oxygen necessary for combustion resulting in lower HC emissons [42]. Biodiesel has greater oxygen content, which can accomplish improved combustion and lessen the HC emissions [43]. The occurrence of oxygen in the blends increases oxidation process of UBHCs (Unburnt hydrocarbons) post flame in the combustion chamber. Similar results were obtained by 19 & 40.

3.8 Smoke Opacity

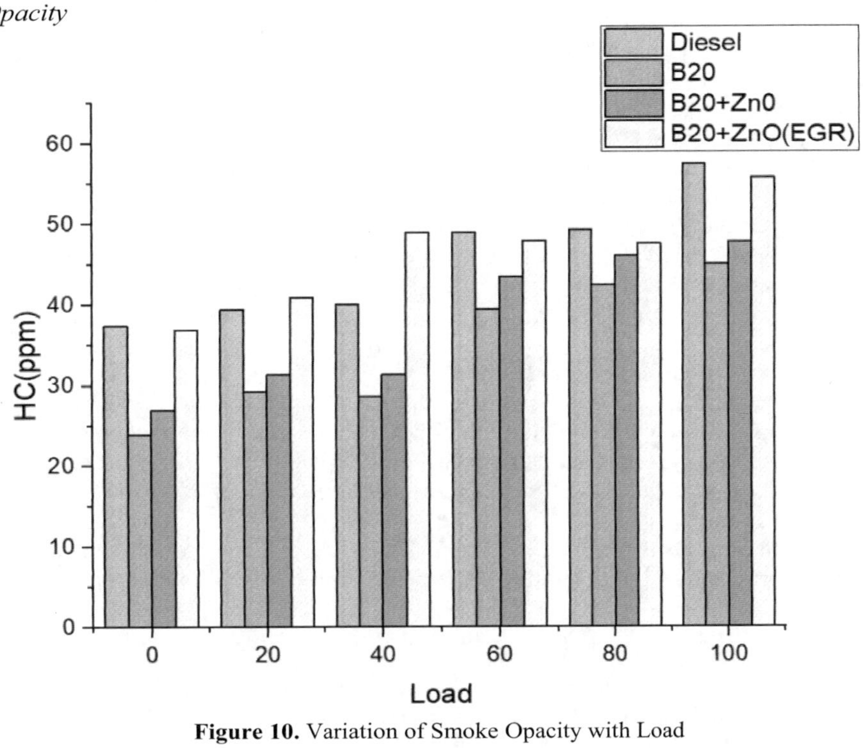

Figure 10. Variation of Smoke Opacity with Load

From the Fig. 10 it can be seen that Smoke opacity falls by 8% for B20 and 3% for B20+ZnO (when compared to diesel). There is decrease of Smoke Opacity because of the superior volume to surface area ratio of Zinc Oxide and also the superior cetane Number of the biodiesel blend, so both these factors result in better more complete combustion. [47] The reduction of Smoke Opacity can be caused by the higher Sulphur and oxygen content in the blend [44]. Smoke Opacity is reduced because of the bonded oxygen which is higher in biodiesel that tends to reduce the soot formation [44, 45]. It is seen that smoke opacity decreases for EGR because of drop in the available oxygen level in the combustion chamber which is unfavorable for development of Soot. Also this reduction in oxygen causes decline of the flame temperature which again is negative for soot formation [46).Also the decline of aromatic compounds (which are considered soot precursors) because biodiesel does not provide the preliminary radicals which are required for development of aromatic rings [47]. This type

Smoke opacity trend was also noticed by Prabakaran and Vijayabalan [34] & Karthikeyan and Elango [43].

4. Conclusions

The superior properties of Zinc Oxide have already been demonstrated in this experimentation. Zinc oxide nanoparticle has superior Volume to Surface area ratio and also high thermal efficiency, thus when ZnO nanoparticle is added to a fuel, it considerably enhances the combustion reaction by increasing the volume available for combustion in the combustion chamber. Furthermore, metal nanoparticle additives in general have more reactive surfaces than conventional metal additives thus being more effective and efficient as an additive.

1) The Brake Thermal efficiency reduces because of oxygen buffer of ZnO which lowers requirement of oxygen complete combustion process.
2) BSFC reduces by 12% in comparison with diesel at 100% load because Zinc Oxide acts as an oxidizing agent which results in better combustion. EGR when applied to the mixture rises the BSFC by 9% because oxygen deficit and the dwindling burn rate, thus making it harder to attain stable state for combustion.
3) Heat Release Rate for the Zinc Oxide blend reduces by 11% caused by the poorer calorific value of the blend. P-θ diagram shows that there is a reduction of peak pressure value by 10%. EGR affects the P-θ by reducing the maximum peak pressure value because of the different composition which EGR introduces mainly Carbon Dioxide.
4) Smoke Opacity of the Zinc Oxide blend marginally decreases due because of the superior volume to surface area ratio of Zinc Oxide and also the superior cetane Number of the blend, so both these factors result in better more complete combustion. When EGR is applied to the blend smoke opacity reduces further because of drop in the available oxygen level for combustion which is unfavorable for development of soot.
5) Nitrogen oxides/NOx emissions of the Zinc Oxide blend reduces by about 15% as Zinc Oxide increases the combustion temperature which results in greater oxygen in the blend to react giving lower NOx emissions. EGR when applied to the ZnO blend reduces the NOx emissions by about 20% because of decrease in oxygen content due to it being replaced in EGR also leads to poorer flame temperature and thereby reducing NOx emissons, as formation of NOx is a highly temperature dependent phenomenon.
6) CO and HC emissions decrease by 20% because more fuel is being burnt in the chamber and this is explained by the presence of ignition delay as depicted in the P-θ diagram and this decreases the CO and HC emissions.

References

[1] Mittelbach M and Tritthart P, "Diesel fuel derived from vegetable oils, III. Emission tests using methyl esters of used frying oil," J American Oil Chemists' Soc 1988;65(7):1185–7.
[2] Payri F, Macián V, Arregle J, Tormos B and Martínez JL, "Heavy-duty diesel engine performance and emission measurements for biodiesel (from cooking oil) blends used in the ECOBUS project," SAE paper 2005-01-2205, 2005. Doi: 10.4271/2005-01-2205
[3] Kelvin S and Gerpan J., "The effect of biodiesel fuel composition on diesel combustion and emission," SAE paper 961086. Doi: 10.4271/961086
[4] Çetinkaya M, Ulusay Y, Tekìn Y and Karaosmanoglu F., "Engine and winter road test performances of used cooking oil originated biodiesel," Energy Convers Manage 2005; 46:1279–91. Doi: 10.1016/j.enconman.2004.06.022
[5] Murillo S, Míguez JL, Porteiro J, Granada E and Morán JC. , "Performance and exhaust emissions in the use of biodiesel in outboard diesel engines," Fuel 2007; 86:1765–71. Doi: 10.1016/j.fuel.2006.11.031
[6] Dorado MP, Ballesteros E, Arnal JM, Gómez J and López F., "Exhaust emissions form a Diesel engine fueled with transesterified waste olive oil," Fuel 2003; 82:1311–5. Doi: 10.1016/S0016-2361(03)00034-6
[7] Tat ME, "Investigation of oxides of nitrogen emissions from biodiesel-fueled engines," PhD thesis, Iowa State University, 2003.
[8] Canakci M and Van Gerpen JH. , "Comparison of engine performance and emissions for petroleum diesel fuel, yellow grease biodiesel, and soybean oil biodiesel," Trans ASAE 2003;46(4):937–44. Doi: 10.13031/2013.13948

[9] Yetter RA, Risha GA and Son SF., "Metal particle combustion and nanotechnology," Proc Combust Inst 2009;32(2):1819–38. Doi: 10.1016/j.proci.2008.08.013
[10] Tyagi H, Patrick EP, Ravi P, Robert P, Taewoo L, Jose RP and Paul A. , "Increased hot plate ignition probability for nanoparticle laden diesel fuel," NanoLetters 8:1410–1416. Doi: 10.1021/nl080277d
[11] Mu-Jung Kao, Chen-Ching Ting, Bai-Fu Lin and Tsing-TshihTsung, "Aqueous aluminum nanofluid combustion in diesel fuel," J.Testing and Evaluation 36 (2) (2008). Doi: 10.1520/JTE100579
[12] Lai S L, Guo J Y, Petrova V, Ramanath G and Allen L H, "Size - dependent melting properties of small tin particle – Nano calorimetric measurement,"Phys Rev Lett, 77 (1996) 99 -102. Doi: 10.1103/PhysRevLett.77.99
[13] Bajpai S, Sahoo PK and Das LM, "Feasibility of blending karanja vegetable oil in Petrol-diesel and utilization in a direct injection diesel engine," Fuel 2009; 88:705-11. Doi: 10.1016/j.fuel.2008.09.011
[14] Yanan Gan and Li Qiao Combustion characteristics of fuel droplets with addition of nano and micron-sized aluminium particles Combustion and Flame 2011; 158:354-368. Doi: 10.1016/j.combustflame.2010.09.005
[15] Murugesan A, Umarani C, Subramanian R and Nedunchezian N., "Bio-diesel as an alternative fuel for diesel engines – A review," Renew Sustain Energy Rev 2009; 13:653-62. Doi: 10.1016/j.rser.2007.10.007
[16] Sahoo PK and Das LM, "Comparative evaluation of performance and emission characteristics of jatropha,karanja and polanga based biodiesel as fuel in a tractor engine," Fuel 2009; 88:1698-707. Doi: 10.1016/j.fuel.2009.02.015
[17] Justin LS, Daniel MD, Richard AY, Frederick LD and Ilhan AA , "Functionalized graphene sheet colloids for enhanced fuel/ propellant combustion," ACS Nano 3:3945–3954. Doi: 10.1021/nn901006w
[18] Lobo, Ashley, "Influence of Zinc oxide on Direct Injection CI Engine Fueled with Waste Cooking Oil Biodiesel," International Journal of Scientific and Engineering Research. 6. 32-40. Doi: 10.17148/IARJSET.2019.61106
[19] Ashok, B.Nanthagopal, K. Mohan, AravindJohny and AjithTamilarasu , "Comparative analysis on the effect of zinc oxide and ethanox as additives with biodiesel in CI engine," Doi: 10.1016/j.energy.2017.09.021
[20] S. K. Das, S. U. S. Choi, W. Yu, and T. Pradeep, Nanofluids: Science and Technology; Wiley, Hoboken, NJ, 2007.
[21] Soudagar,MKalam, Abul Badruddin, Banapurmath, Akram and Naveed, "The effect of nano-additives in diesel-biodiesel fuel blends : A comprehensive review on stability, engine performance and emission," Energy Conversion and Management, doi: 10.1016/j.enconman.2018.10.019
[22] Nicos Ladommatos, RazmikBalian ,Roy Horrocks and Laurence Cooper, "The Effect of Exhaust Gas Recirculation on Soot Formation in a High-Speed Direct-injection Diesel Engine," SAE Technical Paper No. 960841. Doi: 10.4271/960841
[23] Selvan VA, Anand RB and Udayakumar M, "Effects of cerium oxide nanoparticle addition in diesel and diesel–biodiesel–ethanol blends on the performance and emission characteristics of a CI engine. J EngAppl Sci. 2009 (7):1819-6608.
[24] May WR and Hirs EA, "Catalyst for improving the combustion efficiency of petroleum fuels in diesel engines," 11th diesel engine emissions reduction conference, August 21–25; 2005. Chicago Il
[25] Jung H, Kittelson DB and Zachariah MR., "The influence of a cerium additive on ultrafine diesel particle emissions and kinetics of oxidation," Combust Flame 2005;142:276–88. Doi: 10.1016/j.combustflame.2004.11.015
[26] Sayin C, Ilhan M, Canakci M and Gumus M., "Effect of injection timing on the exhaust emissions of a diesel engine using diesel methanol blends," Renew Energy 2009; 34:1261e9. Doi: 10.1016/j.renene.2008.10.010
[27] Puhan S, Jegan R, Balasubbramanian K and Nagarajan G. , " Effect of injection pressure on performance, emission and combustion characteristics of high linolenic linseed oil methyl ester in a DI diesel engine," Renew Energy 2009;34: 1227e33. Doi: 10.1016/j.renene.2008.10.001
[28] Sadhik Basha, J. and Anand, R.B., "An experimental study in a CI engine using nanoadditive blended water-diesel emulsion fuel," J. of Green Energy, Vol.8, 332-348, 2011. Doi: 10.1080/15435075.2011.557844
[29] Sabourin, J.L., Daniel, M.D., Yetter, R.A., Frederick, L.D. and Ilhan, A.A., "Functionalized graphene sheet colloids for enhanced fuel/propellant combustion," ACS Nano, Vol.3, 3945-3954, 2009. Doi: 10.1021/nn901006w
[30] Sajith, V., Sobhan, C.B., and Peterson G.P., "Experimental investigations on the effects of cerium oxide nanoparticle fuel additives on biodiesel,"Advances in Mechanical Engineering, Vol. 2010, ID 581407, 2010. Doi: 10.1155/2010/581407
[31] N. Ladommatos, S. Abdelhalim, H. Zhao and Z. Hu, "The effects of carbon dioxide in exhaust gas recirculation on diesel engine emissions," Proc. Inst. Mech. Eng., Part D: J. Automobile Eng. 212 (1998) 25–42. Doi: 10.1243/0954407981525777
[32] Murlidharan, K. and.Vasudevan, and D., "Performance, emission and combustion characteristics of a variable compression ratio engine using methyl esters of waste cooking oil and diesel blends.Doi:10.1016/ j.apenergy. 2011. 04.014
[33] Alla GH., "Using exhaust gas recirculation in internal combustion engines: a review," Energy Convers Manage 2002;43:1027–42
[34] Prabakaran and B.Vijayabalan., "Influence of zinc oxide nano particles on performance, combustion and emission characteristics of butanol-diesel-ethanol blends in di CI engine," IOP Conference Series: Materials Science and Engineering. Doi: 10.1088/1757-899X/377/1/012069
[35] Shaafi, T. Sairam, K. Gopinath, A. Kumaresan and G.Velraj, "Effect of dispersion of various nanoadditives on the performance and emission characteristics of a CI engine fuelled with diesel," Renewable and Sustainable Energy Reviews, 0016-2361. Doi: 10.1016/j.rser.2015.04.086
[36] Murphy MJ, Taylor JD and McCormick RL. Compendium of experimental cetane number data, NREL Report 2004: NREL/SR-540-36805.
[37] Graboski MS, Alvarez JR, McCormick R. NOx solution for biodiesel, NREL/ SR-510-31465.
[38] Ban-Weiss GA, Chen JY, Buchholz BA Dibble RW, "A numerical investigation into the anomalous slight NOx increase when burning biodiesel: a new (old) theory," Fuel Process Technol 2007; 88(7):659–7. Doi: 10.1016/j.fuproc.2007.01.007
[39] Ladommatos, N Balian, Razmik, and Roy Cooper, "The effect of exhaust gas recirculation on soot formation in a high-speed direct-injection diesel engine," SAE Technical Paper No. 960841. Doi: 10.4271/960841

[40] Venu, Harish Madhavan, and Venkataramanan, "Effect of Al2O3 nanoparticles in biodiesel-diesel-ethanol blends at various injection strategies," Doi: 10.1016/j.fuel.2016.08.046
[41] Deepak Agarwal ,Shailendra Sinha, and Avinash Kumar, "Experimental investigation of control of NOx emissions in biodiesel-fueled compression ignition engine," Doi: 10.1016/j.renene.2005.12.003
[42] Abedin MJ, Masjuki HH, Kalam MA, Sanjid A, Rahman SMA, and Fattah IMR, "Performance, emissions, and heat losses of palm and jatropha biodiesel blends in a diesel engine," Ind Crops Prod 2014; 59:96–104. Doi:
[43] Karthikeyan, and S Elang, "An environmental effect of GSO methyl ester with ZnO additive fuelled marine engine," Indian Journal of Geo-Marine Sciences 564 Vol. 43(4), April 2014, pp. 564-570. Doi:
[44] Ladommatos, N., Balian, R., Horrocks, R., and Cooper, L., "The Effect of Exhaust Gas Recirculation on Combustion and NOx Emissions in a High-Speed Direct-injection Diesel Engine," SAE Technical Paper 960840, 1996, doi: 10.4271/960840
[45] Wang WG, Lyons DW, Clark NN, Gautam M, Norton PM., "Emissions from nine heavy trucks fuelled by diesel and biodiesel blend without engine modification," Environ Sci Technol 2000;34(6):933–9. Doi: 10.1021/es981329b
[46] Boehman AL, Song J, and Alam M., "Impact of biodiesel blending on diesel soot and the regeneration of particulate filters," Energy Fuels 2005; 19:1857–64. Doi: 10.1021/ef0500585
[47] Wilson, R.P., Muir, E.B., and Pellicciotti, F.A., "Emissions study of a single-cylinder diesel engine, SAE Papers 740123, 1974.
[48] Sadhik Basha J, and Anand RB. , "An experimental investigation in a diesel engine using CNT blended water-diesel emulsion fuel," Proceedings of I Mech E. Part A. J Power Energy 225:279–288. Doi: 10.1177/2041296710394247
[49] Kegl B, and Hribernik A., "Experimental analysis of injection characteristics using biodiesel fuel," Energy Fuels 2006; 20:2239e48. Doi: 10.1021/ef060285m
[50] Basha JS, and Anand RB. , "Performance, emission and combustion characteristics of a diesel engine using carbon nanotubes blended jatropha methyl ester emulsions," Alexandria Engineering Journal, 2014; 53(2):259-73. Doi: 10.1016/j.aej.2014.04.001
[51] Vijay KumarA., Veeresh Babu, and P.Ravi Kumar, "The impacts on combustion, performance and emissions of biodiesel by using additives in direct injection diesel engine," Alexandria Engineering Journal. Doi: 10.1016/j.aej.2016.12.016
[52] Kim SH, Fletcher RA, and Zachariah MR., "Understanding the difference in oxidative properties between flame and diesel soot nanoparticles: the role of metals." Environ Sci Technol 2005; 39:4021. Doi: 10.1021/es048828z
[53] Jung H, Kittelson DB, and Zachariah MR., "The influence of a cerium additive on ultrafine diesel particle emissions and kinetics of oxidation," Combust Flame 2005; 142:276–88. Doi: 10.1016/j.combustflame.2004.11.015
[54] Metin Gürü, Atilla Koca, özer Can, Can Cinar, and Faith Sahin, "Biodiesel production from waste chicken fat based sources and evaluation with Mg based additive in a diesel engine," Renewable Energy 2010; 35:637–43. Doi: 10.1016/j.renene.2009.08.011
[55] Knothe G, Matheaus AC, Ryan TW. , "Cetane numbers of branched and straight chain fatty esters determined in an ignition quality tester," Fuel 2003; 82:971–5. Doi: 10.1016/S0016-2361(02)00382-4
[56] Ramos MJ, Fernandez CM, Casas A, Rodriguez L, and Perez A. , "Influence of fatty acid composition of raw materials on biodiesel properties," Bioresour Technol 2009; 100:261–8. Doi:
[57] Mittelbach M, Remschmidt, Biodiesel, The comprehensive handbook. Boersedruck Ges.m.b.H., Viena; 2004. Doi:
[58] Lapuerta Magín, Rodríguez-Fernández José, and Fontde Mora Emilio. , "Correlation for the estimation of the cetane number of biodiesel fuels and implications on the iodine number," Energy Policy 2009; 37:4337–44. Doi: 10.1016/j.enpol.2009.05.049
[59] Chang DY, and Van Gerpan JH. , "Fuel properties and engine performance for biodiesel prepared from modified feed stocks," SAE Paper 2005:2005-01- 2200. Doi: 10.4271/971684.
[60] Prabakaran, B Udhoji, and Anurag , "Experimental investigation into effects of addition of zinc oxide on performance , combustion and emission characteristics of diesel-biodiesel-ethanol blends in CI engine," Alexandria Engineering Journal, Volume 55, Issue 4, December 2016, Pages 3355-3362. Doi: 10.1016/j.aej.2016.08.022

IOP Conf. Series: Materials Science and Engineering 1013 (2021) 012038 doi:10.1088/1757-899X/1013/1/012038

Comparative analysis of Machine Learning algorithms for Intrusion Detection

Vasudeva Pai [*] Devidas [*] Adesh N. D. [+]
*NMAM Institute of Technology, Nitte, Udupi, Karnataka, India
[+]Manipal Institute of Technology, Manipal Academy of Higher Education, Manipal, Udupi, Karnataka, India
[*1]paivasudeva@nitte.edu.in, [*2]devidasbhat@nitte.edu.in, [+]adesh.nd@manipal.edu

***Abstract* - In this modern era, the network related applications, programs and services are growing enormously but the network security issues also grow along with them. Keeping the network secure is a challenging and a crucial task. To maintain the secure network there must be some system which can detect and identify any malicious activity happening in network. This system is called as Intrusion Detection System. There are many traditional network security tools and techniques of preventing intrusion like firewalls, anti-virus, encryption-decryption, access control etc. But all are not effective in protecting network from increasing attacks. The network traffic can be categories into normal and intrusive traffic using Machine Learning (ML) algorithms. Here, the preliminary comparative study regarding which type of machine learning algorithm performs better in identifying the attacks namely Denial of Service, Probe, User to Root and Remote to Local. The NSL-KDD dataset is used to study features and behavior of malicious attacker using machine learning techniques. This study can be taken as reference for mechanical engineers for developing a safe automation in industrial atmosphere and automation in automobile.**

***Index Terms*—Intrusion Detection System, Machine Learning algorithms, NSL-KDD dataset, feature selection.**

I. Introduction

The advancement of internet in this era has brought a positive impact to a wide variety of purpose. This improvement also has impact on emerging large amounts of data and information. Along with the progress of internet there are number of challenges to be faced to make it more reliable, stable and secure system. To make system more secure there are lot of alternatives like firewalls, some dynamic mechanisms, software etc., can be used. Out of these, Intrusion Detection System (IDS) is one among the most powerful dynamic mechanism that determines and detects the specific attacks in the network.

Intrusion detection system mainly monitors the process of network and analyzes them to detect any deviation from the normal operation or any sort of abnormalities. The software that scans the activity in the network to detect the suspicious and policy breaching. The IDS is classified into five categories such as Network Intrusion Detection System, Host Intrusion Detection System, Protocol-based Intrusion Detection System, Application Protocol-based Intrusion Detection System, and Hybrid Intrusion Detection System. There are two main methods of detection: misuse detection (signature-based detection) and anomaly detection.

Now a day, few researchers are looking towards machine learning algorithms to detect anomalies in the network. They are using large amount of training dataset to train the machine learning model and test the model by applying the input data that detect attacks or not [3]. It is also observed that some study has already been done [1-4] and only few algorithms are considered for the study purpose. Many of the research works were related to performance and evaluation of algorithms such as Support Vector Machine (SVM) [4], Artificial Neural Network (ANN) etc. But some ML techniques are more suitable for analyzing huge data for intrusion detection of network and information system.

In this paper, the research work is more intend to study the best algorithm, which is more prominently detects any anomaly in the network. Selecting the best possible algorithm for detection of anomaly in network plays a very important role in coming up with good solution to network security issues. The ML algorithms which are selected for the study are SVM, Decision Tree, Logistic Regression, Naive Bayes, and Random Forest. Here the NSL-KDD dataset is used to train and test the ML model. The dataset is preprocessed and features are extracted from it. These features are then applied to ML algorithms and the performance of these algorithms over different intrusions like Probe, Denial of Service (DoS), Remote to Local (R2L), and User to Root (U2R) are compared. The study gives a brief idea about which machine learning algorithm should be used by the Intrusion Detection System that will best identify the deviation in network. It also helps to build a more accurate and efficient detection system for identifying each attack.

The rest of the section is as follows, Section II gives the research work of other researchers pertaining to the machine learning and its application in network security. The Section III describes the methodology of the proposed study and Section IV explains the different machine learning algorithms. The brief description of performance metric is explained in Section V and the outcomes of experiments are discussed in Section VI. Lastly the highlights of the research work are drawn in section VII.

II. Literature Survey

The author [1] has studied the network security issues and conducted the experiment using Naive Bayes, Random Forest Support Vector Machine and K-means ML algorithms to detect four types of attacks like DOS, PROBE, U2R and R2L. They concluded that the Random Forest Classifier (RFC) surpasses the other methods and also stated that hierarchical clustering method can be used to improve the performance of the system. In paper [2] the author has done a comparison using supervised machine learning classifiers namely, Random Forest, Support Vector Machine, Gaussian Naive Bayes, and Logistic Regression are compared for an intrusion detection in network. Effective classifying algorithm is identified based on performance matrix namely F1-Score, accuracy, precision, and recall. Based on the observed results they have concluded that the Random forest classifier outperforms other classifiers for the considered data-set and parameters.

A light weight IDS method is proposed here [3] mainly concerned on pre-processing of the data, so that they can use important features of online data. The main step is to remove the redundant data from dataset to standardize the data. This helps the machine learning algorithms to give the unbiased and accurate result. In paper [4] the author proposed intrusion detection system (IDS) using supervised machine learning techniques to detect the online network data as normal or anomaly. The proposed method only identifies the Denial of Service (DOS) and probe attacks, but the other attacks are not taken into consideration.

The author proposed Intrusion detection method using Support Vector Machine (SVM) [5]. They also used feature removal method to improve the efficiency of the algorithm. Using the proposed feature removal method, they selected best nineteen features from the KDD-CUP99 data-set. The authors have proposed [6] anomaly intrusion detection using improved Self Adaptive Bayesian Algorithm to process the large amount of data. In this paper [7] authors proposed a novel idea to reduce the dimensionality of the data by using triangle-based K-NN approach. An Intrusion Detection system using fuzzy logic is tested [8]. This technique uses a set of fuzzy rules which are obtained from the definite rules using frequent items. The classification accuracy of this approach is above 90% for all types of attacks.

G. Meera Gandhi et al., [9] examined the performance measure of four supervised machine learning algorithms in detecting the four types of attack such as DoS, R2L, Probe, and U2R. The result shows that the C4.5 decision tree classifier performs best in prediction accuracy compared to Naıve Bayes. The authors [10] have compared the performance of the three machine learning algorithm namely Neural Network, Support Vector Machine and Decision Tree. The algorithms were measured based on false alarm rate, accuracy and detection rate of four categories of attacks classes. From these experiments they found that the Decision tree (J48) algorithm outperformed the other two algorithms.

The author [11] suggested using a collective of, Support Vector Machines (SVMs), Multivariate Adaptive Regression Splines (MARS), and Artificial Neural Networks (ANNs). Whereas [12] proposed a hybrid approach in which they have used a Support Vector Machine (SVM) and Radial Basis Function (RBF).

The sequential search strategy for feature selection or feature extraction through determining the importance of a given attribute by simply removing it and recording the performance [13]. If performance of the algorithm is increased, then the feature is unimportant and thus shall be removed. The author [14] suggested that every attribute in the dataset is not much important and it will not give the accurate result as expected. It is very important to reduce the no of features using feature selection technique and also they concluded that simple Cart algorithms gives accurate result than other five algorithm J4.8 Naive Bayes, NBTree, Multi-Layer Perceptron, and SVM.

III. Methodology

In this paper, the research work is more intend to study the best machine learning algorithm, which is more prominently detects any anomaly in the network. Selecting the best possible algorithm for detection of anomaly in network plays a very important role in coming up with good solution to network security issues. There are various steps to be carried out in order to build a ML model such as data preparation, data preprocessing, feature selection, data visualization, clustering, model training and model validation. Different ML algorithms are selected for the study purposes are SVM, Decision Tree, Logistic Regression, Naive Bayes, J48, and Random Forest. The NSL-KDD dataset [15] is used for training the machine learning model to detect the intrusion in network. NSL-KDD dataset consist of several attributes which gives the features of network-based intrusion detection systems. The dataset consist of 42 attributes out of which 41 attributes are features and one attribute is a class label as normal traffic or anomaly traffic is as shown in Figure 1. This makes the detection more accurate and realistic as the ML model is also verified for the unknown attacks. NSL KDD dataset is divided into train and test data as shown in Table 1. The train data set is used to train the ML model and test dataset is given as inputs to trained model for validation purpose. The advantage of this dataset is that it does not include any redundant records in the training data, so the classifiers are accurate and not biased towards more frequent records.

F#	Feature name	F#	Feature name	F#	Feature name
F1	Duration	F15	Su attempted	F29	Same srv rate
F2	Protocol type	F16	Num root	F30	Diff srv rate
F3	Service	F17	Num file creations	F31	Srv diff host rate
F4	Flag	F18	Num shells	F32	Dst host count
F5	Source bytes	F19	Num access files	F33	Dst host srv count
F6	Destination bytes	F20	Num outbound cmds	F34	Dst host same srv rate
F7	Land	F21	Is host login	F35	Dst host diff srv rate
F8	Wrong fragment	F22	Is guest login	F36	Dst host same src port rate
F9	Urgent	F23	Count	F37	Dst host srv diff host rate
F10	Hot	F24	Srv count	F38	Dst host serror rate
F11	Number failed logins	F25	Serror rate	F39	Dst host srv serror rate
F12	Logged in	F26	Srv serror rate	F40	Dst host rerror rate
F13	Num compromised	F27	Rerror rate	F41	Dst host srv rerror rate
F14	Root shell	F28	Srv rerror rate	F42	Class label

Figure 1: Attributes of NSL – KDD dataset [15]

Table 1: Distribution of instance in NSL-KDD dataset for training and testing

Types of Attack	Number of Records	
	Training dataset	**Test dataset**
Normal	67343	9711
Denial of Service	45927	7456
Probe	11656	2421
Remote to Local	995	2756
User to Root	52	200
Total	**125973**	**22544**

A. Data Preparation

Data preparation is the process of cleaning and transforming the raw data prior to data preprocessing and analysis. It is a mandatory step prior to processing and often involves making corrections to data, reformatting data and the combining of data sets to enrich data. The first step of data preparation is dividing connections into normal and anomaly classes based on 'labels' column. Then attacks are further divided into four main categories: DoS, Probe, U2R and R2L are as shown in Figure 2. After this step, all the categories are indexed and furthermore, column ID is included to simplify the work with clustered data.

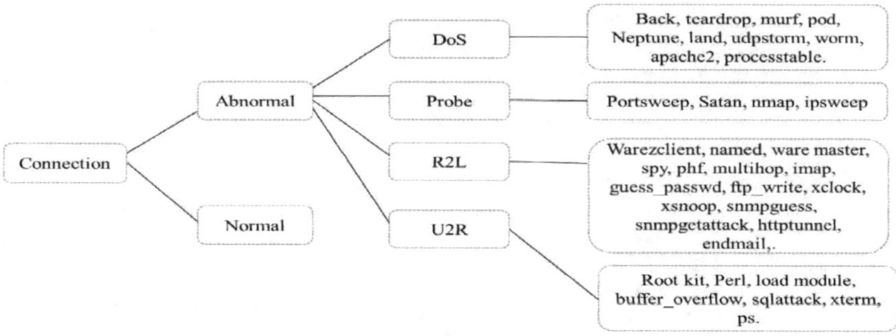

Figure. 2: Classification of intrusion attacks in NSL-KDD dataset

B. Data Preprocessing

Data preprocessing is an integral step in Machine Learning as the quality of data and the useful information that can be derived from it directly affects the ability of our model to learn. Some of the concepts that handles the data preprocessing are Handling Null values, Standardization, Handling Categorical Variables, One-Hot Encoding, and Multicol linearity. The real-world data is incomplete, inconsistent, lacking in certain behavior, and likely to contain many errors. Data pre-processing is proven to resolve these errors.

The technique of data pre-processing will get rid of irrelevant data and missing values. Here, the One Hot Encoding technique is used to convert nominal values into binary values in dataset. So, that the machine learning algorithms can use these data values to training a model.

C. Feature Selection

Feature selection in machine learning is one of the core concepts which largely impact the performance of the model. The features that are used to train machine learning models have a huge influence on the performance because partially relevant or irrelevant features can negatively impact the performance. The process of selecting the features manually or automatically from dataset mainly contributes to the prediction or output of model which are interested in. The technique used here for feature selection is attribute ratio. After feature selection data standardization is done which is the process that converts the structure of disparate datasets into a Common Data Format. Standardization is required as many distance based algorithms are used.

D. Data Visualization

Data visualization is used to understand how actually the data looks like and what type of correlation is held by the attributes of data. It is the best and quickest way to see if the features correspond properly to the output. In this study, the Principal Component Analysis (PCA) algorithm is used on NSL-KDD dataset for data visualization is as shown in Figure 3. One of the advantages of using PCA algorithm is for speeding up machine learning algorithms and also used in reducing the dimension of data. It is also used later on for preprocessing in Gaussian Mixture clustering.

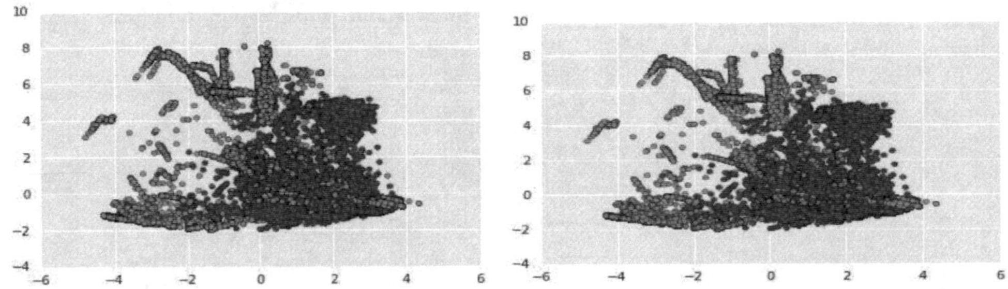

a. Attack vs Normal labels b. Different types of attacks vs normal connections

Figure 3: Data visualization using PCA algorithm on NSL – KDD dataset.

E. Clustering

Clustering is the process of dividing the data points into several groups such that the data points in the group are more similar to each other than data points in other groups. In simple words, the task is to segregate groups with similar properties and assign them into clusters. In this study, the K-Means and Gaussian Mixture are used for clustering purpose because the output of K Means is different from Gaussian Mixture is as shown in the Figure 4. So, that the results of both algorithm are combined to achieve high performance. Here, it is divide into two categories, Frist category include clusters that contain both 'attack' and 'normal' connections and also have more than 25 connections. The second category includes all other clusters and maps these cluster to 'attack' or 'normal' based on the majority. The clusters that include less than or equal to 25 connections are considered as outliers.

cluster_labels2	attack	normal	count
0	6659	46448	53107
1	9125	2266	11391
2	626	61	687
3	27742	101	27843
4	2670	5073	7743
5	1	0	1
6	0	24	24
7	2	42	44

a. K-Means

cluster_labels2	attack	normal	count
0	22895	0	22895
1	7548	19846	27394
2	30	26900	26930
3	4024	0	4024
4	4154	2212	6366
5	1136	1007	2143
6	5479	0	5479
7	1559	4050	5609

b. Gaussian Mixture

Figure 4: Process of clustering the NSL – KDD dataset.

F. Training

After all the previous process for NSL-KDD dataset, now the ML model can be trained using various different classification algorithms namely Random Forest, Logistic Regression, Naive Bayes, Decision Tree, SVM and J48 with NSL-KDD dataset.

IV. Classification Algorithms

A. *Random Forest*

Random Forests are troupe learning strategy for classification and regression strategies. It builds various decision trees at preparing time and yields the class that is the method of the classes yield by individual trees. It likewise endeavors to limit the issues of high change and high predisposition by averaging to locate a characteristic harmony between the two boundaries.

B. *Naive Bayes*

Naive Bayes is an upgraded version of Bayes Theorem as it considers strong independence among the attributes. It is based on Bayes probability theory. It is a classification technique which is based on an assumption of independence between predictors. This implies that the probability of one attribute does not affect the probability of the other attribute.

C. *Support Vector Machine*

SVM is a supervised learning technique which classifies data into two classes over a hyper plane. In SVM, each data item is plot as a point in n-dimensional space where n represents the number of features with the value of each feature being the value of a particular coordinate

D. *J48*

In this algorithm every aspect of the information is split into minor subsets based on decision. J48 look at the standardized data gain and split the information by choosing an attribute. To summarize, the attribute extreme standardized data gained is utilized. The minor subsets are returned by the algorithm. The split strategies stop if a subset has a place with a similar class in all the instances. J48 develops a decision node utilizing the expected estimations of the class. J48 decision tree can deal with particular characteristics, lost or missing attribute estimations of the data and varying attribute costs.

E. *Logistic Regression*

Logistic regression is a classification algorithm used to assign observations to a discrete set of classes. Unlike linear regression which outputs continuous number values, logistic regression transforms its output using the logistic sigmoid function to return a probability value which can then be mapped to two or more discrete classes.

F. *Decision Tree*

A decision tree classifier is one of the most known supervised machine learning technique. A decision tree is composed of three basic elements. A decision node represents test or condition on a data item, and an edge or a branch which corresponds to the one of the possible attribute values which means one of the test attribute outcomes. A leaf determines the class to which the object belongs.

V. Performance Metrics

A. *Accuracy*

It is the number of correctly predicted data points out of all the data points and it is represented in terms of percentage. The accuracy is calculated as shown in Equation 1.

$$Accuracy = \frac{TP + TN}{TP + TN + FP + FN} \quad (1)$$

Where False positive (FP): It defined as the number of detected attacks which are actually normal, and False negative (FN): means the wrong prediction i.e. it detects the instances as normal but in actual its an attack. True positive (TP): is an instances that are correctly predicted as normal, and True negative (TN): is an instances that are correctly classified or detected as attack.

B. *Precision*

It is a measure which estimates the probability that a positive prediction is correct, and the formula is as shown in Equation 2.

$$Precision = \frac{TP}{TP + FP} \quad (2)$$

C. *Recall*

It is the proportion of instances belonging to the positive class that are correctly predicted as positive, and the formula is as shown in Equation 3.

$$Recall = \frac{TP}{TP + FN} \tag{3}$$

D. Kappa

Its value ranges from 0 to 1. 0 means totally disagreement and full agreement. It checks the reliability of classifying algorithm on dataset.

E. Receiver Operating Characteristics (ROC)

It is used to design the curve between true positive rate and false positive rate, and the Area Under Curve (AUC) gives the value of ROC. More the area under curve and more will be the value of ROC.

F. F1 Score

The F1 score or F-measure is a measure of the test's accuracy. It considers both the precision P and the recall R of the test to compute the score.

VI. Results

A. DoS attack

Table 2: Performance of the selected machine learning classifiers against the DoS attack

Machine Learning algorithms	Accuracy (%)	Precision	Recall	F1-Score	ROC
Random Forest	**99.83**	**1**	**0.999**	0.999	**0.999**
J48	99.76	0.99	**0.999**	0.999	0.658
Logistic Regression	99.47	0.99	0.998	0.999	0.727
Decision Tree	99.66	0.99	0.999	**1.0**	0.618
SVM	99.19	0.98	0.999	0.999	0.5
Naïve Bayes	94.19	0.97	0.903	0.948	0.844

B. Probe attack

Table 3: Performance of the selected machine learning classifiers against the Probe attack

Machine Learning algorithms	Accuracy (%)	Precision	Recall	F1-Score	ROC
Random Forest	**99.94**	**0.999**	**1.0**	**1.0**	**1.0**
J48	**99.94**	0.994	0.999	0.999	0.999
Logistic Regression	99.47	0.995	0.998	0.998	0.996
Decision Tree	99.91	0.993	0.999	0.998	**1.0**
SVM	99.90	0.994	**1.0**	**1.0**	**1.0**
Naïve Bayes	90.31	0.986	0.966	0.956	0.974

C. Remote to Local attack

Table 4: Performance of the selected machine learning classifiers against the Remote to Local attack

Machine Learning algorithms	Accuracy (%)	Precision	Recall	F1-Score	ROC
Random Forest	**99.99**	**0.999**	**1.0**	**0.999**	**1.0**
J48	99.89	0.998	0.999	0.998	0.994
Logistic Regression	99.81	0.997	0.992	0.992	0.997
Decision Tree	99.74	0.997	0.993	0.993	0.999
SVM	98.92	0.995	0.990	0.990	0.979
Naïve Bayes	98.92	**0.999**	0.882	0.931	0.953

D. User to Root attack

Table 5: Performance of the selected machine learning classifiers against the User to Root attack

Machine Learning algorithms	Accuracy (%)	Precision	Recall	F1-Score	ROC
Random Forest	**99.85**	0.999	**0.998**	**0.998**	0.997
J48	99.67	**1**	**0.998**	**0.998**	0.936
Logistic Regression	99.16	**1**	0.995	0.995	0.982
Decision Tree	99.29	0.999	0.993	0.993	**0.999**
SVM	98.97	0.999	0.992	0.992	0.826
Naïve Bayes	88.85	0.999	0.942	0.960	0.948

VII. Conclusion and Future Work

In this study, the machine learning technique that can efficiently identify the anomalies in the network such as DoS, Probe, R2L and U2R. The machine learning algorithm considered for the study are Random forest, Naive Bayes, J48, Logistic Regression, Decision Tree, and SVM. The NSL-KDD dataset is used for training and testing the machine learning model. The dataset needs to be normalized before feeding the model. The performance of classifiers on the test data set and compute a classification metrics for each model. To check the performance of ML model is evaluated in terms of accuracy, precision, recall, F1-score, and ROC. The best classifier was the random forest classifier with an accuracy of 99.997% for the DoS attack and similarly J48 is the best classifier with an accuracy of 99.947% for U2R attack. Likewise, the performance of R2L and U2R attack is computed and do a comparative study.

There is an increasing demand for a reliable and real-world attacks dataset among research community. As a future work, it is possible to work in several directions, at one direction enhancing the current work by implementing various balancing algorithm and investigate its effects on model learning. Study the trouble with the two attacks (Web Attack-Brute Force Web Attack-XSS). In addition, applying deep learning algorithms on this data set such as convolution and recurrent neural networks.

References

[1] A. T. Yousef El Mourabit, AnouarBouirden and N. E. Moussaid, "*Intrusion Detection Techniques in Wireless Sensor Network Using Data Mining Algorithms: Comparative Evaluation Based on Attacks Detection*", International Journal of Advanced Computer Science and Applications, vol. 6, no. 9, pp. 164–172, (2015).

[2] J. Manjula C. Belavagi and Balachandra Muniyal, "*Performance Evaluation of Supervised Machine Learning Algorithms for Intrusion Detection*" Twelfth International Multi-Conference on Information Processing- 2016.

[3] S. S. S. Sindhu, S. Geetha and A. Kannan, "*Decision Tree Based Light Weight Intrusion Detection Using a Wrapper Approach, Expert Systems with Applications*", vol. 39, no. 1, pp. 129–141, (2012).

[4] P.Sangkatsanee, N. Wattanapongsakorn and C. Charnsripinyo,, "*Practical Real-Time Intrusion Detection Using Machine Learning Approaches, Computer Communications*" , vol. 34, no. 18, pp. 2227–2235, (2011).

[5] Y.Li, J. Xia, S. Zhang, J. Yan, X. Ai and K. Dai, "*An EfficientIntrusion Detection System Based on Support Vector Machines and Gradually Feature Removal Method, Expert Systems with Applications*", vol. 39, no. 1, pp. 424430, (2012).

[6] D. M. Farid and M. Z. Rahman, "*Anomaly network intrusion detection based on improved self-adaptive Bayesian algorithm*" Journal of computers, vol. 5, no. 1, pp. 23–31, 2010.

[7] JB. Luo and J. Xia, "*A novel intrusion detection system based on feature generation with visualization strategy*", Expert Systems with Applications, vol. 41, no. 9, pp. 4139– 4147,2014.

[8] R. Shanmugavadivu and Dr. N. Nagarajan, "*Network intrusion detection system using fuzzy logic*", Indian Journal of Computer Science and Engineering, Vol. 2, pp. 101-111.

[9] G. Meera Gandhi, "*Machine learning approach for attack prediction and classification using supervised learning algorithms*", International Journal of Computer Science & Communication, Vol. 1, No. 2, pp. 247-250. July-December 2010.

[10] K. AbdJalil, and S. Mara, "*Comparison of machine learning algorithms performance in detecting network intrusion*", In Proceedings of Networking and Information Technology (ICNIT), pp. 221 – 226, Manila 2010.

[11] S. Mukkamala, A. H. Sung, and A. Abraham, "*Intrusion detection using an ensemble of intelligent paradigms*" J. Netw. Comput. Appl., vol. 28, no. 2, pp. 167–182, 2005.

[12] M. Govindarajan and R. Chandrasekaran, "*Intrusion Detection using an Ensemble of Classification Methods,*" In Proceedings of World Congr. Eng. Computer. Sci., vol. I, no. October, 2012

[13] Karan Bajaj, Amit Arora, "*Improving the Intrusion Detection using Discriminative Machine Learning Approach and Improve the Time Complexity by Data Mining Feature Selection Methods*", International Journal of Computer Applications (0975 – 8887) Volume 76– No.1, August 2013.

[14] C.Ingre B. and Yadav A., "*Performance analysis of NSL-KDD dataset using ANN,*" 2015 International Conference on Signal Processing and Communication Engineering Systems, Guntur, 2015, pp. 92-96.

[15] NSL-KDD dataset [online] available: http://nsl.cs.unb.ca/nsl-kdd/. Last accessed on 15-7-2020.

AUTHOR INDEX